Microwave Semiconductor Devices:

Fundamentals and Radiation Effects

Microwave Semiconductor Devices:
Fundamentals and Radiation Effects

R. J. CHAFFIN
Sandia Laboratories
Albuquerque, New Mexico

A Wiley–Interscience Publication
JOHN WILEY & SONS
New York London Sydney Toronto

Library of Congress Cataloging in Publication Data

Chaffin, Roger J. 1941–
Microwave semiconductor devices.

Includes bibliographical references.
1. Semiconductors. 2. Semiconductors, Effect of radiation on. 3. Microwave circuits. I. Title.
QC611.C48 537.6'22 73–7581
ISBN 0–471–14311–1

Printed in the United States of America

10 9 8 7 6 5 4 3 2 1

Acknowledgments

I would like to thank the U.S. Atomic Energy Commission and Sandia Laboratories for allowing me the time required to write this book. Further, I would like to acknowledge helpful discussions held with E. P. EerNisse, D. G. Schueler, C. E. Barnes, C. B. Norris, W. E. Wilson, J. M. McKenzie, and B. L. Gregory, from Sandia Laboratories. The microwave device measurements described in the book were performed at Sandia Laboratories with the aid of T. E. Gerst. Special thanks go to E. D. Graham for his thorough proofreading of and helpful comments on the original manuscript. The manuscript could not have been produced without the assistance of Judy Roberts, helped by Sharon Erickson and Jan Hawkins, whose careful typing, retyping, and organizational efforts are appreciated. The drawings used in this book were prepared by the Technical Publications and Art Department of Sandia Laboratories.

Albuquerque, New Mexico

Roger Chaffin

Contents

Microwave Semiconductor Devices:
Fundamentals and Radiation Effects

1

Introduction

1.1 Purpose of the Book

The main purpose of this book is to provide electronic circuit designers and system planners with the understanding needed to design radiation-tolerant microwave semiconductor circuits. The basic format used in the book is as follows: the physics of device and circuit operation are discussed in moderate detail, and the macroscopic semiconductor material parameters (conductivity, mobility, etc.) critical to device operation are identified. A discussion of how radiation affects these macroscopic materials' parameters is provided in Chapter 5. This approach enables the reader to predict failure modes and levels as well as to generate circuit designs that are radiation tolerant. The discussions of each device type (Chapters 6–13) include experimental measurements as well as theoretical predictions of radiation damage to microwave semiconductor devices.

The book is designed so that it can be used by engineers with varying levels of understanding. If the reader does not have a background in the device physics area, he is encouraged to start reading at the beginning of Chapter 2. If he understands device physics but not radiation effects, he may begin with Chapter 5. If he possesses a good knowledge of both device physics and radiation effects, he may wish to go directly to the chapter on the specific device of interest.

1.2 Why an Interest in Radiation Effects?

An interest in radiation effects studies has grown in response to several needs. Radiation damage occurs when a semiconductor device is irradiated with energetic electrons, protons, neutrons, or gamma rays. The exact fluence† required to degrade the performance of a given microwave semi-

† Fluence ≡ number incident particles per unit area.

conductor device depends on the type and energy of the incident radiation. Radiation effects are usually classed in one of two ways, either permanent effects or transient effects. Permanent radiation effects are those effects which remain (at room temperature) long after the radiation source is removed. Permanent effects are usually attributed to displacements of lattice atoms from their initial equilibrium position at a lattice site. Transient radiation effects are those effects which disappear shortly after the radiation source (usually a γ-ray pulse) has been removed. Transient effects are usually attributed to ionization effects in the semiconductor crystal.

Several types of electronic systems must be capable of operation in a radiation environment or be capable of operation after radiation exposure.

The greatest amount of research has been carried out on nuclear weapons environments. Electronic systems in ballistic and antiballistic missiles must remain operational even after exposure to radiation from friendly or unfriendly nuclear detonations. For example, at a range of one mile from a 1 kiloton airburst near sea level one would expect a total energetic neutron fluence of 7×10^8 n/cm^2† and an absorbed gamma dose of 4 rads [1] At higher altitudes (where atmospheric shielding is less effective) or for larger yield weapons, considerably larger radiation doses are to be expected. Since most of the energy output from a nuclear detonation occurs in the first 100 nsec [2], the radiation dose rates are quite high (6×10^{15} n/cm^{-2} sec^{-1} and 4×10^7 rad·sec^{-1} for the above example). These high dose rates can produce the transient ionizing effects described earlier.

The increasing number of earth satellites and interplanetary probes has created a need for electronic circuits which can withstand the space radiation environment. The most severe radiation problem occurs when a satellite or probe spends a substantial period of time in the magnetic field near a planet. These planetary magnetic fields are known to trap high-energy protons and electrons which produce regions of fairly high radiation fluences. The van Allen radiation belt of the earth is typical of these radiation belts. For example, a satellite passing through the most intense region of the van Allen belt receives a daily fluence of electrons (with energies >1.2 MeV) of 1.8×10^{13}/cm^2 and a daily proton fluence (energy greater than 14 MeV) of 3.3×10^{10}/cm^2 [3]. Prolonged exposure to radiation fluences of this magnitude can certainly cause system degradation and eventual failure. The life of the system can be extended considerably by use of a radiation-tolerant design. A second source of radiation in space satellites and probes is caused by the use of radioisotopic-fueled thermoelectric generators to provide electrical power for systems operation during the mission [4]. In this case we are concerned with a fairly low flux‡ but an exposure of several years.

† See Chapter 5 for a discussion of units of radiation dose.

‡ Flux ≡ fluence per second.

1.3 Quick-Reference Table

This section presents, in abbreviated form, some of the major results of the book. Each device type is listed in Table 1.1 together with either a number or an equation for neutron failure fluence. (The techniques discussed in Chapter 5 can be used to extend Table 1.1 to other radiation types.) The

TABLE 1-1

Device	Neutron Failure Fluence (n/cm²)	Failure Mechanism
Schottky-Barrier Diode	10^{16} See Section 6.7	Carrier removal in epitaxial layer increases R_s
p-i-n Switch	$\Phi_f = \frac{4 \times 10^{16}}{W^2}$ See Figure 7.15	$W = I$ region width; failure due to reduced lifetime in I region
IMPATT TRAPATT	$\Phi_f = 3 \times 10^6 \cdot f_0$ or 10^{16} max See Figure 8.22	f_0 = TRAPATT design frequency; failure due to recombination and/or radiation-induced dc differential negative resistance
Gunn	$\Phi_f = 1.2 \times 10^8 (f_0)^{0.6}$ See Eq. (9.25)	f_0 = design frequency; failure due to carrier removal effects changing the $n \cdot L$ product
LSA	$\Phi_f = 5.5 \times 10^7 (f_0)^{0.6}$ See Eq. (9.26)	f_0 = design frequency; failure due to changes in low field resistance due to carrier removal effects
BARITT	$\Phi_f = 3.0 \times 10^5 \cdot f_0$ See Eq. (10.16)	f_0 = design frequency; failure due to avalanche breakdown caused by trapped charge effects
SRD	$\Phi_f = 2.7 \times 10^7 \cdot f_i$ See Eq. (11.21)	f_i = input frequency to multiplier; failure due to recombination of injected charge
Microwave Transistors	10^{15}–10^{16} See Chapter 12	Failure due to onset of Kirk effect; exact level depends on operating current, voltage
Tunnel Diodes	Si, Ge – 10^{15} GaAs – 10^{16}	Failure due to tunneling from radiation-produced states in forbidden gap

number or equation representing the failure fluence is for a radiation-hardened device and circuit (i.e., one that has been designed for maximum radiation tolerance) and the reader is cautioned that these limits may not necessarily be achieved by off-the-shelf circuits or devices. The table also states the predominant failure mechanism for each device type.

REFERENCES

[1] S. Glasstone, ed., *The Effects of Nuclear Weapons*, U.S. Atomic Energy Commission, U.S. Government Printing Office, Washington, D.C., April 1962.

[2] R. J. Boucher and J. B. Pearson, "Transient Radiation Environment of Nuclear Weapon Detonations," *J. Appl. Phys.*, **36**, 2722 (1965).

[3] W. Rosenzweig, "Space Radiation Effects in Silicon Devices," *IEEE Trans. Nuc. Sci.*, **NS-12**, 18 (October 1965).

[4] P. G. Kase, "The Radiation Environments of Outer Planets Missions," presented at IEEE Conference on Nuclear and Space Radiation Effects, Seattle, Washington, July 1972.

2

Characteristic Properties of Bulk Semiconductors

2.1 Introduction

The microwave semiconductor devices to be discussed in this book will be made primarily from silicon, germanium, or gallium arsenide. These materials will be used in the form of very pure (unless deliberately doped with impurity atoms) crystals of varying size. The bulk electrical properties of crystalline semiconductors have been extensively studied by many authors and very elegant treatments are available (see [1], for example). Only those properties of crystalline semiconductors important to the operation of microwave solid-state devices will be discussed here. Section 2 of this chapter is devoted to a discussion of crystal lattice structures and the nomenclature used to discuss them. Section 3 discusses the motion of electrons in a semiconductor crystal, including the concepts of energy band structure and effective mass. Section 4 is devoted to calculation of the population of the energy bands in the intrinsic case. Section 5 considers the effect of impurity (dopant) atoms on the population of the energy bands. Section 6 introduces the concepts of mobility and conductivity and velocity saturation. Section 7 examines the behavior of carriers under nonequilibrium conditions and discusses diffusion and recombination, and Section 8 is devoted to a discussion of the current continuity equation as it applies to semiconductors. Detailed treatments of these subjects are beyond the scope of this book. Therefore, in each case, the discussions will be brief, but there are references to additional literature if further study is desired.

2.2 The Crystalline State

The discussion in this section will be limited primarily to crystalline Si, Ge, and GaAs. A crystal is defined as a regular arrangement of atoms. The arrangement may be quite complex, but nonetheless that arrangement is periodic. A small subgroup of atoms from which the crystal array can be constructed by simple translation of the subgroup is called a *unit cell.* Crystalline materials are usually described in terms of the unit cell. The deterministic arrangement of the atoms in the unit cell is due to the fact that the system seeks to attain a minimum energy configuration. Different chemical elements form different types of unit cells. The structure of the unit cell for Si and Ge is called the diamond unit cell (since its placement of atoms is similar to that found in diamond). The diamond unit cell, depicted in Figure 2.1, contains fourteen atoms, some of which are shared with other adjacent cells (see [2]).

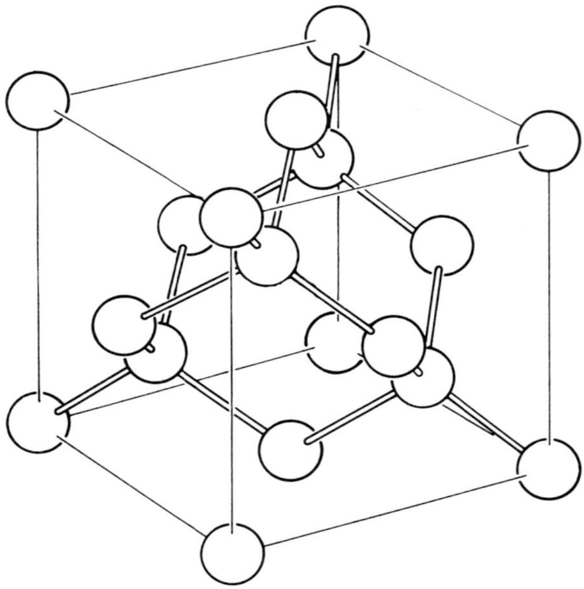

FIGURE 2.1
Diamond unit cell, the basic structure of Si and Ge crystals.

The structure for GaAs, called the zincblende structure [2], is basically the same as the diamond structure except that it contains two atom types (i.e., Ga and As atoms are both contained in the same unit cell). Description of various symmetry planes in a crystal structure is usually accomplished with

Miller indices. Assume that the edges of the unit cell form a set of coordinate axes, and a plane to be described intersects these axes at a, b, and c. The Miller indices then are defined by taking the reciprocal of a, b, and c and multiplying them by a common multiplier which reduces them to their lowest integer values, i.e., if the intercepts of the plane are 1, 2, and 4, then the Miller indices are 4, 2, and 1, respectively. For example, in cubic crystals (not shown here) the y-z plane† which intersects the x axis at $x = 1$ has Miller indices 1, 0, and 0 usually written (100). If a bar is placed over one of the numbers it implies the intercept is on the negative axis. For example, $(\bar{1}00)$ refers to the y-z plane which has an intercept at $x = -1$.

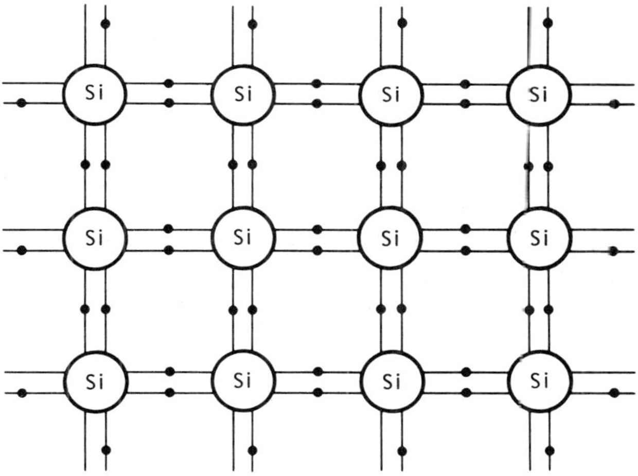

FIGURE 2.2
Covalent bonding in the silicon lattice.

The atoms in the crystals of Si, Ge, and GaAs are held together by covalent bonding. This means that the atoms share valence electrons with one another. Consider Si as an example. Silicon has four electrons in its outermost unfilled shell. This shell will be filled when it contains eight electrons, and this is a minimum energy condition. In the diamond lattice, the Si atoms are arranged so that they have four nearest neighbors (and each of these neighbors has four nearest neighbors, etc.). The Si atom shares its four outer-shell electrons with its four nearest neighbors, and the result is that the outer shell of each atom now contains eight electrons (all of which are shared). Figure 2.2 depicts this situation; however, the figure does not show the diamond lattice structure but only schematically represents the covalent bonding of the silicon atoms. The

† The intercepts are assumed to be at $z = \infty$, $y = \infty$.

actual diamond crystal structure was depicted in Figure 2.1. The study of crystalline structures has had a long history and several books have been written on this subject (see, for example, [3]).

2.3 Motion of Electrons in a Crystal

The motion of electrons in a periodic crystal is considerably more complex than that of a free electron. To make the subject more understandable to the reader, we will endeavor to describe the motion in terms of classical mechanics concepts. According to the classical electrodynamic motion of an electron, the potential energy of a free electron is zero, and the total energy consists of only kinetic energy, $\mathscr{E}_k$, which is given by

$$\mathscr{E}_k = \tfrac{1}{2}mv^2 = \frac{p^2}{2m} \tag{2.1}$$

where v = velocity, m = mass, and $p = mv$ = momentum of the electron. A plot of $\mathscr{E}_k$ versus p is shown in Figure 2.3A for a free particle. The right-hand side of Figure 2.3A represents the continuum of allowed values of energy for a free particle. The mass may be found from:

$$\frac{d^2\mathscr{E}_k}{dp^2} = \frac{d^2}{dp^2}\left(\frac{p^2}{2m}\right) = \frac{1}{m} \tag{2.2}$$

which may be written as

$$m = \frac{1}{d^2\mathscr{E}_k/dp^2} \, . \tag{2.3}$$

The fact that the second derivative of the kinetic energy with respect to momentum yields the particle's mass will be of use later. Now consider the motion of an electron in orbit around a single hydrogen atom. Studies of the hydrogen atom around the turn of the century led Bohr to postulate that the angular momentum of the electron could only have values of $n\hbar$ where n = an integer and $\hbar$ = Planck's constant/2π. The kinetic energy of the electron orbiting the nucleus at a radius r may be found from the fact that the centrifugal and Coulomb forces must balance, i.e.,

$$\frac{mv^2}{r} = \frac{e^2}{4\pi\epsilon_0 r^2} \, . \tag{2.4}$$

Hence, the kinetic energy of the electron is

$$\mathscr{E}_k = \tfrac{1}{2}mv^2 = \frac{e^2}{8\pi\epsilon_0 r} \tag{2.5}$$

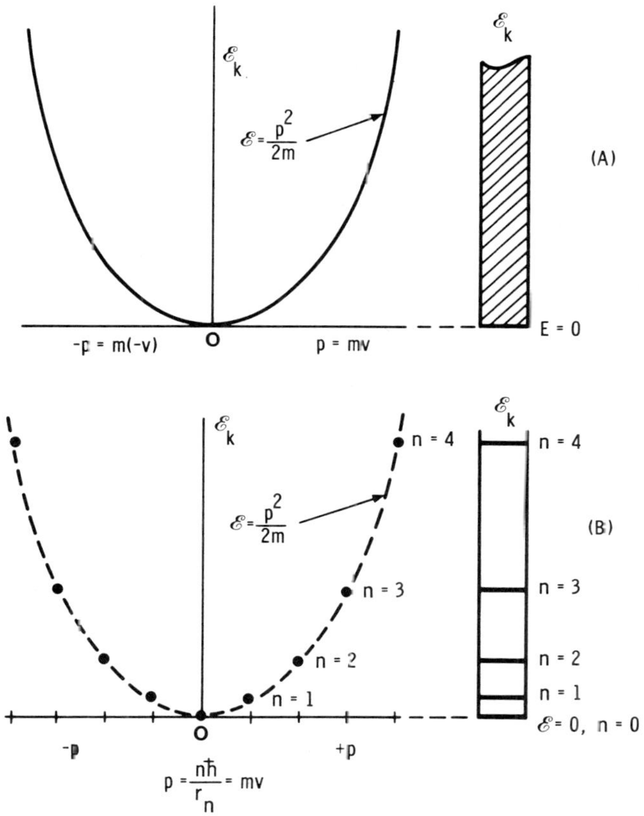

FIGURE 2.3
(A) Energy versus momentum for a free electron. (B) Energy versus momentum for an electron bound to a hydrogen atom.

by combining Eq. (2.5) with the Bohr postulate, $mvr = n\hbar$, one can show that the kinetic energy of the electron is quantized (it can have only certain allowed values) to:

$$\mathscr{E}_k(n) = \frac{me^4}{32\pi^2\epsilon_0^2 n^2\hbar^2} \,. \tag{2.6}$$

The radius of the nth Bohr orbit can be derived as

$$r_n = \frac{4\pi\epsilon_0 n^2\hbar^2}{me^2} \,. \tag{2.7}$$

The allowed values of momentum are

$$p_n = mv = \frac{n\hbar}{r_n} \cdot \tag{2.8}$$

A comparison of Eq. (2.8) and (2.6) shows that the classical relation between $\mathscr{E}_k$ and $p(\mathscr{E}_k = p^2/2m)$ still applies for the Bohr model of the hydrogen atom. From these results, the kinetic energy versus momentum for an electron bound to a hydrogen atom is plotted in Figure 2.3B. From Figure 2.3B it can be seen that the energy-momentum plot of the bound electron consists of discrete points rather than the continuous curve obtained for the free particle shown in Figure 2.3A. This means that the kinetic energy can only have certain allowed energies (shown on the right-hand side of Figure 2.3B and all other energies are forbidden). Thus, when a bound electron moves in the presence of an atom, the motion becomes more complex than that of a free particle.

When many atoms join together in a crystal lattice, the problem becomes even more complex, and quantum mechanics must be used to obtain the solution. One simple observation is that when electrons move in the presence of a periodic potential (an array of atoms in a crystal), the quantum-mechanical solutions indicate that the discrete energy levels in Figure 2.3B broaden into bands, and hence there are bands of allowable energy separated by bands of forbidden energy. The quantum-mechanical solution of the problem is accomplished by considering the wavelike behavior of electrons. deBroglie postulated that all particles have a characteristic wavelength given by $\lambda = h/p$ where h = Planck's constant and $p = mv$ = momentum. In terms of Bohr's postulate, this implies that there must be an integral number of deBroglie wavelengths around the circumference of the electron's classical orbit. The Schrödinger wave equation is the basic differential equation of quantum mechanics, and the motion of electrons in a crystal can be solved for in terms of the periodic potential of the atoms when appropriate boundary conditions are supplied. One of the parameters of motion is called the crystal momentum, k. k is essentially a propagation constant of the electron wave packet and is given by $k = 2\pi/\lambda$, which is analogous to the transmission line propagation constant. Using the definition of the deBroglie wavelength, we can write $p = \hbar k$; hence there is a relation between crystal momentum and classical momentum (this is a crude attempt to put quantum-mechanical solutions into the classical framework). The energy band structures for Ge, GaAs, and Si are shown in Figure 2.4. These band structures are obtained by a combination of experimental and theoretical techniques [13]. From Figure 2.4 it can be seen that the energy versus momentum (crystal momentum) plots are quite different from those obtained for a free particle or a single atom (Figure 2.3). The lines on the $\mathscr{E}$ versus k diagrams in Figure 2.4 represent borders of forbidden and allowed regions rather than single-valued functions of $\mathscr{E}$ versus k as in Figure 2.3. The right-hand side of Figure 2.4 attempts to illustrate this by showing a portion of the conduction band, C.B., and valence band, V.B., for the crystal. The C.B. extends upward much further than shown; similarly the valence band extends below the levels shown. However,

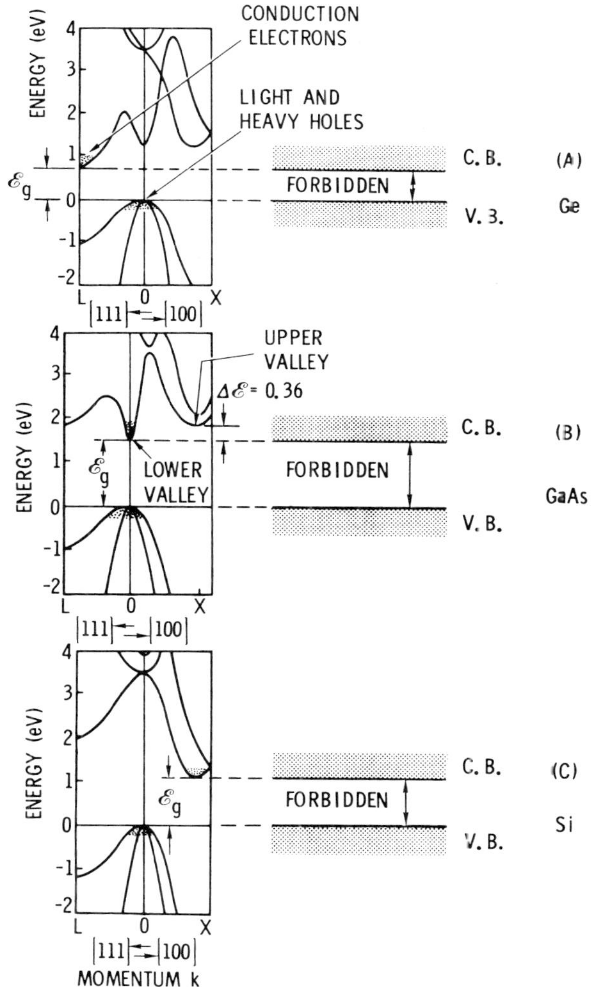

FIGURE 2.4
Energy band structures of Ge, GaAs, and Si (after Cohen and Bergstresser [13]).

it will be shown later that we are interested only in a relatively narrow strip of energies near the bottom of the conduction band and near the top of the valence band. There are several more allowed energy bands above and below those shown on Figure 2.4. The conduction mechanism in semiconductors depends only on the uppermost nearly filled energy band (called the valence band) and the next (almost empty) energy band above it (called the conduction band). At a finite temperature it is known that thermal agitation can impart sufficient energy to some of the electrons in the valence band to allow them to jump across the forbidden region to the bottom of the conduction band. The motion of electrons near the bottom of the conduction band is one of the two important conduction mechanisms in semiconductors.

To put the motion of electron wave packets into a classical framework, refer to the $\mathscr{E}$ versus k diagrams of Figure 2.4 and to Eq. (2.3). Define the effective mass of the electrons near the bottom of the conduction band as

$$m^* = \frac{1}{d^2\mathscr{E}/dp^2} = \frac{\hbar^2}{d^2\mathscr{E}/dk^2}. \tag{2.9}$$

From Figure 2.4 it can be seen that the $\mathscr{E}$-k relation depends on direction; hence, m^* will be a tensor quantity. For example, the conduction band in silicon really has identical minima in the (100), (010), (001), ($\bar{1}$00), (0$\bar{1}$0), and (00$\bar{1}$) directions, for a total of six minima. The constant energy surfaces of the minimum take the form of ellipsoids of revolution in k space [2]. The result is that the effective mass of an electron in the lower part of the C.B. in silicon and germanium can be expressed as [1]

$$m_{de}^* = (m_\perp^{*2} m_\parallel^*)^{1/3} \tag{2.10}$$

and

$$m_{ce}^* = 3\left(\frac{2}{m_\perp^*} + \frac{1}{m_\parallel^*}\right)^{-1} \tag{2.11}$$

where m_{de}^* is the effective electron mass to be used for density-of-states calculations, m_{ce}^* is the effective electron mass to be used for conductivity calculations, $m_\parallel^*$ is the effective mass in Eq. (2.9) calculated along the (111) direction for Ge and the (100) direction for silicon, and $m_\perp^*$ is the effective mass calculated from Eq. (2.9) perpendicular to these directions. For GaAs the conduction band has only one minimum which occurs at $k = 0$ (see Figure 2.4); hence, $m_{de}^* = m_{ce}^* = m_e^*$ which can be simply calculated from Eq. (2.9). (The effect of the upper valley is only important for high applied fields and will be elaborated on in Chapter 9 on Gunn devices.)

So far we have discussed only the motion of the electrons in the bottom of the conduction band. When the electrons were excited from the valence band, they left behind a vacancy in the valence band. Therefore, we now have a nearly filled band to consider. This vacancy can move, in that another electron in the V.B. can fill it, thus creating a different vacancy. The problem of electron motion in a nearly filled band is extremely complex since the Pauli exclusion principle states that no two electrons can occupy the same state, so we must consider the coordinated motion of all of the electrons in the nearly filled valence band. (This is unlike the problem of a few electrons in a nearly empty conduction band.) The easiest solution to the problem is to introduce the concept of the *hole*. The hole is a carrier which has positive effective mass and positive charge. There are only a few holes in a nearly filled valence band. Thus, the problem of the motion of electrons in the nearly filled valence band reduces to that of the motion of holes in any essentially empty (of holes) valence band. The effective mass of the holes may be found from Eq. (2.9) (use $\mathscr{E} = -\mathscr{E}$ for holes since their opposite charge causes increased hole energy to be downward in Figure 2.4). As a result of the double minima that occur at the top of the V.B. for Si, Ge, and GaAs (at $k = 0$ on Figure 2.4), there are really two masses of holes present. These are termed heavy (m_{hh}^*) and light (m_{lh}^*) holes [depending on the value of Eq. (2.9)]. It can be shown [1] that this leads to the following effective masses for holes:

$$m_{dh}^* = (m_{lh}^{*3/2} + m_{hh}^{*3/2})^{2/3} \tag{2.12}$$

$$m_{ch}^* \simeq \frac{m_{lh}^{*5/2} + m_{hh}^{*5/2}}{m_{lh}^{*3/2} + m_{hh}^{*3/2}} \tag{2.13†}$$

where m_{dh}^* = density-of-states effective hole mass and m_{ce}^* = conductivity effective hole mass. Smith [1] states that in Si and Ge the contribution of the light holes is usually negligible except in cyclotron resonance experiments. The values of $m_\perp^*$, $m_\parallel^*$, m_{lh}^*, and m_{hh}^* for Si, Ge, and GaAs may be found in Appendix A. For the remainder of the book we will be concerned primarily with the conductivity effective masses for holes and electrons; these will be denoted as $m_h^* = m_{ch}^*$ and $m_e^* = m_{ce}^*$, respectively. The values of m_h^* and m_e^* for Si, Ge, and GaAs are also given in Appendix A. Another simplification is that we will usually discuss energy bands, such as those shown on the right side of Figure 2.4, rather than energy-momentum diagrams, such as those shown on the left side of Figure 2.4.

Later we will also mention direct and indirect semiconductor materials. A direct semiconductor is one which has the conduction band minimum located at the same k (momentum) value as the valence band maxima. In direct semiconductors, an electron may make a transition from the conduction band to the valence band radiating a photon of energy $h\nu = \mathscr{E}_g$ (to conserve energy).

† This expression is a result of the fact that the ratio of the numbers of light to heavy holes is equal to $(m_{lh}^*/m_{hh}^*)^{3/2}$ [5] and is only a crude approximation since the constant energy surfaces are not spherical [10].

Figure 2.4 shows that GaAs is a direct semiconductor. In an indirect semiconductor such a transition cannot occur without the additional introduction of a phonon to conserve the momentum. Figure 2.4 shows that Si and Ge are indirect semiconductors.

2.4 Intrinsic Carrier Densities

The previous sections showed that electrons and holes can move in certain allowed energy bands (the C.B. and V.B.) in a crystal. Two important problems that must be solved are the values of the density of electrons in the conduction band and the density of holes in the valence band. To solve these problems we need to answer two basic questions: (1) what is the energy distribution function for the electrons? and (2) what is the distribution of available states for the electrons to occupy?

The answer to the first question involves the application of statistical mechanics to a system where the Pauli exclusion principle holds (i.e., an energy level can be occupied by two electrons at most). The resulting distribution function is called the Fermi-Dirac distribution function (see [6]) and is written as follows:

$$f(\mathscr{E}) = \frac{1}{1 + \exp[(\mathscr{E} - \mathscr{E}_f)/kT]} \tag{2.14}$$

where $\mathscr{E}$ = energy of interest and $\mathscr{E}_f$ = the energy at which $f(\mathscr{E}) = \frac{1}{2}$ ($\mathscr{E}_f$ is called the Fermi energy). $f(\mathscr{E})$ is the probability that a state at the energy $\mathscr{E}$ will be occupied by an electron. Figure 2.5C shows a plot of $f(\mathscr{E})$ versus $\mathscr{E}$ at a finite temperature. The fact that $f(\mathscr{E}) = 0.5$ at $\mathscr{E} = \mathscr{E}_f$ can be seen clearly on Figure 2.5C.

The answer to the second question, the distribution of available states, is well-known (see [2]). For the bottom of the conduction band the result is

$$N_c(\mathscr{E}) = M\frac{\sqrt{2}(m_{de}^*)^{3/2}}{\pi^2\hbar^3}(\mathscr{E} - \mathscr{E}_c)^{1/2} \tag{2.15}$$

where M = number of conduction band minima (6 for Si), and $\mathscr{E}_c$ = lower limit of conduction band, see Figure 2.5A. The density of available hole states near the top of the valence band is given by

$$N_v(\mathscr{E}) = \frac{\sqrt{2}(m_{dh}^*)^{3/2}}{\pi^2\hbar^3}(\mathscr{E}_v - \mathscr{E})^{1/2} \tag{2.16}$$

where $\mathscr{E}_v$ = upper limit of valence band (see Figure 2.5A). A plot of $N_c(\mathscr{E})$ and $N_v(\mathscr{E})$ versus $\mathscr{E}$ is shown in Figure 2.5B. The density of electrons at the energy $\mathscr{E}$ is $N_c(\mathscr{E})\cdot f(\mathscr{E})$ for the conduction band and $N_v(\mathscr{E})\cdot[1 - f(\mathscr{E})]$ for the valence band. Since a hole is the absence of an electron, the probability of a state not being occupied is $1 - f(\mathscr{E})$. The resulting distribution of electrons and holes is shown in Figure 2.5D.

$$n_0 = \int_{E_c}^{\infty} f(\mathscr{E}) N_c(\mathscr{E})\, d\mathscr{E} \,. \tag{2.17}$$

(The zero subscript implies thermal equilibrium values.) The upper limit of the integral is not important due to the nature of $f(\mathscr{E})$. If the Fermi level is more than $3kT$ below the conduction band edge (which is almost always the

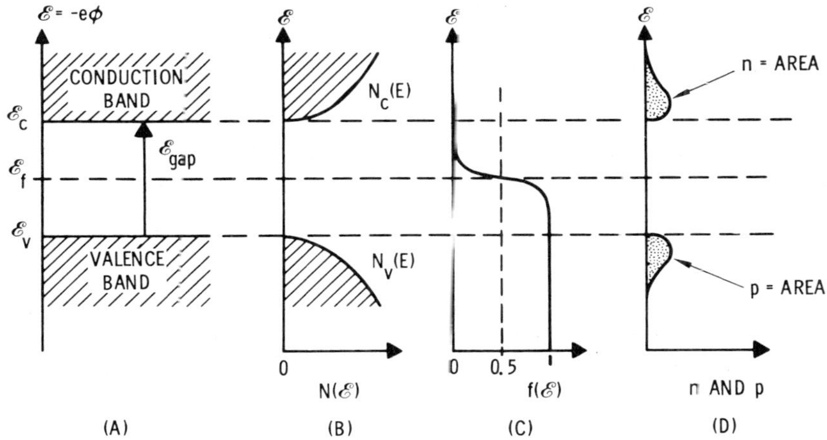

FIGURE 2.5
Intrinsic semiconductor: (A) band structure, (B) density-of-states function, (C) Fermi-Dirac function, and (D) density of holes and electrons.

case for intrinsic semiconductors), then Eq. (2.17) may be simply integrated to obtain

$$n_o = N_c \exp\left(\frac{\mathscr{E}_f - \mathscr{E}_c}{kT}\right) \tag{2.18}$$

where

$$N_c = 2\left(\frac{2\pi m_{de}^* kT}{h^2}\right)^{3/2} \cdot M \,. \tag{2.19}$$

Similarly, for holes,

$$p_0 = \int_{-\infty}^{\mathscr{E}_v} N_v(\mathscr{E})(1 - f(\mathscr{E}))\, d\mathscr{E} \tag{2.20}$$

and again, if $\mathscr{E}_f - \mathscr{E}_v > 3kT$, Eq. (2.20) may be integrated to

$$p_0 = N_v \exp\left(\frac{\mathscr{E}_v - \mathscr{E}_f}{kT}\right) \tag{2.21}$$

where

$$N_v = 2\left(\frac{2\pi m_{dh}^* kT}{h^2}\right)^{3/2}. \tag{2.22}$$

Since we are dealing with an intrinsic semiconductor, the only carriers present are due to the electrons which were excited across the forbidden gap into the conduction band leaving a hole behind in the valence band. Therefore, $n_0 = p_0$. The product

$$p_0 n_0 = n_i^2 \tag{2.23}$$

is frequently discussed in semiconductor physics where n_i is called the intrinsic density. We may calculate n_i from Eq. (2.18) and (2.21), i.e.,

$$n_i^2 = p_0 n_0 = N_c N_v \exp\left(\frac{\mathscr{E}_v - \mathscr{E}_c}{kT}\right) = N_c N_v \exp\left(\frac{-\mathscr{E}_g}{kT}\right). \tag{2.24}$$

For example, in silicon $n_i \cong 1.4 \times 10^{10}$ per cm^3 at 300°K. n_i is a constant for a given temperature and semiconductor. We may also use Eq. (2.18) and (2.21) to calculate the position of the Fermi level in the intrinsic case (using the fact $n_0/p_0 = 1$). The result is

$$\mathscr{E}_f = \frac{\mathscr{E}_c + \mathscr{E}_v}{2} + \frac{kT}{2} \ln \frac{N_c}{N_v}. \tag{2.25}$$

Since $N_c \cong N_v$, the Fermi level for the intrinsic case is located essentially at the center of the forbidden gap.

2.5 Extrinsic Carrier Densities

In the practical case, we would usually like higher carrier densities than can be obtained from intrinsic generation, and we would like $n \gg p$ or $p \gg n$ rather than the $n = p$ obtained in the intrinsic case. This can be accomplished by "doping" the semiconductor crystal. Doping simply means that atoms of a different type are substituted for some of the host atoms in the crystal. Such a doped crystal is called extrinsic. For example, if a group V atom such as phosphorous is substituted for one of the silicon atoms shown in Figure 2.2, the phosphorous atom has five electrons in its outer shell, and, after sharing four of these with its silicon neighbors, the fifth electron has no covalent bond and is rather easily broken free from the phosphorous atom to wander freely through the crystal. The density of impurity atoms ranges typically from one impurity atom for every 10^5 to 10^{10} silicon atoms. Since there are approximately 10^{22} silicon atoms per cm^3, there are about 10^{12} to 10^{17} impurity atoms per cm^3. This means that the density of the electrons donated by the phosphorous atoms could vary from 10^{12} to 10^{17} per cm^3, which is considerably higher than the 1.4×10^{10} per cm^3 intrinsic density discussed in the previous section. The fact that group V atoms (P, As, and Sb) add extra electrons when substituted into the lattice has resulted in their being termed donor atoms. The density of donor (or group V impurity) atoms is usually called N_D, per cm^3. The energy levels of the donor impurities lie in the forbidden gap, usually very close to the edge of the conduction band, as is depicted in Figure 2.6B. Hence, it takes very little energy to excite

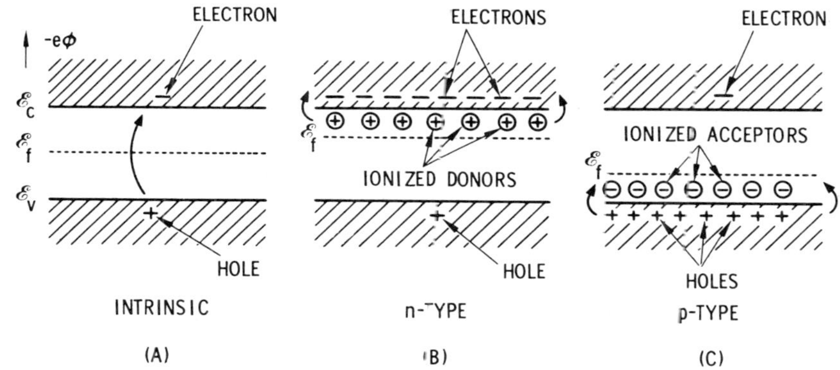

FIGURE 2.6
(A) Intrinsic semiconductor. (B) *n*-type semiconductor. (C) *p*-type semiconductor.

an electron from a donor atom into the conduction band, and this leaves behind a positively ionized, immobile donor atom. It is usually assumed that at room temperature all of the donors are ionized; hence we have N_D extra electrons in the conduction band in addition to those present due to intrinsic excitation. Semiconductor material which is doped with donor atoms is usually called *n*-type. Similarly a group III (B, Al, Ga) atom substituted for a silicon atom (see Figure 2.2) is called an acceptor impurity. The density of acceptor impurities is called N_A, per cm^3. The group III atoms have only three electrons in their outer shell to form covalent bonds with the neighboring atoms, and one more electron is needed to complete the bonds. The energy level of the group III atoms lies very close to the valence band edge; consequently, it is very easy for the acceptor to capture an electron from the valence band and leave behind a hole near the top of the valence band. This also creates a negatively ionized immobile acceptor atom (see Figure 2.6C).

The acceptor level is usually close enough to the valence band so that all of the acceptor atoms are ionized at room temperature; thus the density of holes is N_A plus the intrinsically excited holes. Material which is doped with acceptor atoms is usually called *p*-type. The doping of germanium is accomplished in the same way as described for silicon. Gallium-arsenide, on the other hand, is doped somewhat differently. One can still make *n*-type or *p*-type material, but it can be accomplished by substituting group II, IV, or VI for the host atoms, see [7].

In *n*-type material (doped with donors), there are many more electrons than holes, and therefore, the electrons are called the majority carriers and the holes are called the minority carriers. In *p*-type material (doped with acceptors), there are many more holes than electrons, and, therefore, the holes are the majority carriers and the electrons are the minority carriers. One convenient notation that is often used is a subscript *p* or *n* to indicate *p*-type or *n*-type material. For example, p_n = density of holes in an *n*-type region. The density of electrons and holes in intrinsic (pure) or extrinsic (doped) semiconductors are still given by Eq. (2.18) and (2.21). The difference will be that Fermi energy level will change with doping. The crystal must be electrically neutral; hence,

$$n_0 + N_A^- = p_0 + N_D^+$$

where n_0, p_0 are the thermal equilibrium electron and hole densities, respectively, and N_A^- and N_D^+ are the ionized acceptor and donor densities. For most cases either

$$\left.\begin{aligned} N_D &\gg N_A \\ N_D &\gg n_i \end{aligned}\right\} \textit{n-type} \tag{2.26A}$$

or

$$\left.\begin{aligned} N_A &\gg N_D \\ N_A &\gg n_i \end{aligned}\right\} p\text{-type} . \tag{2.26B}$$

For the conditions in Eq. (2.26A), and from Eq. (2.18)

$$n_{n0} \cong N_D = N_c \exp\left(\frac{\mathscr{E}_f - \mathscr{E}_c}{kT}\right) \tag{2.27A}$$

or

$$\mathscr{E}_c - \mathscr{E}_f = kT \ln \frac{N_c}{N_D} \cdot \tag{2.27B}$$

Equation (2.27B) shows that as N_D gets larger and approaches N_c ($N_c \cong 10^{19}$ per cm^3 in Si) $\mathscr{E}_f$ approaches $\mathscr{E}_c$, this is also shown in Figure 2.6B. When $N_D = N_c$, $\ln (N_c/N_D) = 0$ and $\mathscr{E}_f = \mathscr{E}_c$. For the *p*-type conditions given in

Eq. (2.26B) it can also be shown that $\mathscr{E}_f$ approaches $\mathscr{E}_v$ as $N_A \to N_v$.† The value of the $p \cdot n$ product in doped semiconductors can be found through Eq. (2.27B) and (2.21) (this will be for the n-type case). In particular,

$$p_{n0} = N_v \exp\left(\frac{\mathscr{E}_v - \mathscr{E}_f}{kT}\right) \tag{2.21}$$

where

$$\mathscr{E}_v - \mathscr{E}_f = \mathscr{E}_c - \mathscr{E}_f - \mathscr{E}_g \tag{2.28}$$

where $\mathscr{E}_g$ = energy gap (see Figure 2.5A). Substituting from Eq. (2.27B), Eq. (2.21) can be reduced to

$$p_{n0} = N_v \exp\left[\frac{kT \ln (N_c/N_D) - \mathscr{E}_g}{kT}\right]$$

or

$$p_{n0} = \frac{N_v N_c}{N_D} \exp\left(\frac{-\mathscr{E}_g}{kT}\right) \cdot \tag{2.29}$$

A comparison of this equation with Eq. (2.24) shows that for the n-type case (Eq. 2.26A) the hole density is given by

$$p_{n0} = \frac{n_i^2}{N_D} \tag{2.30}$$

or since $n_{n0} \cong N_D$ we find that $p_{n0} \cdot n_{n0} = n_i^2$ still holds in the extrinsic case. Similar arguments show that for the p-type case (Eq. 2.26B), $p_{n0} \cong N_A$ and

$$n_{n0} = \frac{n_i^2}{N_A} \cdot \tag{2.31}$$

It was stated earlier that n_i is a constant for a given semiconductor and temperature. The fact that the hole density decreases with increasing donor doping (Eq. 2.30) and that electron density decreases with increasing acceptor density (Eq. 2.31) is referred to as minority carrier suppression. Equations (2.30) and (2.31), and the fact that $p_0 n_0 = n_i^2$, make it very simple to calculate the carrier densities in a doped semiconductor. As an example, take silicon ($n_i \cong 1.4 \times 10^{10} \cdot \text{cm}^{-3}$ at 300°K) and assume 10^{14} phosphorous atoms$\cdot$cm^{-3} are substituted for silicon lattice atoms. Phosphorous is a donor and, since $n_{n0} \cong N_D$, there will be 10^{14} electrons$\cdot$cm^{-3}. From Eq. (2.30) we know $p_{n0} = (1.4 \times 10^{10})^2/10^{14}$ or $p_{n0} \cong 2 \times 10^6$ holes$\cdot$cm^{-3}.

† If the Fermi level is closer than $2kT$ to $\mathscr{E}_c$ or $\mathscr{E}_v$, the material is said to be degenerately doped.

2.6 Mobility, Conductivity, and Velocity Saturation

The preceding section discussed the concentration of electrons and holes that are free to move in the semiconductor crystal. This section examines the behavior of these charge carriers under the influence of an applied electric field. First, consider the behavior of electrons (holes will behave in similar fashion) in a lattice at room temperature with no applied field. The electrons move through the crystal suffering random collisions with vibrating lattice atoms.† The process keeps the electrons in thermal equilibrium with the lattice; consequently, the electrons' average thermal velocity can be estimated from

$$v = \sqrt{\frac{3kT}{m}} \cong 10^7 \text{ cm/sec} \tag{2.32}$$

at 300°K. The mean free path between collisions in silicon at room temperature is approximately 10^{-6} cm. A microscopic picture of the motion without an applied electric field is depicted in Figure 2.7A. If a small‡ electric field, $\bar{E}$, is applied to the semiconductor, it will apply a force to the electrons, $\bar{F} = -e\bar{E}$, and a net motion such as that shown in Figure 2.7B will occur. The average velocity attained between collisions will be $\frac{1}{2}$ the maximum velocity. Hence,

$$\bar{v}_e = \tfrac{1}{2}\bar{a}\tau = \frac{-e\bar{E}}{2m_{ce}^*}\tau = -\mu_n\bar{E} \tag{2.33}$$

where $\bar{a}$ = acceleration $= \bar{F}/m_{ce}^*$, τ = average time between collisions, and the minus sign is a result of the fact that the electron drifts in the opposite direction of the field. The constant μ is referred to as the mobility and is given by

$$\mu_n = \frac{e\tau}{2m_{ce}^*} \text{ cm}^2 \text{ volt}^{-1} \text{ sec}^{-1} . \tag{2.34}$$

A similar argument may be applied to the motion of holes in the presence of an electric field with the result being

$$\bar{v}_p = \mu_p\bar{E} \tag{2.35}$$

where

$$\mu_p = \frac{e\tau}{2m_{ch}^*} \text{ cm}^2 \text{ volt}^{-1} \text{ sec}^{-1} . \tag{2.36}$$

† The thermal energy of a crystal is primarily in the form of small amplitude vibrations of lattice atoms around their equilibrium positions.

‡ Small enough so that drift velocity is much lower than the thermal velocity.

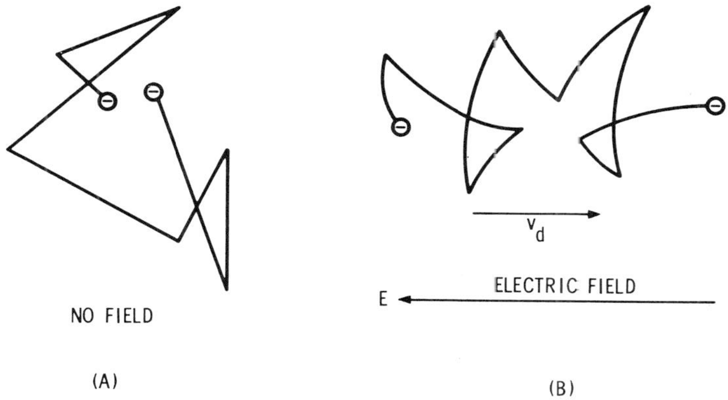

FIGURE 2.7
Microscopic motion of electrons in a crystal (A) without and (B) with an applied electric field.

Generally, rather than discuss the microscopic behavior shown in Figure 2.7, we will consider the macroscopic motion of electrons and holes in terms of their average velocities given by Eq. (2.33) and (2.35). Equations (2.34) and (2.36) are the result of a simple intuitive argument. A more rigorous analysis [2] yields similar results. The low field mobilities expressed by Eq. (2.34) and (2.36) are approximately correct for lightly doped semiconductors. For more heavily doped semiconductors, the electrons and holes also suffer collisions with the impurity atoms; this reduces the time between collisions, and hence reduces the mobilities. Figure 2.8 shows this effect quite clearly for Si, Ge, and GaAs. From Figure 2.8 it can be seen that low field mobilities are not affected until the impurity concentration reaches approximately 10^{15} cm^{-3}. Chapter 5 will discuss similar behavior caused by the effects of radiation induced defects on mobility. Mobility is also dependent on temperature, and it is found to decrease with increasing temperature (see [4]) because the ampitude of the lattice vibrations (which are responsible for the collisions) increases with temperature.

The net drift of holes and electrons that occurs in the presence of a small applied electric field causes a particle current density to flow, which is given by

$$\bar{j} = -n_0\mu_n\bar{E} + p_0\mu_p\bar{E}\,. \tag{2.37}$$

The electrons and holes drift in opposite directions. The electric current density that flows as a result of this particle current density is

$$\bar{J} = (n_0e\mu_n + p_0e\mu_p)\bar{E}\ \text{amps/cm}^2\,, \tag{2.38}$$

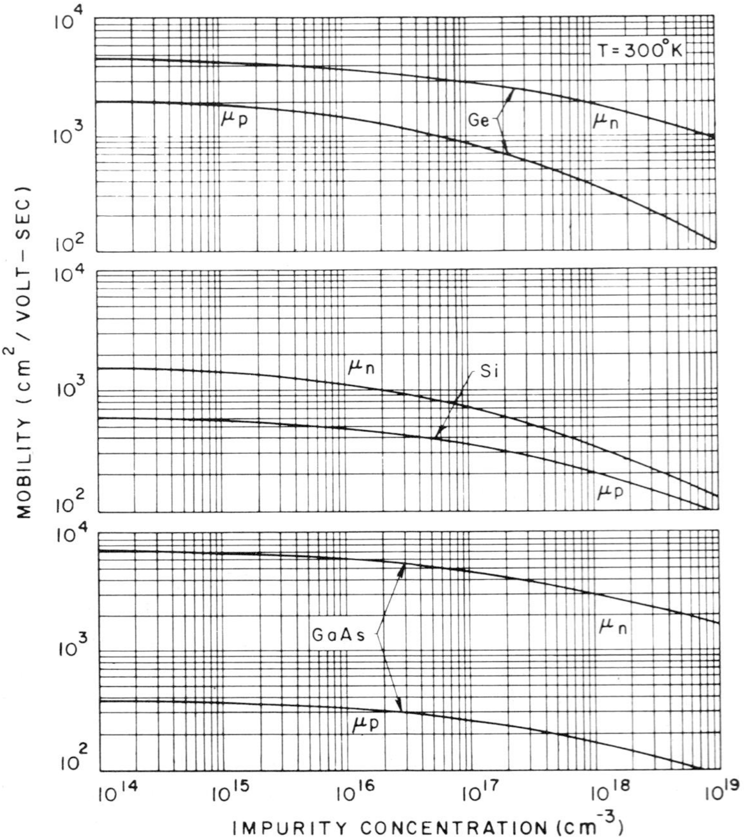

FIGURE 2.8
Drift mobility in Ge and Si and Hall mobility of GaAs at 300°K (after Sze and Irvin [8]).

because of the opposite signs of the electron and hole charges their currents add. The point form of Ohm's law is $\bar{J} = \sigma\bar{E}$, hence an inspection of Eq. (2.38) shows that σ, the conductivity of the semiconductor, is given by

$$\sigma = (n_0 e\mu_n + p_0 e\mu_p) \text{ ohm}^{-1}\text{ cm}^{-1} . \tag{2.39}$$

Usually $p_0 \gg n_0$ or $n_0 \gg p_0$, so that there is only one type of carrier to consider in Eq. (2.39). From Eq. (2.39) it can be seen that bulk conductivity, σ, or resistivity, $\rho = 1/\sigma$, depends on both the majority carrier density (which is approximately equal to doping density) and on the mobility (which decreases slowly with doping density); therefore, the conductivity (or resistivity) is not a simple function of doping concentration. Sze and Irvin [8] have published a very useful graph giving the resistivity of both *n*- and *p*-type Si, Ge, and GaAs as a function of doping density at 300°K. This graph is reproduced in Figure 2.9. In the semiconductor industry, the terms *doping density* and *resistivity* are used interchangeably; thus a graph of the type shown in Figure 2.9 is extremely useful.

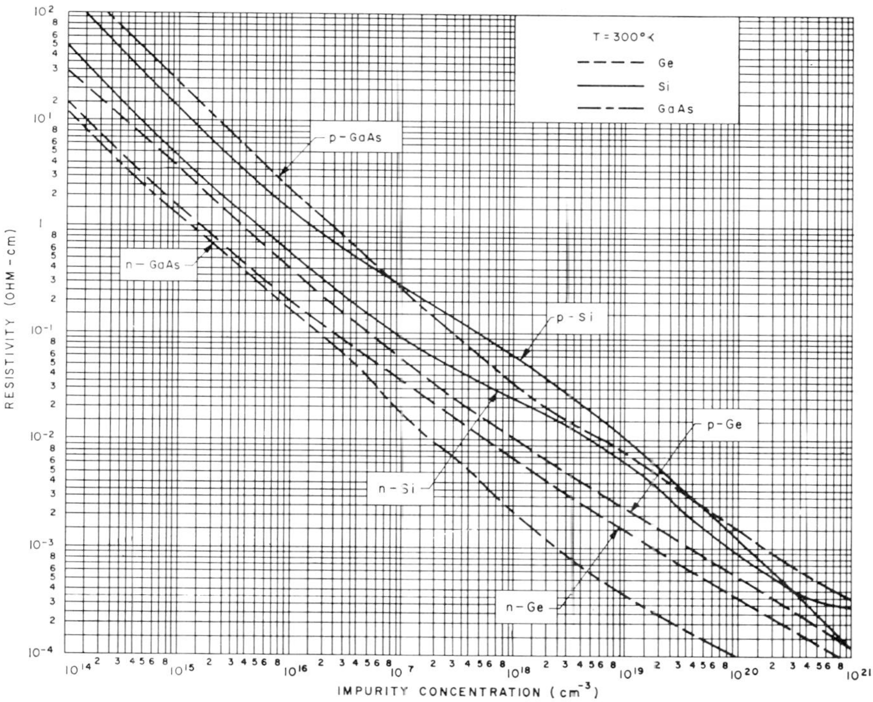

FIGURE 2.9
Resistivity versus impurity concentration for Si, Ge, and GaAs at 300°K (after Sze and Irvin [8]).

The linear relation between carrier velocity and field [Eq. (2.33) and (2.35)] only holds for low values of electric field (up to a few kV per cm). In many microwave semiconductor devices, the field does not go above this value; consequently, the carrier velocities are adequately described by Eq. (2.33) and (2.35). However, several microwave semiconductor devices utilize very high fields; for example, fields of the order of 300 kV/cm are found in IMPATT devices. For these high fields, one can no longer treat μ_p or μ_n as a constant. When the electric field is increased, the carrier velocity eventually reaches an upper limit called the scatter-limited, or saturated, drift velocity. The value of the saturated drift velocity is of the order of the thermal velocity (10^7 cm sec^{-1}). The reason for the saturation of the velocity can be explained in thermal equilibrium terms. At low fields the carriers are in equilibrium with the lattice, and, on the average, they pick up as much energy per collision as they lose. As the carrier drift velocity approaches the thermal velocity, the carriers, on the average, give more energy to the lattice than they receive per collision, and the velocity saturates. A more elegant explanation is given in Ref. [9]. Experimental measurements of carrier velocity versus electric field have been made for most semiconductor materials. Figure 2.10 shows the

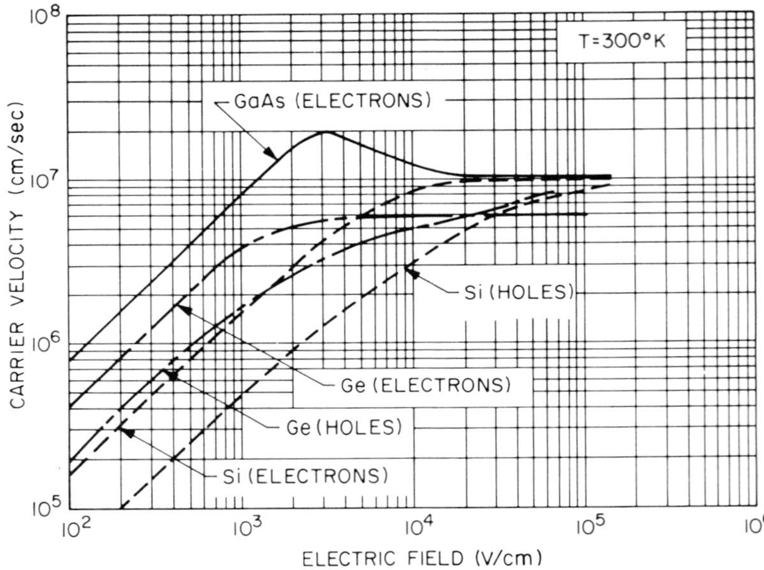

FIGURE 2.10
Velocity-field characteristics for Si, Ge, and GaAs (after Sze [4]).

results for Si, Ge, and GaAs. From Figure 2.10 it can be seen that at low fields the relation between the drift velocity and the electric field is linear, as predicted by Eq. (2.33) and (2.35). However, at higher electric field strengths the velocity approaches a saturated value of approximately 10^7 cm sec^{-1}. The peculiar behavior of electrons in GaAs near 3 kV cm^{-1} (i.e., a decrease in velocity with an increase in field) will be explained in Chapter 9 on Gunn devices.

If the electric field were increased further, a point eventually would be reached where the carrier concentration would increase rapidly due to avalanche multiplication. This phenomenon will be discussed in Chapter 3 on *p-n* junctions.

2.7 Diffusion and Recombination

The preceding sections of this chapter have been concerned with thermal-equilibrium carrier concentrations p_0 and n_0. (The zero subscript implies thermal equilibrium.) Quite often, however, we are interested in non-thermal-equilibrium conditions, which can occur, for example, when short wavelength light is used to increase the local hole-electron pair generation rate, or when excess holes or electrons are injected from a contact. There are mechanisms that exist which try to return the system to thermal equilibrium. The first of these that will be discussed is diffusion. It was shown earlier that electrons (and holes) are constantly bouncing around inside the lattice, suffering one collision approximately every 10^{-13} seconds (see Figure 2.7A). These collisions will cause a localized concentration of charge to spread out (diffuse). Consider a region in a semiconductor that has an electron concentration n (lack of the zero subscript means that n is not necessarily a thermal-equilibrium value), and assume that the concentration of electrons a distance Δx away is $n\,(x + \Delta x)$; then, according to Fick's law [10], there will be a net flow of electrons from the higher-concentration region to the lower-concentration region. This flow is a particle current given by

$$\bar{j}_n = -D_n \frac{n(x) - n(x + \Delta x)}{\Delta x} \bar{a}_x = -D_n \frac{dn(x)}{dx} \bar{a}_x \qquad (2.40)$$

$$\Delta x \to 0$$

where $\bar{a}_x$ is the unit vector in the x direction and D_n is a constant of proportionality called the *diffusion coefficient*, given in units of cm^2 sec^{-1}. Similar reasoning may be applied for hole diffusion with the resulting particle current being given as

$$\bar{j}_p = -D_p \frac{dp(x)}{dx} \bar{a}_x \,. \qquad (2.41)$$

In the three-dimensional case, the d/dx term is replaced by the gradient operator, i.e., $\bar{j}_n = -D_n \cdot \nabla n(x)$. The particle currents give rise to electric currents, which may be expressed as

$$\bar{J}_n = +eD_n \frac{dn}{dx} \bar{a}_x \qquad (2.42\text{A})$$

$$\bar{J}_p = -eD_p \frac{dp}{dx} \bar{a}_x \,. \qquad (2.42\text{B})$$

The total conduction current (due to both drift and diffusion) may be written with the aid of Eq. (2.38) and (2.42A, B). The result is

$$\bar{J}_n = ne\mu_n\bar{E} + eD_n\frac{dn}{dx}\,\bar{a}_x \tag{2.43A}$$

$$\bar{J}_p = pe\mu_p\bar{E} - eD_p\frac{dn}{dx}\,\bar{a}_x\,. \tag{2.43B}$$

In an inhomogeneously doped n-type semiconductor at thermal equilibrium, $n_{n0}(x) = N_D(x)$, see Figure 2.11. If no external fields are applied, then $J_n = J_p = 0$. In this case there will still be a built-in electric field, E, due to

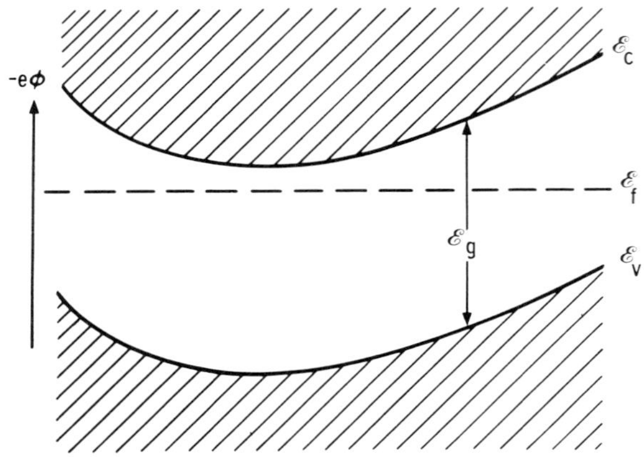

FIGURE 2.11
Energy band diagram of an inhomogeneously doped semiconductor.

the gradient of the potential in the device. This is necessary to maintain thermal equilibrium. For the inhomogeneously doped case,

$$n_{n0}(x) = N_D(x) = N_c \exp\left(\frac{\mathscr{E}_f - \mathscr{E}_c(x)}{kT}\right)\,. \tag{2.44}$$

The variation in density implies that the energy level of the conduction band edge and valence band edge must also be a function of x since at thermal equilibrium and for no applied bias $\mathscr{E}_f$ must be constant. (See Figure 2.11). The energy level $\mathscr{E}_c(x)$ is equal to $-e\phi(x)$ where $\phi(x)$ is the electronic potential. Eq. (2.44) may be rewritten as

$$e\phi(x) = -\mathscr{E}_c(x) = kT\ln\frac{n_{n0}(x)}{N_c} - \mathscr{E}_f\,. \tag{2.45}$$

According to the laws of electrostatics, the electric field is defined as $\bar{E} = -\nabla\phi$. Hence, the built-in electric field (at zero bias) is given by

$$\bar{E} = -\nabla\phi = \frac{-kT}{e}\frac{d}{dx}\left[\ln n_{no}(x) - \ln N_c - \frac{\mathscr{E}_f}{kT}\right]\bar{a}_x$$

or

$$\bar{E} = \frac{-kT}{e}\frac{1}{n_{no}(x)}\frac{dn_0(x)}{dx}\bar{a}_x . \tag{2.46}$$

If Eq. (2.46) is put into (2.43A) (for $J_n = 0$), then it can be shown that

$$\mu_n\left(\frac{kT}{e}\right) = D_n$$

or

$$\frac{D_n}{\mu_n} = \frac{kT}{e} . \tag{2.47}$$

It can also be shown that for holes

$$\frac{D_p}{\mu_p} = \frac{kT}{e} . \tag{2.48}$$

Equations (2.47) and (2.48) are the so-called Einstein relations and hold even in non-thermal-equilibrium cases as long as the semiconductor is nondegenerate ($\mathscr{E}_f$ is greater than $2kT$ away from either band edge). The diffusion coefficients are usually calculated from the mobility through the Einstein relations, Eq. (2.47) and (2.48). A typical value of D_n for silicon is 40 cm^2 sec^{-1}. The direct relation between D and μ [Eq. (2.47), (2.48)] should not be surprising since the mechanism that dominates both diffusion and drift is the same, i.e., collisions.

The second means of reducing nonequilibrium carrier densities is *recombination.* The problem of recombination shall be treated in some detail here since it will be important later in considering radiation effects on carrier lifetime. When one talks about recombination of carriers in a semiconductor one usually thinks about recombination of *excess* carriers. It should be pointed out, however, that recombination of carriers is occurring constantly within a semiconductor, even at thermal equilibrium. At equilibrium, however, the recombination is completely balanced by the reverse process, *carrier generation.* Therefore, recombination equations must be interpreted carefully. The reader is referred to Ref. [2], [4], and [11] for supplemental reading on this subject.

The simplest form of recombination, called *band-to-band recombination*, is shown in Figure 2.12A. In this case an electron in the conduction band recombines directly with a hole in the valence band and radiates a photon of energy of $h\nu = \mathscr{E}_g$. Conservation of crystal momentum, k, requires that $k_n + k_p = 0$. Hence, this process is likely to occur only in a direct band gap semiconductor such as GaAs and is not the dominant mechanism in Si and Ge (see Figure 2.4). In thermal equilibrium, the rate of recombination is expected to be directly proportional to the equilibrium density of electrons in the conduction band times the equilibrium density of holes in the valence band, i.e.,

$$R_{cv_0} = rn_0p_0 = R_{vc_0} \tag{2.49}$$

where r = proportionality constant. If we assume that the material is n-type

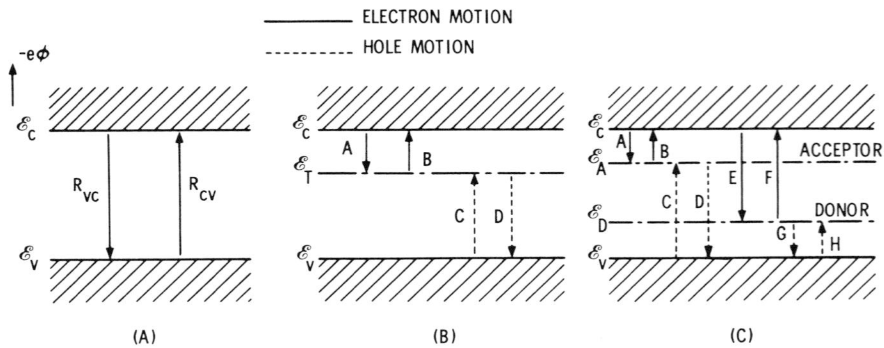

FIGURE 2.12
Carrier recombination models: (A) direct recombination, (B) recombination through a single-level trap, and (C) recombination through a two-level trap.

($n_0 \gg p_0$) and that a small excess density of carriers $\Delta n = \Delta p$ (for neutrality) is injected, then we find $R_{cv} = r(n_0 + \Delta n)(p_0 + \Delta p)$ and

$$R'_{cv} = R_{cv} - R_{cv_0} = r(n_0 + \Delta n)(p_0 + \Delta p) - rn_0p_0 \tag{2.50}$$

where R'_{cv} is net increase in carrier recombination above the thermal equilibrium value. For $n_0 \gg p_0$ Eq. (2.50) becomes

$$\frac{d\,\Delta p}{dt} = R'_{cv} \cong rn_0\,\Delta p\,. \tag{2.51}$$

The average lifetime before recombination of an excess hole in the n-type material is simply the density of excess holes divided by the recombination rate, i.e.,

$$\tau_p = \frac{\Delta p}{R'_{cv}} = \frac{1}{rn_0}\,. \tag{2.52}$$

Should the excitation generating the excess holes suddenly stop, then the excess hole density will decay exponentially as

$$\Delta p = \Delta p|_{t=0}\, e^{-t/\tau_p} . \tag{2.53}$$

It was found that the band-to-band recombination did not fit the experimentally observed lifetimes in Si and Ge. Since these are indirect semiconductors, it was felt that recombination was much more likely to occur at the sites of some types of impurity atoms or crystal defects. This is because these lattice disturbances create allowed energy levels within the forbidden energy gap which can trap and hold one type of carrier until the opposite type comes along and annihilation occurs. This process is much more likely to occur for defect energy levels near the center of the energy gap because there is a much higher probability of having both types of carriers with this energy. The SRH (Shockley-Read-Hall) model [2] for a single level recombination center is used to analyze this situation. The defect energy level can either be deep (not near either band edge) or shallow (near $\mathscr{E}_c$ or $\mathscr{E}_v$). As mentioned earlier, the deep levels are much more effective as recombination centers and thus this discussion will center on them. Several types of impurity atoms introduce at least one deep energy level into the crystal. Some of these are gold, copper, and iron. For a complete tabulation of the energy levels introduced by impurities the reader is directed to Sze [8]. Chapter 5 will show that radiation damage to semiconductors also introduces deep energy levels which serve as recombination centers and which may be treated analogously by the methods in the following outline. The deep impurity level (for the single level SRH model) can either be a donor type or an acceptor type (with behavior similar to that of donors and acceptors as discussed in Section 2.5). As an example of the SRH model, consider a deep acceptor level trap† such as is shown in Figure 2.12B. By definition, an acceptor level can have one of two charge states, either neutral or negative. The four processes that can occur are [2]

1. Electron capture on a neutral acceptor (i.e., an electron is captured and the acceptor becomes negative)
2. Electron emission from a negative acceptor (the acceptor becomes un-ionized and neutral)
3. Hole capture on a negative acceptor (hole-electron annihilation, acceptor becomes neutral)
4. Hole emission from a neutral acceptor (electron from V.B. is captured by the acceptor, acceptor becomes negative)

† These levels are often referred to as *traps* and *recombination centers* interchangeably. It is usually called a trap if the carrier is more likely to be re-emitted before it recombines and a recombination center if it is more likely to recombine before being re-emitted.

These processes are depicted in Figure 2.12B. Assume that the electrons and holes travel through the lattice at a velocity $= v_t$ (thermal) and that electrons must come within a capture cross-section, σ_n, to be captured by the trap. If there are n electrons per cm^3 and N_T^0 neutral traps per cm^3, then the rate of electron capture is (Process 1)

$$R_n = v_t \sigma_n n_0 N_T^0 \text{ cm}^{-3} \text{ sec}^{-1}. \tag{2.54}$$

Similarly the capture rate for holes (Process 3) is

$$R_p = v_t \sigma_p p_0 N_T^- \tag{2.55}$$

where N_T^- is the density of negatively charged traps. At thermal equilibrium the average free time of an electron before being captured is

$$t_{1n} = \frac{n}{R_n} = \frac{1}{v_t \sigma_n N_T^0} \tag{2.56}$$

similarly for holes

$$t_{1p} = \frac{p}{R_p} = \frac{1}{v_t \sigma_n N_T^-} \cdot \tag{2.57}$$

In thermal equilibrium the electron and hole re-emission rates from the traps (Processes 2 and 4) are equal to the capture rates. This is called *detailed balance*. In the case of non-thermal equilibrium, $p = p_0 + \Delta p$ and $n = n_0 + \Delta n$ for neutrality. Using the application of detailed balance, and the fact that $N_T = N_T^0 + N_T^-$ it can be shown [2] that the occupation probabilities of neutral and negative traps are

$$\frac{N_T^0}{N_T} = \frac{\sigma_p p + \sigma_n n_1}{\sigma_n(n + n_1) + \sigma_p(p + p_1)} \tag{2.58}$$

$$\frac{N_T^-}{N_T} = \frac{\sigma_n n + \sigma_p p_1}{\sigma_n(n + n_1) + \sigma_p(p + p_1)} \tag{2.59}$$

where n_1 and p_1 are the electron and hole densities that would be obtained if the Fermi level were at the trap level [i.e., Eq. (2.18) and (2.21) with $\mathscr{E}_f$ replaced by $\mathscr{E}_T$]. For the above case the recombination rates for the SRH model may be shown to be [2] given by

$$\frac{d\,\Delta n}{dt} = \frac{d\,\Delta p}{dt} = -\frac{np - n_i^2}{(n + n_1)\tau_{p0} + (p + p_1)\tau_{n0}} \tag{2.60}$$

where

$$\tau_{n0} = \frac{1}{N_T v_t \sigma_n} \tag{2.61A}$$

$$\tau_{p0} = \frac{1}{N_T v_n \sigma_p} \cdot \tag{2.61B}$$

As an example, consider low level injection, $p = p_0 + \Delta p$, $n = n_0 + \Delta n$, in n-type material $n_0 \gg p_0$, $\Delta n \ll n_0$. For a deep trap ($\mathscr{E}_T$ near midgap) $n_1 \ll n_0$ and Eq. (2.60) reduces to

$$\frac{d\,\Delta n}{dt} = \frac{d\,\Delta p}{dt} \cong \frac{\Delta p n_0}{n_0 \tau_{p0}} = \frac{\Delta p}{\tau_{p0}}. \tag{2.62}$$

Physically, this case corresponds to all traps filled with electrons and waiting to capture holes. τ_{p0} [Eq. (2.61B)] is just the number of times per second each hole comes within the capture cross section of a trap. Equation (2.60) is useful in calculating lifetimes of the carriers under varying injection and doping conditions. Equations (2.58) and (2.59) are important in that they allow the calculation of the density of trapped charge in the semiconductor, which will be very significant later in attempting to solve Poisson's equation for material containing traps. Equations (2.58), (2.59), and (2.60) were derived for a single-level acceptor-type deep trap. Similar equations can be derived for single-level deep donor traps as well.

The single-level SRH recombination model is not adequate to explain recombination through impurity or defect energy levels in all cases. In many cases the crystal defect or impurity atom introduces multiple levels in the forbidden energy gap. Some levels are acceptor traps and some are donor traps. Shockley and Last [12] have analyzed this situation. One important subcase of the multilevel trap, which will be important later, is the case where there are two deep-trap levels, an acceptor and a donor, within the forbidden energy gap. For example, this situation occurs when gold is present in a silicon crystal (the gold introduces an acceptor level at $\mathscr{E}_c - \mathscr{E}_a = 0.54$ eV and a donor level at $\mathscr{E}_D - \mathscr{E}_v = 0.35$ eV [4]). This situation is also similar to the radiation-induced divacancy defect in silicon to be discussed in Chapter 5. The two-level trap is shown in Figure 2.12C. The following eight basic processes can occur in the situation depicted in Figure 2.12C:

1. Electron capture on a neutral acceptor (cross section σ_n^0, acceptor becomes negative)
2. Electron emission from a negative acceptor (acceptor becomes neutral)
3. Hole capture on a negative acceptor (cross section σ_p^-, acceptor becomes neutral)
4. Hole emission from a neutral acceptor (electron from V.B. captured, acceptor becomes negative)
5. Electron capture on a positive donor (cross section σ_n^+, donor becomes neutral)
6. Electron emission from a neutral donor (donor becomes positive)
7. Hole emission from a positive donor (electron from V.B. captured, donor becomes neutral)

8. Hole capture on neutral donor (cross section σ_p^0, donor becomes positive)

Moll [2], using the S-L (Shockley-Last) model, has shown that occupation probabilities for the two-level trap, as described here and in Figure 2.12C, can be written as:

$$\eta^+ = \frac{N_T^+}{N_T} = \frac{(p\sigma_p^0 + n_{1D}\sigma_n^+)(p\sigma_n^- + n_{1A}\sigma_n^0)}{D} \tag{2.63}$$

$$\eta^0 = \frac{N_T^0}{N_T} = \frac{(n\sigma_n^+ + p_{1D}\sigma_p^0)(p\sigma_p^- + n_{1A}\sigma_n^0)}{D} \tag{2.64}$$

$$\eta^- = \frac{N_T^-}{N_T} = \frac{(n\sigma_n^+ + p_{1D}\sigma_p^0)(n\sigma_n^0 + p_{1A}\sigma_p^-)}{D} \tag{2.65}$$

where

$$D = (p\sigma_p^0 + n_{1D}\sigma_n^+)(p\sigma_p^- + n_{1A}\sigma_n^0) + (n\sigma_n^+ + p_{1D}\sigma_p^0)(p\sigma_p^- + n_{1A}\sigma_n^0) + (n\sigma_n^+ + p_{1D}\sigma_p^0)(n\sigma_0^n + p_{1A}\sigma_p^-)$$

p_{1D}, n_{1D} = hole, electron concentration if the Fermi level were at the donor-trap level, $\mathscr{E}_D$

p_{1A}, n_{1A} = hole, electron concentration if the Fermi level were at the acceptor-trap level, $\mathscr{E}_A$.

The actual energy levels, $\mathscr{E}_A$ and $\mathscr{E}_D$, may differ from the ionization energies by a few kT [2], but for simplicity one usually uses $\mathscr{E}_A$ and $\mathscr{E}_D$ directly. From detailed balance considerations the recombination rate for the S-L model may be written [2] as

$$\frac{d\,\Delta n}{dt} = \frac{d\,\Delta p}{dt} = \frac{(n_{1A}\tau_{pA} + p\tau_{nA} + p_{1D}\tau_{nD} + n\tau_{pD})(pn - n_i^2)}{A + B + C} \tag{2.66}$$

where

$$A = (p\tau_{nD} + n_{1D}\tau p_D)(p\tau_{nA} + n_{1A}\tau p_A)$$
$$B = (n\tau_{pD} + p_{1D}\tau_{nD})(p\tau_{nA} + n_{1A}\tau_{pA})$$
$$C = (n\tau_{pD} + p_{1D}\tau_{1D})(n\tau_{pA} + p_{1A}\tau_{nA})$$

and

$$\tau_{nD} = \frac{1}{N_T v_t \sigma_n^+} \qquad \tau_{pD} = \frac{1}{N_T v_t \sigma_p^0}$$

$$\tau_{nA} = \frac{1}{N_T v_t \sigma_n^0} \qquad \tau_{pA} = \frac{1}{N_T v_t \sigma_p^-}.$$

As an example, consider n-type material and low-level injection, $n_0 \gg p_0$, $n = n_0 + \Delta n$, $p = p_0 + \Delta p$, $\Delta n \ll n_0$, $n \gg n_{1D}, n_{1A}, p_{1A}, p_{1D}$. In this case, Eq. (2.65) reduces to

$$\frac{d\,\Delta p}{dt} = \frac{\Delta p}{\tau_{pA}} \,. \tag{2.67}$$

This corresponds to all of the acceptor traps being negatively charged, and recombination occurs as soon as a hole comes within the capture cross-section of an acceptor trap. Equation (2.65) is useful in calculating the carrier lifetimes under varying conditions of doping and injection. Equations (2.63), (2.64), and (2.65) allow the calculation of the trapped charge density in the semiconductor and will be important later for solving Poisson's equation for material containing traps produced by radiation damage.

2.8 Continuity Equation

It is a well-known fact that electronic charge must be conserved. This is expressed mathematically as

$$\nabla \cdot \bar{J} + \frac{\partial \rho}{\partial t} = 0 \tag{2.68}$$

where $\bar{J}$ is the total current density and ρ is the total electronic charge density. Just as in the case of recombination, it is possible to consider detailed balance, i.e., Eq. (2.68) holds for electrons and holes separately. Using Eq. (2.43A and B) the continuity equation may be written as

$$\frac{\partial n}{\partial t} = g - R + \mu_n \nabla \cdot (n\bar{E}) + D_n \nabla \cdot \nabla n \tag{2.69}$$

and

$$\frac{\partial p}{\partial t} = g - R - \mu_p \nabla \cdot (pE) + D_p \nabla \cdot \nabla p \,. \tag{2.70}$$

In Eq. (2.69) and (2.70) the g term is the generation rate of carriers (by light or ionizing radiation),† and the R term is the recombination rate of carriers which was discussed in the previous section. The last two terms on the R.H.S. of Eq. (2.69) and (2.70) are just an expansion of the $\nabla \cdot \bar{J}_n$ and $\nabla \cdot \bar{J}_p$ terms. From time to time in this book, various limiting cases of Eq. (2.69) and (2.70) will be used to solve carrier transport problems in semiconductor devices. Jonscher [11] discusses these equations in detail.

As an example of the use of Eq. (2.69) and (2.70), consider the following

† If a photon of energy $h\nu > \mathscr{E}_g$ is absorbed, it can excite an electron across the forbidden gap creating an electron-hole pair.

boundary value problem. A bar of uniform n-type semiconductor material of infinite length is illuminated by light ($h\nu > \mathscr{E}_g$) in the region ($-\infty \leq x \leq 0$), and the hole density in this region is assumed to be $p_n = p_{n0} + \Delta p$. The remainder of the bar ($0 \leq x \leq +\infty$) is assumed to be unilluminated. The problem to be solved is what is the density of holes in the bar? Figure 2.13 illustrates this problem. Assume that there are steady-state conditions and that the electric field within the sample is negligible; hence Eq. (2.70) becomes

$$\frac{\partial p}{\partial t} = 0 = g - R + D_p \frac{d^2 p_n}{dx^2} \, . \tag{2.71}$$

It was shown in the previous section that for low injection we may write $R = \Delta p/\tau_p$. Let us further assume that, in region A, p_n is independent of x. Hence, Eq. (2.71) becomes

$$\frac{\Delta p}{\tau_p} = g \quad \text{or} \quad \Delta p = g\tau_p \tag{2.72}$$

or the excess hole density is proportional to the carrier generation rate (which is in turn related to the intensity and frequency of the light) times the minority-carrier lifetime. In region B, $g = 0$ since there is no incident light. Then, in the steady-state, Eq. (2.70) becomes

$$\frac{\partial p}{\partial t} = 0 = -R + D_p \frac{d^2 p_n}{dx^2} \, . \tag{2.73}$$

If we again assume $R = \Delta p/\tau_p$, Eq. (2.73) can be written

$$\frac{d^2 p_n}{dx^2} - \frac{p_n - p_{n0}}{D_p \tau_p} = 0 \, . \tag{2.74}$$

This is a linear second-order differential equation which can easily be solved. Apply the boundary conditions that $p_n(x) = p_{n0}$ at $x = +\infty$ and $p_n(x) = p_{n0} + \Delta p$ at $x = 0$. The solution to Eq. (2.74) for $x > 0$ can then be shown to be

$$p_n(x) = \Delta p \exp\left(-\frac{x}{L_p}\right) + p_{n0} \tag{2.75}$$

where

$$L_p = \sqrt{D_p \tau_p} \text{ cm} \, , \tag{2.76}$$

and L_p is called the hole diffusion length. The interpretation of L_p is the average distance an excess minority hole will diffuse before recombining (actually the injected density is down to $1/e$ of its maximum value at $x = L_p$).

A similar technique may be used to show that the diffusion length for electrons in p-type material is

$$L_n = \sqrt{D_n \tau_n} \text{ cm} . \tag{2.77}$$

An important variation of the above problem occurs if the generation rate is not constant. Assume that the light shining on the sample (Figure 2.13) has two components, an ac one and a dc one; and, therefore, $g = g_0 + g_1 e^{j\omega t}$ (where $g_1 \ll g_0$). This implies that hole density in the region $-\infty < x \leq 0$

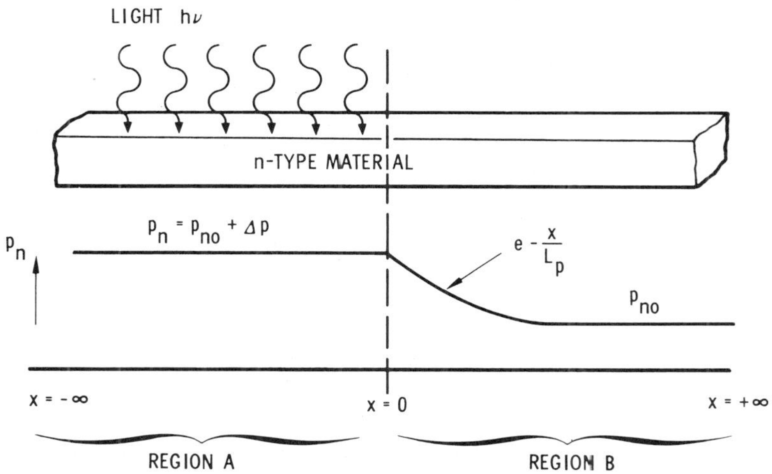

FIGURE 2.13
Recombination of excess injected holes.

(see Figure 2.13) has an ac component as well as a dc component. Therefore, at $x = 0$,

$$p_n(0) = p_{no} + \Delta p_0 + \Delta p_{ac} e^{j\omega t} \tag{2.78}$$

where $\Delta p_{ac} \ll \Delta p_0$. If this density is used as a boundary condition at $x = 0$ and the boundary condition at $x = \infty$ is $p_n = p_{no}$, Eq. (2.73) becomes

$$\frac{\partial p_n(x, t)}{\partial t} = \frac{p_n(x, t) - p_{no}}{\tau_p} + D_p \frac{\partial^2 p_n(x, t)}{\partial x^2} . \tag{2.79}$$

The solution of Eq. (2.79) is assumed to be of the form

$$p_n(x, t) - p_n(x) + \tilde{p}_n(x) e^{j\omega t} \tag{2.80}$$

where the tilde (~) indicates the ac part of the solution. If Eq. (2.80) is substituted into Eq. (2.79) and the time-dependent and time-independent components are separated out, the two resulting equations are:

$$j\omega\tilde{p}_n(x) = \frac{-\tilde{p}_n(x)}{\tau_p} + D_p \frac{d^2\tilde{p}_n(x)}{dx^2} \tag{2.81}$$

$$0 = \frac{p_n(x) - p_{n0}}{D_p\tau_p} + \frac{d^2p_n(x)}{dx^2} \,. \tag{2.82}$$

Equation (2.82) has been solved before, and the result is that the injected dc excess density falls off exponentially with a characteristic length given by Eq. (2.76). Equation (2.81), for the ac component, may be rewritten as

$$\frac{d^2\tilde{p}_n(x)}{dx^2} - \left(\frac{1 + j\omega\tau_p}{D_p\tau_p}\right)\tilde{p}_n = 0 \,. \tag{2.83}$$

The solution to Eq. (2.83), with the boundary conditions that $\tilde{p}_n(0) = \Delta p_{\text{ac}}e^{j\omega t}$ and $\tilde{p}_n(\infty) = 0$ may be found to be

$$\tilde{p}_n(x) = \Delta p_{\text{ac}} \exp\left(-\frac{x}{\lambda_p}\right)e^{j\omega t} \tag{2.84}$$

where

$$\lambda_p = \frac{\sqrt{D_p\tau_p}}{\sqrt{1 + j\omega\tau_p}} = \frac{L_p}{\sqrt{1 + j\omega\tau_p}} \,. \tag{2.85}$$

Hence, the ac excess density also falls off exponentially with distance from the injection region and has a characteristic diffusion length given by Eq. (2.85). A similar argument for an ac component of excess electrons in *p*-type material gives

$$\lambda_n = \frac{L_n}{\sqrt{1 + j\omega\tau_p}} \,. \tag{2.86}$$

λ_n and λ_p are called the ac diffusion lengths and can be seen to reduce to the dc lengths as $\omega \to 0$. The concept of the ac diffusion length will be important when we consider high frequency *p-n* junction behavior.

REFERENCES

[1] R. A. Smith, *Semiconductors*, Cambridge at the University Press, London, 1959.

[2] J. L. Moll, *Physics of Semiconductors*, McGraw-Hill, New York, 1964.

[3] F. D. Bloss, *Crystallography and Crystal Chemistry*, Holt, Rinehart, and Winston, New York, 1971.

[4] S. M. Sze, *Physics of Semiconductor Devices*, John Wiley & Sons, New York, 1969.

[5] E. S. Rittner, "Simultaneous Transport of Heavy and Light Holes in Semiconductors with a Degenerate Valence Band," *Phys. Rev.*, **101**, 1291 (April 1956).

[6] R. A. Griener, *Semiconductor Devices and Applications*, McGraw-Hill, New York, 1961.

[7] O. Madelung, *Physics of III–V Compounds*, John Wiley & Sons, New York, 1964.

[8] S. M. Sze and J. C. Irvin, "Resistivity, Mobility, and Impurity Levels in GaAs, Ge and Si at 300°K," *Solid-State Electronics*, **11**, 599 (1968).

[9] H. A. Watson, *Microwave Semiconductor Devices and Their Circuit Applications*, McGraw-Hill, New York, 1969.

[10] N. B. Hannay, *Semiconductors*, Reinhold Publishing Corp., New York, 1959.

[11] A. K. Jonscher, *Principles of Semiconductor Devices*, John Wiley & Sons, New York, 1960.

[12] W. Shockley and J. T. Last, "Statistics of the Charge Distribution for a Localized Flaw in a Semiconductor," *Phys. Rev.*, **107**, 392 (February 1957).

[13] M. L. Cohen and T. K. Bergstresser, "Band Structures and Pseudopotential Form Factors for Fourtum Semiconductors," *Phys. Rev.*, **141**, 789 (1966).

3

p-n Junction Diodes and Transistors

3.1 Introduction

Nearly all of the microwave semiconductor devices to be discussed in this book are *p-n* junction devices. A *p-n* junction is formed when the doping as a function of position in the semiconductor crystal changes rapidly from *p*-type to *n*-type over a short distance. Practical methods for achieving this are discussed in Chapter 4. *p-n* junction devices have found a great many useful electronic applications at dc and low frequencies as well as at microwave frequencies; consequently, they have been studied and discussed extensively. Shockley [1] was the first to present a thorough study of the subject. There are many excellent texts discussing basic *p-n* junction device theory (see for example Ref. [2], [3], [4], [5], and [6]). Therefore, only a brief treatment will be given here to serve as a foundation for later work in this book. Section 3.2 of this chapter will consider energy bands, potential barriers, equilibrium currents, and the depletion region in a zero-biased *p-n* junction. Section 3.3 will consider the effects on these quantities when a dc bias is applied to the device and will derive the *V-I* relation for the device. Section 3.4 is devoted to dynamic junction properties, depletion capacitance, diffusion capacitance, and junction conductance; and it concludes with a presentation of the equivalent circuit of a *p-n* diode. Section 3.5 discusses nonideal effects, such as space charge generation, avalanche effects, and others. A simplified dc treatment of the junction transistor is given in Section 3.6. Section 3.7 discusses a simple equivalent circuit of a transistor.

3.2 The *p-n* Junction

The simplest model of a *p-n* junction is called the *abrupt model.* In this model the transition from *n*-type to *p*-type material is assumed to be at $x = 0$. The doping for $x < 0$ is constant at N_D donors cm^{-3}, and for $x > 0$ the doping is constant at N_A acceptors cm^{-3}. This is depicted in Figure 3.1*A*. While perfectly abrupt doping changes cannot be accomplished in practice, it will be found that this model gives results which are in reasonable agreement with practical devices. The energy-band diagram that results from the abrupt doping profile is shown in Figure 3.1B (for no applied bias). At thermal equilibrium we require $\mathscr{E}_f$ to be constant (straight line) throughout the crystal, as shown in Figure 3.1B. In Chapter 2, Section 2.5, it was shown that $\mathscr{E}_f$ approaches $\mathscr{E}_c$ in *n*-type material and $\mathscr{E}_v$ in *p*-type material. This can be seen in Figure 3.1B. This implies that a potential barrier (of height eV_D on Figure 3.1B) must exist between the *n*-region and *p*-region. An alternate way of stating the need for the existence of the potential barrier follows. Assume that the *n*- and *p*-regions are initially separated in space and then brought into intimate contact. Due to the large concentration gradients, large numbers of electrons would diffuse into the *p*-type material, and large numbers of holes would diffuse into the *n*-type region. These carriers would leave behind immobile, unneutralized, ionized impurities (+ charge in *n*-region and − charge in *p*-region, see Figure 3.1C). This space charge sets up an electric field which retards the diffusive flow of electrons and holes. Eventually this process will reach the equilibrium condition shown in Figure 3.1B. The integral of the electric field at equilibrium is the barrier potential V_D. The region in which the donor and acceptor atoms are unneutralized by mobile charge carriers is called the *depletion, space-charge,* or *transition region* (see Figure 3.1C).

Assume that no recombination occurs in the depletion region, that far outside of the depletion region the carrier densities have their thermal-equilibrium values, and that the electric field is zero outside the depletion region. If we also assume that the holes at both edges of the depletion region (at $x = -W_D$ and $x = W_A$) are in thermal equilibrium with one another, then the ratio of their numbers can be found from Boltzmann statistics. The same assumption is also made for electrons [4].

Poisson's equation in one dimension is

$$\frac{d^2V(x)}{dx^2} = \frac{-\rho(x)}{\epsilon_s} \tag{3.1}$$

where $V(x)$ = electric potential, ϵ_s = dielectric constant, and $\rho(x)$ = charge density. Let us assume the following values for $\rho(x)$

$$
\begin{array}{lll}
x < -W_D, \rho(x) = 0 & \text{neutral material} \\
-W_D \leq x \leq 0, \rho(x) = +eN_D & \text{ionized donors} \\
0 \leq x \leq W_A, \rho(x) = -eN_A & \text{ionized acceptors} \\
x > W_A, \rho(x) = 0 & \text{neutral material}
\end{array}
\tag{3.2}
$$

For the sake of overall charge neutrality, we require $N_D W_D = N_A W_A$ (see Figure 3.1C). The solution to Poisson's equation proceeds by using these values of $\rho(x)$ and applying the boundary conditions that $dV(x)/dx$ is zero at both $x = W_A$ and $x = -W_D$, and $V(0) = 0$. The solution is

$$V(x) = -\frac{eN_D}{\epsilon_s}\left(\frac{x^2}{2} + W_D x\right) \qquad \text{for } (-W_D \leq x < 0) \tag{3.3}$$

and

$$V(x) = -\frac{eN_A}{\epsilon_s}\left(-\frac{x^2}{2} + W_A x\right) \qquad \text{for } (0 < x \leq W_A)\,. \tag{3.4}$$

The variation in potential across the depletion region [Eq. (3.3) and (3.4)] is plotted in Figure 3.1D. The total barrier height is $V(-W_D) - V(W_A) = V_D$. From Eq. (3.3) and (3.4) we find

$$V_D = V(-W_D) - V(W_A) = \frac{eN_A}{2\epsilon_s}(W_A^2 + W_D^2) \tag{3.5}$$

or, using the fact that $N_A W_A = N_D W_D$, we may solve for W_A and W_D as

$$W_A = \left[\frac{2\epsilon_s V_D}{eN_A(1 + N_A/N_D)}\right]^{1/2} \tag{3.6}$$

$$W_D = \left[\frac{2\epsilon_s V_D}{eN_D(1 + N_D/N_A)}\right]^{1/2}. \tag{3.7}$$

The electric field within the depletion region may be found by differentiating Eq. (3.3) and (3.4) since $E = -dV/dx$. The electric field is plotted on Figure 3.1E. The maximum value is seen to occur at $x = 0$ and is given by:

$$E_{\max} = \frac{eN_D W_D}{\epsilon_s} = \frac{eN_A W_A}{\epsilon_s}\,. \tag{3.8}$$

Therefore, when a *p-n* junction is formed, an electric field, a potential barrier, and a depletion region appear at the junction.

In the energy diagram shown in Figure 3.1B, the electrons tend to move downward (towards lower energy), and the holes tend to move upward (towards lower hole energy). This means that nearly all of those electrons in the *p* material within a diffusion length of the depletion region will "fall down" the potential hill into the *n*-region. This current density is called

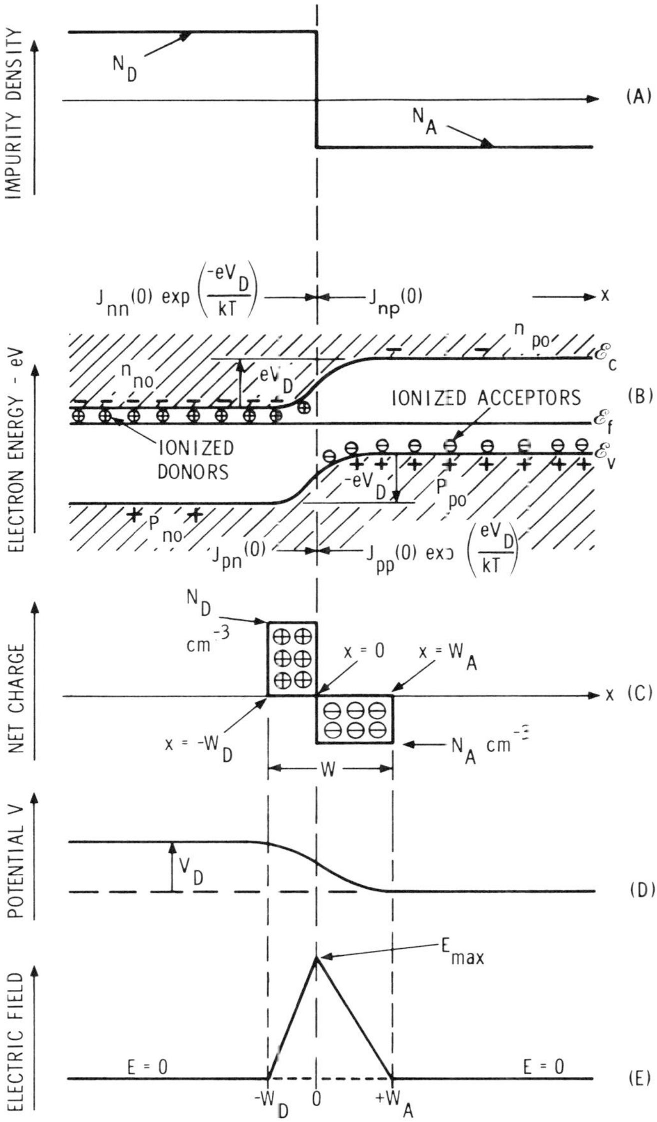

FIGURE 3.1
Zero-biased *p-n* junction: (A) doping density, (B) energy diagram, (C) net charge density, (D) potential, and (E) electric field.

$J_{np}(0)$, which is a constant depending on the lifetime, minority-carrier concentration, and diffusion coefficient. A certain percentage of the electrons in the n-region conduction band possess enough energy to surmount the potential barrier (Figure 3.1B) and are injected into the p-region. This current density is given by $J_{nn}(0) \exp(-eV_D/kT)$. At zero bias and thermal equilibrium, detailed balance must hold; therefore,

$$J_{np}(0) = J_{nn}(0) \exp\left(\frac{-eV_D}{kT}\right). \tag{3.9}$$

Similarly, for holes

$$J_{pn}(0) = J_{pp}(0) \exp \frac{-eV_D}{kT}. \tag{3.10}$$

All of the relations in this section have been derived for the zero-bias, thermal-equilibrium case. The next section considers the effects of nonzero bias.

3.3 *p-n* Junction with Applied Bias

If a dc bias is applied to a p-n junction in the manner shown in Figure 3.2A, the net result will be to shift the Fermi level in p- and n-regions by just the amount of the applied bias. If the p-region is made positive with respect to the n-region, this is called *forward bias*, and the applied voltage *lowers* the diffusion voltage barrier as is shown in Figure 3.2C. If the n-region is made positive with respect to the p-region, this is called *reverse bias*, and the applied voltage *raises* the potential barrier, see Figure 3.2D. All of the equations derived in the previous section can be used in the applied-bias case if the diffusion voltage is replaced by the new barrier height, i.e., if we replace V_D by $V_D - V_A$ (V_A will be positive for forward bias and negative for reverse bias), Eq. (3.5), (3.6), (3.7), (3.9), and (3.10) are still valid. Let us see how the currents discussed in the previous section are changed with applied bias. The currents $J_{np}(0)$ and $J_{pn}(0)$ do not change with barrier height (applied voltage) since they are basically determined by the material parameters. The currents $J_{nn}(0) \exp\{e[(V_A - V_D)/kT]\}$ and $J_{pp}(0) \exp\{e[(V_A - V_D)/kT]\}$ [see Figure 3.1B, and Eq. (3.9) and (3.10)] will obviously change with applied voltage due to a raising or lowering of the potential barrier. The total current density flowing in the device will just be the net hole current flowing from the p- to the n-region plus the net electron current flowing from the n- to the p-region (the opposite sign of the charge carriers makes the electron and hole currents additive). This can be expressed as

$$J = [J_{pp}(0) + J_{nn}(0)] \exp\left[e \frac{V_A - V_D}{kT}\right] - J_{np}(0) - J_{pn}(0). \tag{3.11}$$

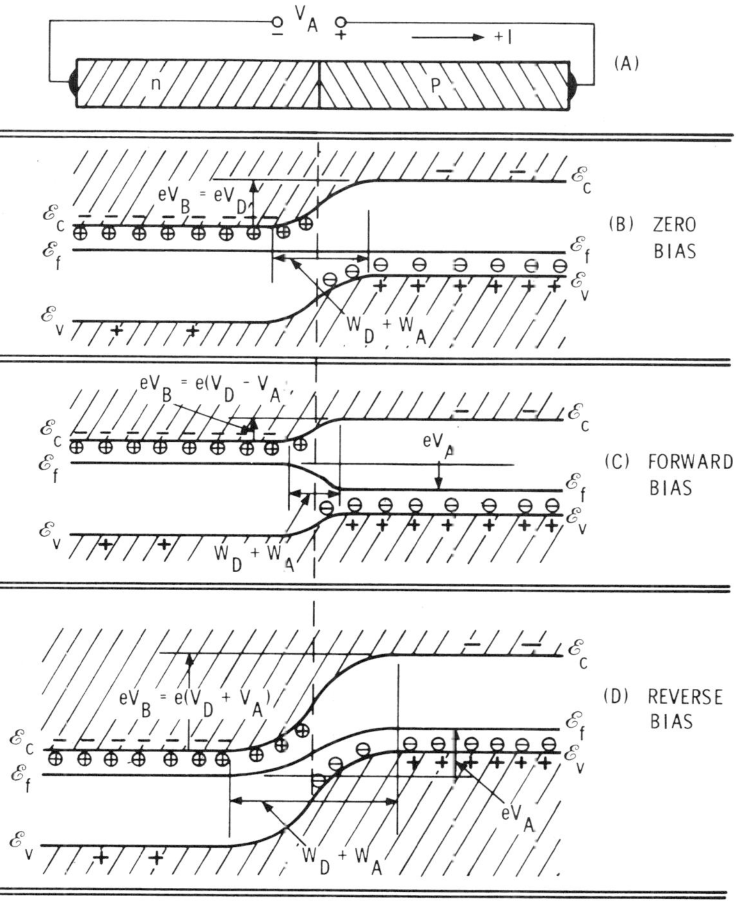

FIGURE 3.2
***p-n* junction: (A) cross-section, (B) zero-bias energy diagram, (C) forward-bias energy diagram, and (D) reverse-bias energy diagram.**

Using Eq. (3.9) and (3.10), we may rewrite Eq. (3.11) as:

$$J = J_0\left(\exp\frac{eV_A}{kT} - 1\right) \tag{3.12}$$

where

$$J_0 = J_{pn}(0) + J_{np}(0)\,. \tag{3.13}$$

Equation (3.12) is the familiar so-called diode equation. From Eq. (3.12) it can be seen that current increases exponentially with forward bias.† In the reverse bias direction (V_A negative) the current quickly saturates at a value $-J_0$ (J_0 is usually quite small). A typical plot of Eq. (3.12) is shown in Figure 3.3. We shall now solve for J_0, the saturation current, which is needed to fully evaluate the diode equation. The current in the device beyond the edge of the depletion region is carried completely by diffusion (since the electric field is assumed to be zero outside the depletion region). The value of

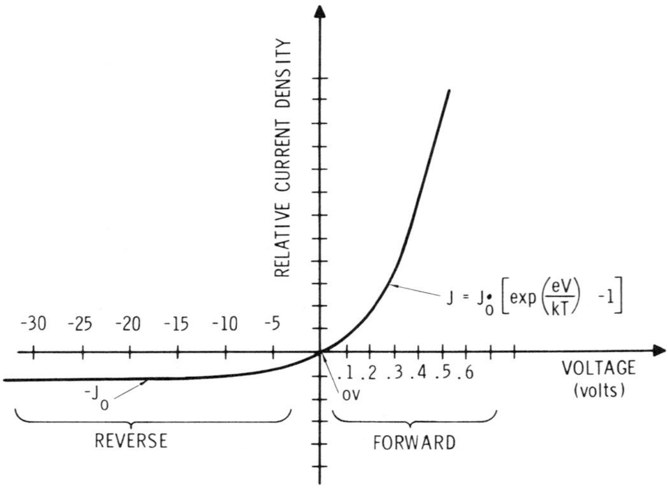

FIGURE 3.3
Ideal diode V-I characteristics (note different forward and reverse voltage scales).

$p_n(x)$ far from the depletion region boundary is the thermal-equilibrium value p_{n0}. The value at $x = -W_D$ is [3] equal to $p_{n0}\exp(eV_A/kT)$ from the Boltzmann relation. Thus, we have a problem similar to that already solved in Chapter 2, Section 2.8. The equation to be solved is Eq. (2.74), and the boundary conditions are given above. The result is:

$$p_n(x) = p_{n0}\left[\exp\left(\frac{eV_A}{kT}\right) - 1\right]e^{+x/L_p} + p_{n0} \tag{3.14}$$

† At high current density, the ohmic voltage drop in the bulk semiconductor material between the device terminals and the *p-n* junction must be subtracted from the terminal voltage to determine the forward-bias voltage.

where the origin is taken at $x = -W_D$ and all values of x are negative. The current $J_{pn}(0)$ at $x = 0$ is given by

$$J_{pn}(0) = +eD_p \left.\frac{dp_n(x)}{dx}\right|_{x=0} \tag{3.15}$$

or

$$J_{pn}(0) = \frac{eD_p p_{n0}}{L_p}\left[\exp\left(\frac{eV_A}{kT}\right) - 1\right]. \tag{3.16}$$

Similarly, at $x = +W_A$ we can show that

$$J_{np}(0) = \frac{eD_n n_{p0}}{L_n}\left[\exp\left(\frac{eV_A}{kT}\right) - 1\right]. \tag{3.17}$$

The total current $J_{np}(0) + J_{pn}(0)$ can now be written as:

$$J = J_{np}(0) + J_{pn}(0) = \left[\frac{eD_n n_{p0}}{L_n} + \frac{eD_p p_{n0}}{L_p}\right]\cdot\left[\exp\left(\frac{eV_A}{kT}\right) - 1\right]. \tag{3.18}$$

Comparison of Eq. (3.18) and (3.12) shows that the reverse saturation current is given by:

$$J_0 = \left[\frac{eD_n n_{p0}}{L_n} + \frac{eD_p p_{n0}}{L_p}\right] \tag{3.19}$$

or, using Eq. (2.30) and (2.31), we may rewrite Eq. (3.19) as

$$J_0 = en_i^2\left[\frac{D_n}{L_n N_A} + \frac{D_p}{L_p N_D}\right]. \tag{3.20}$$

Two assumptions made in the derivation of Eq. (3.20) should be clarified:

1. The only important diffusion current at $x = -W_D$ is due to holes. This is because the injected hole density produces a large change in the equilibrium hole density p_{n0}. Since this region is neutral (just beyond the space charge region), there must also be an excess of majority carriers $\Delta n = \Delta p$; however, the resulting concentration gradient of majority carriers is much smaller than that for minority carriers and is neglected. An analogous statement can be made for the injected electrons at $x = +W_A$.
2. Since we have neglected recombination in the space-charge region, the total current given by $J_{pn}(-W_D) + J_{np}(+W_A)$ is equal to $J_{pn}(0) + J_{np}(0)$.

3.4 Dynamic Junction Parameters

The first parameter to be considered in this section is the depletion-layer capacitance. From Fig. 3.1C it can be seen that there are unneutralized charges exposed in the depletion region, and from Eq. (3.6) and (3.7) (with

V_D replaced by $V_D - V_A$ for nonzero bias), it can be seen that the amount of charge exposed depends on the applied bias. This effect gives rise to a capacitance called the *depletion capacitance*, C_d. The definition of capacitance is [7]

$$C = \frac{\partial Q}{\partial V} \tag{3.21}$$

where V is the applied voltage, Q is the charge on one electrode (in this case the charge in the p-region or the n-region). We know that, on the p-side, $Q = -eN_AW_A$. From Eq. (3.6) (with V_D replaced by $V_D - V_A$) we have

$$C_d = \frac{\partial}{\partial V_A}\left(-eN_A\left[\frac{2\epsilon_s(V_D - V_A)}{eN_A(1 + N_A/N_D)}\right]^{1/2}\right) \tag{3.22}$$

which is equal to

$$C_d = \left[\frac{\epsilon_s N_A N_D e}{2(N_D + N_A)}\right]^{1/2}(V_D - V_A)^{-1/2}\ \text{farad}\cdot\text{cm}^{-2} \tag{3.23}$$

The same result would also be obtained if one considered the n-side of the junction ($Q = -eN_DW_D$). Through a simple manipulation of Eq. (3.6), (3.7) (with $V_D = V_D - V_A$), and Eq. (3.23), it can also be shown

$$C_d = \frac{\epsilon}{W_D + W_A} = \frac{\epsilon}{W}\ \text{farad}\cdot\text{cm}^{-2}. \tag{3.24}$$

This is an interesting result because it says that the depletion capacitance is just the same as would be calculated for a plane parallel capacitor with a plate spacing $W = W_D + W_A$ filled with a dielectric, $\epsilon = \epsilon_s$. For large values of reverse bias ($V_A \gg V_D$), C_d is proportional to $V_A{}^{-1/2}$ [see Eq. (3.23)], and a plot of $1/C^2$ versus V_A has a slope of $2(N_D + N_A)/(\epsilon_s N_A N_D e)$. If $N_A \gg N_D$ or $N_D \gg N_A$ (which is usually the case), the value of the smaller of the two, N_A or N_D, can be obtained from the slope of the $1/C^2$ versus V plot (see Figure 6.3). Equation (3.23) was derived for an abrupt-junction model. In practice, the doping may not vary abruptly as shown in Figure 3.1A but may vary linearly† with x, changing from n-type to p-type at $x = 0$. In this case it is known [8] that the depletion capacitance is given by

$$C_d = \frac{C_0}{(1 - V_A/V_D)^{1/3}}\ \text{farad}\cdot\text{cm}^{-2} \tag{3.25}$$

and

$$C_0 = \left(\frac{e\epsilon_s^2 a}{12V_D}\right)^{1/3} \tag{3.26}$$

where a is the slope of the linearly changing doping function, i.e., $\rho(x) = eax$.

† This is called a graded junction.

However, it can be shown [4] that Eq. (3.24) is valid regardless of the form of the variation of doping with distance, x. The depletion capacitance is most important at reverse bias.

At forward bias and low frequencies, the diffusion capacitance, C_D, becomes dominant. The basic mechanism for the diffusion capacitance is shown in Figure 3.4. If the barrier height is periodically raised and lowered, the amount of excess injected minority charge will change. This change in

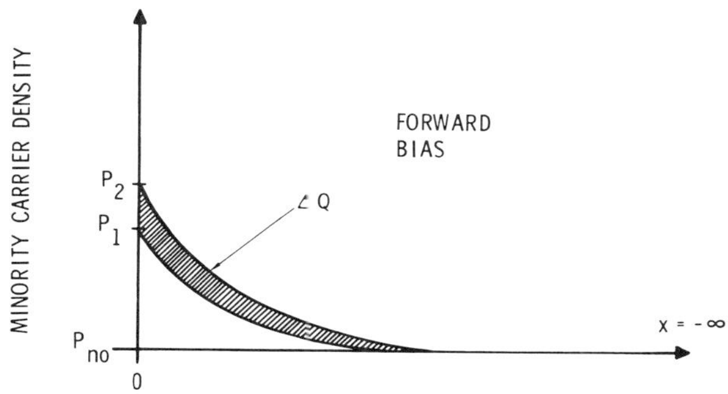

FIGURE 3.4
Stored minority charge which produces forward-bias diffusion capacitance.

stored charge is depicted by the shaded area in Figure 3.4. The excess charge in the p-region may be calculated from

$$Q'_{pn} = \int_{-\infty}^{0} ep'_n(x)\, dx \tag{3.27}$$

where $p'_n(x)$ is the excess injected minority carrier density. $p'_n(x) = p_n(x) - p_{n0}$ where $p_n(x)$ is given by Eq. (3.14). Hence, Eq. (3.27) may be written as

$$Q'_{pn} = \int_{-\infty}^{0} ep_{n0}\left[\exp\left(\frac{eV_A}{kT}\right) - 1\right]e^{+x/L_p}\, dx \tag{3.28}$$

which can be integrated to obtain

$$Q'_{pn} = ep_{n0}L_p\left[\exp\left(\frac{eV_A}{kT}\right) - 1\right]. \tag{3.29}$$

The capacitance due to the stored holes is defined as

$$C_p = \frac{dQ'_{pn}}{dV_A} \tag{3.30}$$

and hence

$$C_p = \frac{e^2 p_{n0} L_p}{kT} \exp\left(\frac{eV_A}{kT}\right) \cdot$$

There is a similar term due to the stored electrons in the *p*-region, and the total diffusion capacitance is given by [3]:

$$C_D = C_n + C_p = \frac{e^2}{kT}(n_{p0}L_n + p_{n0}L_p)\exp\left(\frac{eV_A}{kT}\right) \cdot \tag{3.31}$$

The diffusion capacitance increases exponentially with forward voltage and rapidly swamps out the depletion capacitance [Eq. (3.23)]. Equation (3.31) is the dc capacitance. While this derivation gives excellent insight into the diffusion capacitance, it is only correct at low frequencies since the ac diffusion process (see Chapter 2, Section 2.8) was not taken into account. We shall now derive the ac admittance of the forward-biased abrupt *p-n* junction in a more rigorous (but less informative) way. The diode equation [Eq. (3.18)] is

$$J = \left(\frac{eD_n n_{p0}}{L_n} + \frac{eD_p p_{n0}}{L_p}\right)\left[\exp\left(\frac{eV_A}{kT}\right) - 1\right]; \tag{3.18}$$

The applied voltage, V_A, can be written in the form $V_A' = V_A + \tilde{V}_1 e^{j\omega t}$ where $\tilde{V}_1 \ll V_A$. $\tilde{V}_1$ is the ac term and V_A is the dc term. Assume that the resulting current density may be written in the form $J = J_0 + \tilde{J}_1 e^{j\omega t}$, where J_0 and $\tilde{J}_1$ are the dc and ac currents respectively. The diode equation [Eq. (3.18)] may now be written as

$$J = J_0 + \tilde{J}_1 e^{j\omega t} = \left(\frac{eD_n n_{p0}}{L_n} + \frac{eD_p p_{n0}}{L_p}\right)\left[\exp\left(\frac{eV_A}{kT} + \frac{e\tilde{V}_1 e^{j\omega t}}{kT}\right) - 1\right] \cdot \tag{3.32}$$

By using the series expansion $e^x \cong 1 + x$ for $x \ll 1$ and separating the dc and ac parts of Eq. (3.32), we can write

$$\tilde{J}_1 = \frac{e\tilde{V}_1}{kT}\left(\frac{eD_n n_{p0}}{\lambda_n} + \frac{eD_p p_{n0}}{\lambda_p}\right)\exp\left(\frac{eV_A}{kT}\right) \tag{3.33}$$

where the ac diffusion lengths $\lambda_n = L_n/\sqrt{1 + j\omega\tau_n}$ and $\lambda_p = L_p/\sqrt{1 + j\omega\tau_p}$ [see Eq. (2.85) and (2.86)] have been substituted for the dc diffusion lengths. The ac-junction admittance is simply

$$Y = \frac{\tilde{J}_1}{\tilde{V}_1} = G + j\omega C_D . \tag{3.34}$$

Comparing Eqs. (3.33) and (3.34) we find that at low frequencies ($\omega\tau \ll 1$) the junction conductance G (in mho·cm^{-2}) is given by

$$G = \frac{e}{kT}\left(\frac{eD_n n_{p0}}{L_n} + \frac{eD_p p_{n0}}{L_p}\right)\exp\left(\frac{eV_A}{kT}\right) \text{ mho·cm}^{-2} \tag{3.35}$$

or from a comparison of Eq. (3.18)

$$G = \frac{e(J - J_0)}{kT} \cong \frac{eJ}{kT} \text{mho} \cdot \text{cm}^{-2} \tag{3.36}$$

where $J \gg J_0$. The low frequency ($\omega\tau \ll 1$) capacitance is found to be [using the expansion $(1 + x)^n \cong 1 + nx$ for $x \ll 1$]

$$C_D = \frac{e}{kT}\left(\frac{eL_n n_{p0}}{2} + \frac{eL_p p_{n0}}{2}\right) \exp\left(\frac{eV_A}{kT}\right) \text{farad} \cdot \text{cm}^{-2}. \tag{3.37}$$

The astute reader will notice a difference of a factor of 2 between Eq. (3.37) and (3.31). The discrepancy is a result of the fact that only half of the injected charge is recoverable (see Ref. [11]). Equation (3.37) is the one usually taken by device physicists as the correct one. At high frequencies ($\omega\tau \gg 1$) and $\sqrt{1 + j\omega\tau} \to \sqrt{j}\sqrt{\omega\tau} = [(1 + j)/\sqrt{2}]\sqrt{\omega\tau}$. In this case a comparison of Eq. (3.34) and (3.33) yields

$$G = \frac{e}{kT}\sqrt{\frac{\omega}{2}}\left(e\sqrt{D_n}p_{n0} + e\sqrt{D_p}n_{p0}\right) \exp\left(\frac{eV_A}{kT}\right) \text{mho} \cdot \text{cm}^{-2} \tag{3.38}$$

and

$$C_D = \frac{e}{\sqrt{2\omega}kT}(e\sqrt{D_n}p_{n0} + e\sqrt{D_p}n_{p0}) \exp\left(\frac{eV_A}{kT}\right) \text{farad} \cdot \text{cm}^{-2}. \tag{3.39}$$

From Eq. (3.38) and (3.39) it can be seen that $G = \omega C_D$ at high frequencies. Now consider a model for the *p-n* junction under dynamic conditions. Figure 3.5 shows the diffusion capacitance, C_D [Eq. (3.37) or (3.39)], the

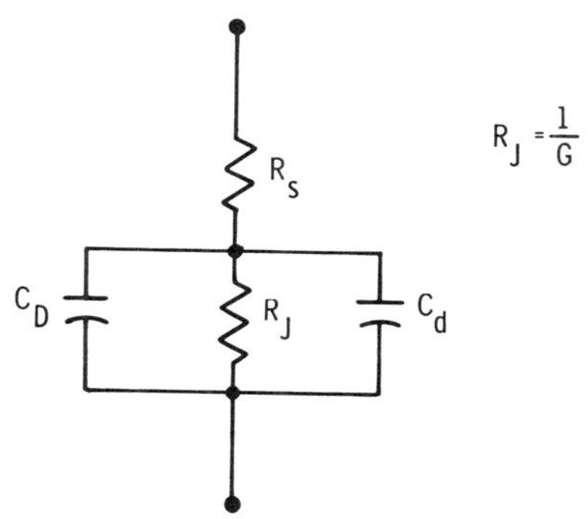

FIGURE 3.5
Equivalent circuit of the *p-n* junction diode.

depletion capacitance, C_d [Eq. (3.24)], and the junction conductance, G [Eq. (3.35) or (3.38)], all connected in parallel. The series resistance R_s is the ohmic resistance of the undepleted p- and n-regions plus any contact resistance. At large forward bias usually only R_J and C_D and R_s need be considered, while at large reverse bias only R_s and C_d are important. The value of R_s may be estimated from

$$R_s = \frac{L_1}{\sigma_n A} + \frac{L_2}{\sigma_p A} \tag{3.40}$$

where σ_n and σ_p are the conductivities of the n-type and p-type regions [see Eq. (2.39) or Figure 2.9], L_1 and L_2 are the lengths of the undepleted n- and p-regions respectively, and A is the cross-sectional area of the device.

3.5 Nonideal *p-n* Junction Effects

The first few sections of this chapter have been devoted to an ideal abrupt p-n junction. This section will cover several important effects which are not included in the ideal model: avalanche breakdown, generation-recombination in the depletion region, high-level injection effects, and the effect of nonplanar geometry. Figure 3.6 shows the V-I characteristic of a nonideal diode and demonstrates some of the effects to be discussed in this section.

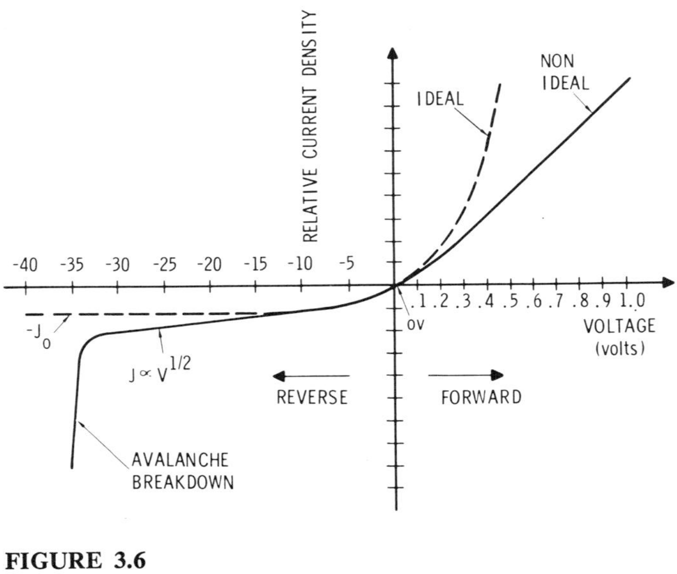

FIGURE 3.6
V-I characteristics of a nonideal diode.

The first point to be considered is *avalanche breakdown*. (A fair amount of detail is required here since the concept will be brought up many times in later chapters.) In Section 6 of Chapter 2 it was mentioned that as the electric field in a semiconductor is increased, the velocity eventually saturates (see Figure 2.10) due to increased collisions with optical phonons (the carriers now lose as much energy as they gain on the average, and the velocity saturates). If the electric field is increased even further (to $\approx 2.8 \times 10^5$ V/cm in Si), there is still a finite probability that a few electrons may evade a collision with optical phonons and gain sufficient energy from the applied field to produce an ionizing collision with a lattice atom. This requires that the incident electron have an energy greater than the bandgap energy. (See Moll [5] for a more thorough discussion of hot electron-phonon interactions.) When this happens, an electron is excited across the forbidden gap, and a hole-electron pair is formed, resulting in the original electron and a new electron plus a hole. This new electron and hole can also go on to generate more pairs. This process, called *avalanche multiplication*, explains the sudden increase in current seen at large reverse voltages in *p-n* junctions, such as is shown in Figure 3.6. Avalanche breakdown can occur in bulk semiconductors as well as *p-n* junctions; however, it is very difficult to maintain a high field in bulk material because of the large currents that would flow ($J = \sigma E$). On the other hand, it is not at all difficult to obtain a sufficiently large electric field in the depletion region of a reverse-biased *p-n* junction to cause avalanche breakdown. This is because in the depletion region there is a high density of unneutralized charge which can give rise to a large field at $x = 0$. (See Figure 3.1E.) The concept of avalanche breakdown is illustrated further in Figure 3.7 which depicts the particle current densities in the depletion region of a reverse-biased *p-n* junction (there is always some current flow due to the reverse saturation current). In Figure 3.7, j_{p0} is the hole particle current density entering the left side of the depletion region and Mj_{p0} is the multiplied hole particle current density exiting the right edge of the depletion region (into the field-free material). Let us define $\alpha(E)$ as the ionization coefficient which is a function of electric field E:

$$\alpha(E) \equiv \text{average number of ionizing events per cm of travel.} \tag{3.41}$$

We will denote $\alpha_p(E)$ for the ionization coefficient for holes and $\alpha_n(E)$ for the ionization coefficient for electrons, since they are not necessarily equal. Referring again to Figure 3.7, we see that the increase in hole particle current density between x and $x + dx$ must be

$$dj_p = \alpha_p(E) j_p \, dx + \alpha_n(E) j_n \, dx \,. \tag{3.42}$$

The first term on the right-hand side of Eq. (3.42) is the contribution due to hole-electron pair generation by holes, and the second term is the contribu-

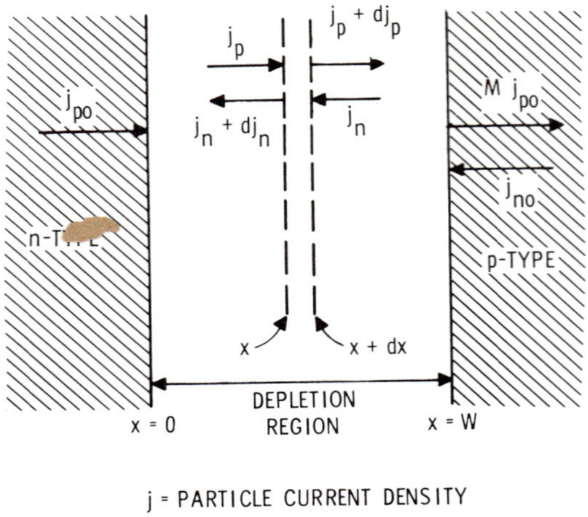

FIGURE 3.7
Avalanche multiplication in the depletion region of a *p-n* junction.

tion to hole-electron pair generation by electrons. Rearranging Eq. (3.42), we may write

$$\frac{dj_p}{dx} = \alpha_p(E)j_p + \alpha_n(E)j_n \,. \tag{3.43}$$

Let us now discuss electronic current densities $J_p = ej_p$ and $J_n = -ej_n$ and $J = J_n + J_p$. We may rewrite Eq. (3.43) as

$$\frac{dJ_p}{dx} = \alpha_n(E)J + [\alpha_p(E) - \alpha_n(E)]J_p \,. \tag{3.44}$$

Equation (3.44) is a differential equation which may be solved by conventional means. If the boundary conditions are taken to be $J_p(0) = J_{p0}$ and $J_p(W) = M_pJ_{p0}$ (where M_p = hole multiplication ratio), Sze [6] has shown the solution to Eq. (3.44) is

$$1 - \frac{1}{M_p} = \int_0^W \alpha_p(E) \exp\left\{-\int_0^x [\alpha_p(E) - \alpha_n(E)]\, dx'\right\} dx \tag{3.45}$$

The condition for avalanche breakdown is $M_p \to \infty$ or

$$\int_0^W \alpha_p(E) \exp\left\{-\int_0^x [\alpha_p(E) - \alpha_n(E)]\, dx'\right\} dx = 1 \,. \tag{3.46}$$

Eq. (3.46) is the avalanche breakdown condition if the breakdown is initiated by holes. If the breakdown is initiated by electrons, the breakdown condition is [6]:

$$\int_0^W \alpha_n(E) \exp\left\{-\int_x^W [\alpha_n(E) - \alpha_p(E)]\, dx'\right\} dx = 1 . \tag{3.47}$$

To solve Eq. (3.46) or (3.47), we must first find the variation of the electric field within the depletion region, $E(x)$. This is done with Poisson's equation [Eq. (3.1], realizing that $E(x) = -dV(x)/dx$]. We must also know the variation of $\alpha(E)$ with E. This can be approximated by [9]:

$$\alpha(E) = A \exp\left[-\left(\frac{b}{E}\right)^m\right] \tag{3.48}$$

where the values of A, b, and m for Si, Ge, GaAs, and GaP are given in Table 3.1.

TABLE 3-1

	Electrons		Holes		
Semiconductor	$A(\text{cm}^{-1})$	$b(\text{V/cm})$	$A(\text{cm}^{-1})$	$b(\text{V/cm})$	m
Ge	1.55×10^7	1.56×10^6	1.0×10^7	1.28×10^6	1
Si	3.8×10^6	1.75×10^6	2.25×10^7	3.26×10^6	1
GaAs	3.5×10^5	6.85×10^5	3.5×10^5	6.85×10^5	2
GaP	4.0×10^5	1.18×10^6	4.0×10^5	1.18×10^6	2

After Sze and Gibbons [9]

A "one-sided" junction is an abrupt junction that has either $N_A \gg N_D$ or $N_D \gg N_A$. This means that the depletion width is approximately $W = W_D$ or $W = W_A$, respectively, [see Eq. (3.6) and (3.7)]. Most of the junction devices considered in this book will be abrupt, one-sided junctions. In the case of the one-sided junction, the depletion width [which forms the integration limit for Eq. (3.46) and (3.47)] will be a function of the smaller of the two dopant densities N_A or N_D and will be called N_B. The depletion width will also be a function of the applied bias voltage V_A [see Eq. (3.6) and (3.7) with V_D replaced by $V_D - V_A$]. The electric field (from Poisson's equation) is a function of position, applied bias, and doping density. Hence,

$$W = W(V_A, N_B)$$

and

$$E = E(x, V_A, N_B) . \tag{3.49}$$

The avalanche breakdown conditions may be calculated by substituting Eq. (3.48) and (3.49) into Eq. (3.46). The problem is extremely complicated, and a computer is usually used to solve the resulting equation. The results of the solutions for breakdown in one-sided abrupt junctions made of Si, Ge, GaAs, or GaP are shown in Figure 3.8. Figure 3.8 illustrates the depletion width at breakdown (see Figure 3.1C) as well as the maximum field, $E_{\max}$, at breakdown (see Figure 3.1E) as a function of the lightest doping density N_B. Most devices of interest have $N_B < 10^{16}$ cm^{-3}, and over this range Figure 3.8 shows

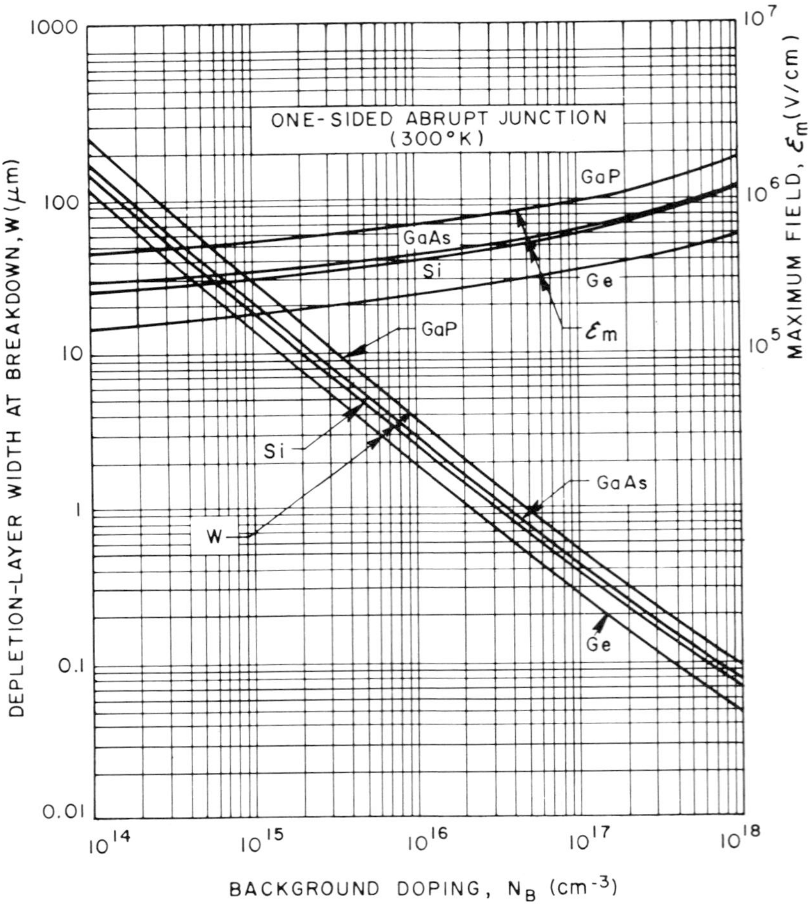

FIGURE 3.8
Maximum field and depletion width at breakdown for a one-sided abrupt junction (after Sze and Gibbons [12]).

that E_{max} is essentially constant, which is to be expected from Eq. (3.8). (E_{max} varies by less than a factor of 2 over this range, and for the remainder of this book we will take it as a constant.) For a one-sided abrupt junction the breakdown voltage† as a function of doping density may be found quite simply as

$$V_B \cong \frac{E_{max} \cdot W}{2} \tag{3.50}$$

where W = depletion width at breakdown and E_{max} = maximum field at breakdown (both W and E_{max} are found from Figure 3.8). Since $E_{max} \cong$ const, Eq. (3.50) and Figure 3.8 show that breakdown voltage increases with decreasing doping density, which is a well-known result.

The effects of *generation and recombination in the space-charge region* were neglected in our original derivation of the diode equation [Eq. (3.12)]. First consider the *reverse-bias* case. The depletion region is assumed to have a single-level, deep trap at $\mathscr{E}_T \cong \mathscr{E}_i$. (See Ref. [4] and [10] for a more thorough discussion.) The effects of deep traps were discussed in Chapter 2, Section 2.7. At thermal equilibrium, the recombination and generation of carriers is balanced. At reverse bias in the depletion region, $n = p \cong 0$, and thus the generation exceeds the recombination. It can be shown [4], [10], that for this case the generation rate is given by

$$g \cong \frac{n_i}{2\tau} \text{ carriers/cm}^3 \tag{3.51}$$

where $\tau = \tau_p = \tau_n$ and n_i = intrinsic density.

The current density due to these generated carriers is simply

$$J_g = eW\left(\frac{n_i}{2\tau}\right) \text{ amp/cm}^2 \tag{3.52}$$

where W = depletion region width. In wide bandgap semiconductors, J_g is usually several orders of magnitude larger [10] than J_0 [Eq. (3.19)], and the reverse characteristics are therefore dominated by generation of carriers in the space-charge region. For an abrupt junction, W is proportional to $V_A^{1/2}$ [see Eq. (3.6) or (3.7) with V_D replaced by $V_D - V_A$]; hence J_g will be proportional to $V_A^{1/2}$. This explains the slope of the reverse current shown on Figure 3.6. Later it will be shown that carrier generation in the depletion region is also important in an ionizing radiation environment. Under *forward bias*, the hole and electron densities in the space-charge region cannot be neglected; consequently, there is a contribution to the current due to recombination of holes and electrons in the space-charge region. At low values of forward bias, it can

† Assuming $V_A \gg V_D$.

be shown (see [4] and [10]) that the recombination current can exceed the normal forward-bias current by a considerable amount. If this happens, then it can also be shown ([4] and [10]) that the forward current is proportional to exp $(eV_A/2kT)$; the factor 2 reverts back to unity as the injection increases sufficiently to swamp the trapping mechanism.

In our original derivation of the diode equation, we assumed that the injected minority densities were much smaller than majority carrier densities. *High-level injection* (large forward-current densities) results in several non-ideal effects. For example, if the minority injected density becomes equal to the majority, the current becomes [4] proportional to exp $(eV_A/2kT)$. Other effects that occur are *conductivity modulation* of the region within a few diffusion lengths of the depletion region, i.e., the injected densities increase n and p considerably above their thermal-equilibrium values and σ [see Eq. (2.39)] increases. Conductivity modulation will be discussed again in Chapter 7 on *p-i-n* diodes. Another effect that occurs is *ohmic drop*. The $I \cdot R$ drop across the bulk series resistance [Figure 3.5 and Eq. (3.40)] can lead to increased voltage across the diode terminals for forward bias as shown in Figure 3.6.

In actual practice the fabrication technique used to form the *p-n* junction (see Appendix B) may result in a *nonplanar junction*. Sze [6] has shown that this can affect the depletion capacitance of the device, and, more importantly, the increase in electric field that occurs at rounded edges leads to a decreased breakdown voltage for nonplanar junctions.

3.6 *p-n* Junction Transistors

The analysis of the *p-n* junction transistor is a simple extension of the work already completed in this chapter. At this point, a greatly simplified† model of a junction transistor will serve as a basis for later work in the chapter on radiation effects in microwave transistors (Chapter 12). The transistor model to be discussed is shown in Figure 3.9, which depicts an *n-p-n* transistor, but the analysis of a *p-n-p* transistor would be analogous to the one following.

The simplifying assumptions that will be made in the dc analysis are:

1. The emitter width, W_E, and base width, W_B, are much shorter than the diffusion lengths of minority carriers in those regions.
2. Carriers of the same type on each side of the emitter-base junction are in equilibrium with each other (the Boltzmann relation holds).
3. Generation and recombination in the emitter-base and collector-base space-charge regions can be neglected.

† For a more detailed analysis see [3] through [6].

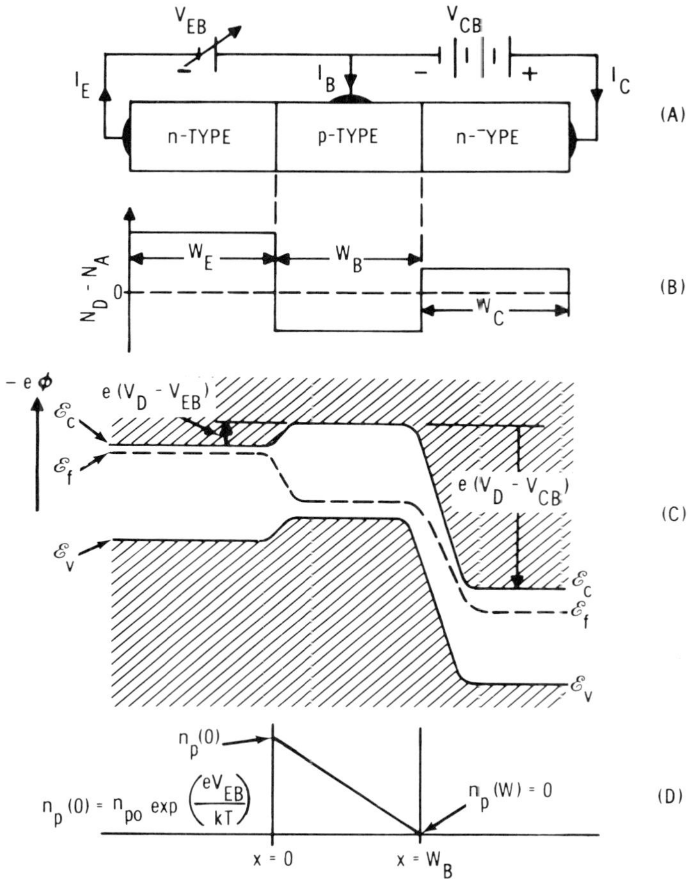

FIGURE 3.9
***n-p-n* transistor: (A) normal operating bias, (B) doping profile, (C) energy band diagram, and (D) minority-carrier distribution in the base.**

4. All ohmic losses are ignored.
5. The emitter is much more heavily doped than is the base ($N_{DE} \gg N_{AB}$).

Figure 3.9A shows an *n-p-n* transistor biased in its normal operating mode. The *E-B* (emitter-base) junction is forward biased and the *C-B* (collector-base) junction is reverse biased. Electrons are injected from the emitter into the base. There they move by diffusion through the base until they reach the *C-B* space charge-region, at which point they are accelerated by the electric field in the depletion region and fall down the potential hill into the collector. This type of device can be seen to be capable of power gain since it takes little energy to lift the electrons over the *E-B* junction barrier, and they can do a lot of work when they fall into the deep potential well at the *C-B* junction.

In practice, the current flow can be controlled by varying the *E-B* bias V_{EB} (Fig. 3.9A); and, therefore, controllable power amplification can be achieved. From the previous sections and from assumption 2, we know that the injected electron density at $x = 0$ is given by

$$n_p(0) = n_{p0} \exp\left(\frac{eV_{EB}}{kT}\right) \tag{3.53}$$

where $n_{p0} = n_i^2/N_A$ = equilibrium electron density in the base.

Since all electrons that reach $X = W$ are immediately swept down into the potential well, $n_p(W) = 0$. If the base is considered to be field free, and carrier generation is neglected, the continuity equation (2.69) in the steady state reduces to

$$D_n \frac{d^2 n_p(x)}{dx^2} - \frac{n_p(x) - n_{p0}}{\tau_n} = 0 \tag{3.54}$$

where the recombination rate R has been assumed to have the form $R = \Delta n/\tau_r$. The solution to Eq. (3.54) with the above mentioned boundary conditions can be shown to be [4]

$$n_p(x) = n_{p0} \exp\left(\frac{eV_{EB}}{kT}\right)\left(1 - \frac{x}{W_B}\right) . \tag{3.55}$$

The electron-diffusion current density in the base resulting from this distribution can be found [Eq. (2.42A)] as

$$J_{nB} = -\frac{D_n e n_i^2}{W_B N_{AB}} \exp\left(\frac{eV_{EB}}{kT}\right) \tag{3.56}$$

where N_{AB} = acceptor doping density in the base. A similar argument for the holes injected from the base into the emitter gives the hole-diffusion current density in the emitter as

$$J_{pE} = -\frac{D_p e n_i^2}{W_E N_{DE}} \exp\left(\frac{eV_{EB}}{kT}\right) . \tag{3.57}$$

The total current density is $J_T = J_{pE} + J_{nB}$ just as in the case of the *p-n* junction. However from Figure 3.9 it can be seen that holes injected from the base to the emitter absorb power and therefore represent an unwanted current component. Let us define the emitter injection ratio as

$$\gamma = \frac{J_{nB}}{J_T} = \frac{J_{nB}}{J_{nB} + J_{pE}} \tag{3.58}$$

Using Eq. (3.56) and (3.57), we may rewrite Eq. (3.58)

$$\gamma = \frac{1}{1 + D_p W_B N_{AB}/D_n W_E N_{DE}} \cdot \tag{3.59}$$

By using the Einstein relation and the fact that $n_{n0} \cong N_{DE}$ and $p_{p0} \cong N_{AB}$ we may rewrite Eq. (3.59) as

$$\gamma = \frac{1}{1 + \sigma_B W_B/\sigma_E W_E} \tag{3.60}$$

where σ_B and σ_E are the conductivities in the base and emitter, respectively [see Eq. (2.39)]. For efficient operation we would prefer $\gamma = 1$ (all the current is due to injected electrons into the base). To achieve this, Eq. (3.60) implies the conductivity of the emitter should be much greater than that of the base (i.e., emitter is more heavily doped than the base). Typical values of the emitter injection ratio of 0.99 are achieved in practice.

Another important transistor parameter is the base-transport factor η. Not all of the electrons injected into the base will reach the collector; some of them will recombine with the majority carriers (holes) in the base. Hence, let us define the base transport factor as

$$\eta = \frac{\text{electron current at } x = W_B}{\text{electron current at } x = 0} \cdot \tag{3.61}$$

It can be shown [4] that if $W_B/L_n \ll 1$, then η is given by

$$\eta = 1 - \frac{1}{2}\left(\frac{W_B}{L_n}\right)^2 \cdot \tag{3.62}$$

Values of η of 0.95 are typically achieved in practice.

One more transport factor to be defined is the collector collection factor, δ. If not all of the electrons at $x = W_B$ reach the collector (due to three-dimensional diffusion), δ will be less than 1. On the other hand, if avalanche multiplication occurs in the base-collector space-charge region, we may have $\delta > 1$.

An overall current transport factor, α, may be defined as

$$\alpha = \frac{\text{total collector current}}{\text{total emitter current}} \tag{3.63}$$

or from the above discussions, it can be seen that

$$\alpha = \gamma\eta\delta = \left(\frac{I_C}{I_E}\right) \cdot \tag{3.64}$$

The mechanisms described here are summarized in Figure 3.10. From Figure

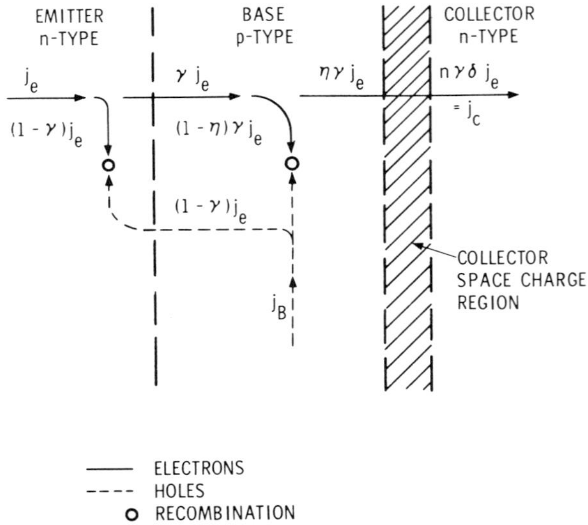

FIGURE 3.10
Currents flowing in the *n-p-n* transistor.

3.10 it can be seen that to maintain current balance there must exist a base current given by

$$I_B = I_E - I_C \tag{3.65}$$

(see Figure 3.9A for the assumed directions of I_B, I_C, and I_E). Using Eq. (3.64) and (3.65), we may write

$$I_B = I_C\left(1 - \frac{1}{\alpha}\right)$$

or

$$\frac{I_C}{I_B} = \frac{\alpha}{1 - \alpha} \equiv \beta \tag{3.66}$$

where β is called the common emitter current gain. Since α typically has values on the order of 0.95, β's of 20 are typical. Eq. (3.66), $I_C = \beta I_B$, shows why the transistor is often thought of as a current-controlled device.

3.7 Transistor Equivalent Circuit

This section briefly describes a simplified small-signal equivalent circuit of a transistor. Figure 3.11A shows a typical application of a transistor as a low frequency ac amplifier. Before a circuit analysis of the amplifier could be performed so that a required design could be achieved, it would be necessary to have an equivalent circuit for the transistor. Figure 3.11B represents a simplified low-frequency equivalent circuit of the *p-n* junction transistor. The series resistance, r_b, in Figure 3.11B represents the contribution due to the finite conductivity of the base bulk region, σ_B. Hence one can expect some ohmic drop across this region (see also Figure 3.9A). Since the emitter bulk

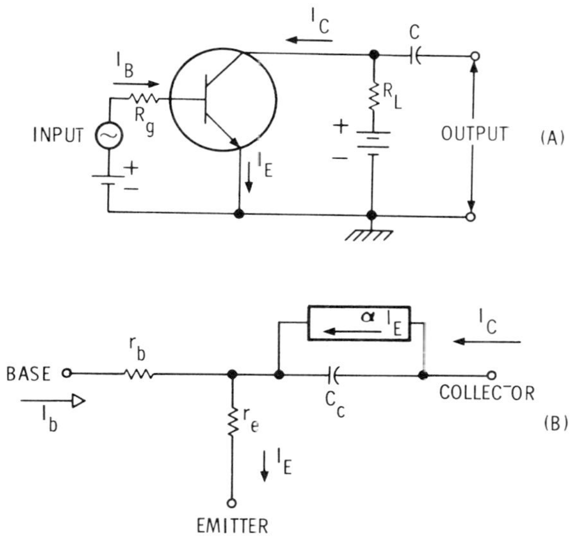

FIGURE 3.11
(A) Typical common-emitter amplifier circuit, and (B) simplified transistor equivalent circuit.

region is very heavily doped, its resistance can generally be ignored. The resistance, r_e, in Figure 3.11 is the inverse of junction conductance given by Eq. (3.36).

The collector-depletion capacitance is represented by C_c in Figure 3.11B. The current transport factor, α, is brought in through the current generator, αI_E, shown in Figure 3.11B. The current-transport factor, α, is frequency-dependent due to the fact that the ac diffusion length [Eq. (2.86)], instead of the dc diffusion length, must be used for ac currents. The well-known [3] form for the frequency dependence of α at low frequencies is

$$\alpha = \frac{\alpha_0}{1 + j(\omega/\omega_\alpha)} \tag{3.67}$$

where ω = operating frequency and

$$\omega_\alpha = \frac{2D_n}{W_B^2} \tag{3.68}$$

is called the alpha cutoff frequency. It should be realized that the equivalent circuit shown in Figure 3.11B is greatly simplified, and more complete equivalent circuit models can be found in the literature ([3] for example).

REFERENCES

[1] W. Shockley, "The Theory of p-n Junctions in Semiconductors and p-n Junction Transistors," *BSTJ*, **28**, 435 (1949).

[2] R. A. Smith, *Semiconductors*, Cambridge at the University Press, London, 1959.

[3] R. A. Griener, *Semiconductor Devices and Applications*, McGraw-Hill, New York, 1961.

[4] A. K. Johnscher, *Principles of Semiconductor Device Operation*, John Wiley & Sons, New York, 1960.

[5] J. L. Moll, *Physics of Semiconductors*, McGraw-Hill, New York, 1964.

[6] S. M. Sze, *Physics of Semiconductor Devices*, John Wiley & Sons, New York, 1969.

[7] E. M. Cherry, "Active Device Capacitances," *IEEE PGED*, **18**, 12, 1166, (1971).

[8] H. A. Watson, *Microwave Semiconductor Devices and Their Circuit Applications*, McGraw-Hill, New York, 1969.

[9] S. M. Sze and G. Gibbons, "Avalanche Breakdown Voltages of Abrupt and Linearly Graded p-n Junctions in Ge, Si, GaAs and GaP," *Appl. Phys. Lett.*, **8**, 5, 111 (1966).

[10] C. T. Sah, R. N. Noyce, and W. Shockley, "Carrier Generation and Recombination in P-N Junctions and P-N Junction Characteristics," *Proc. IRE*, **45**, 9, 1228 (1957).

[11] J. Lindmayer and C. Y. Wrigley, *Fundamentals of Semiconductor Devices*, D. Van Nostrand Co., New York, 1965.

[12] S. M. Sze and G. Gibbons, "Avalanche Breakdown Voltages of Abrupt and Linearly Graded p-n Junctions in Ge, GaAs, Si and GaP," *J. Appl. Phys.*, **8** (1966).

4

Device Fabrication and Parameter Measurement Techniques

4.1 Introduction

Up to this point nothing has been said about how pure single-crystal semiconductors or various doping profiles are obtained. The importance of such bulk parameters as conductivity, mobility, etc., have been discussed without mentioning how these parameters are measured experimentally. This chapter summarizes these subjects so that it will not be necessary to bring them up again in each discussion of a particular device or measurement. Section 4.2 discusses the techniques used in purifying the semiconductor materials and growing single crystals. Section 4.3 describes the alternative methods utilized in fabricating *p-n* junctions. Section 4.4 covers several important techniques used to measure the basic parameters of bulk semiconductor materials. Section 4.5 discusses techniques that are used to measure important bulk parameters in finished devices.

4.2 Fabrication of Pure Single Crystal Semiconductors

Before useful semiconductor crystals can be grown, it is first necessary to obtain very pure semiconductor material. Since most of the devices to be discussed are made from silicon, it shall be used as an example for these discussions. The need for high purity can be seen from the following arguments: In intrinsic silicon at 290°K, the free-carrier concentration is 2×10^{10} carriers/cm^3; if the impurity atoms in the crystal are all electrically active, then it would take only 2 × 10^{10} impurity atoms/cm^3 to double the intrinsic concentration. Since there are approximately 10^{22} atoms/cm^3 in a silicon crystal, intrinsic silicon could have only approximately one active impurity atom for every 10^{12} silicon atoms. In actual practice, intrinsic silicon is not used to make devices, but rather the material is compensated† or doped to arrive at the desired free-carrier concentration. Another important consideration in determining the allowed impurity concentrations can be seen from Figure 2.8. which shows that the mobility of the carriers begins to decrease above impurity concentrations of 10^{15}/cm^3. Minority-carrier lifetime (Chapter 2) is also known to decrease with increased impurity concentration. For these reasons, fabrication of semiconductor crystals is usually started with materials

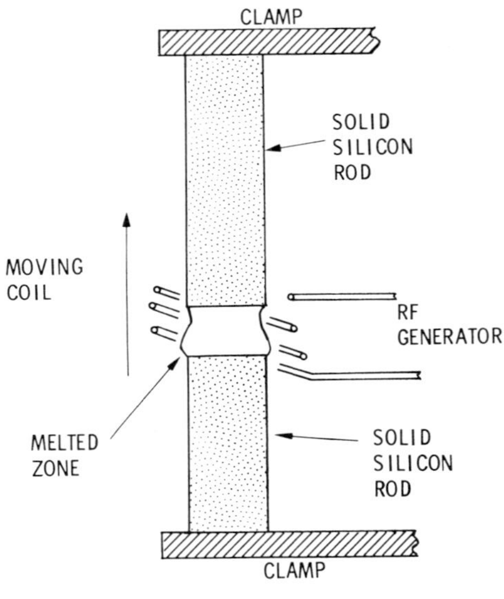

FIGURE 4.1
Apparatus for floating-zone refinement of silicon.

which have impurity concentrations of one part in 10^{10} or less. Obtaining any material to this purity is not an easy task. The problem is complicated by the high melting temperature (1420°C) and high reactivity of molten silicon. One of the most common methods used to purify silicon is the *floating-zone* (FZ) technique depicted in Figure 4.1. In the floating-zone process, a rod of poly-

† Compensation occurs when donors are added to *p*-type material or acceptors are added to *n*-type material to move the Fermi level to the desired level within the energy gap (see Chapter 2).

crystalline silicon (with impurities) is mounted between two clamps, and a small section of the silicon rod is melted by a movable rf induction heating coil. The entire apparatus is enclosed in a controlled atmosphere to prevent contamination. The small melted zone is held in place primarily by surface tension. The coil makes successive passes in one direction, and this has the effect of sweeping most of the impurity atoms to one end of the silicon rod. This end of the rod is then cut off and discarded. Impurities concentrate at one end because they are more soluble in melted silicon than unmelted silicon and hence tend to concentrate the molten layer. Successive sweeps of the molten layer through the rod can reduce purity levels to one part in 10^{11}

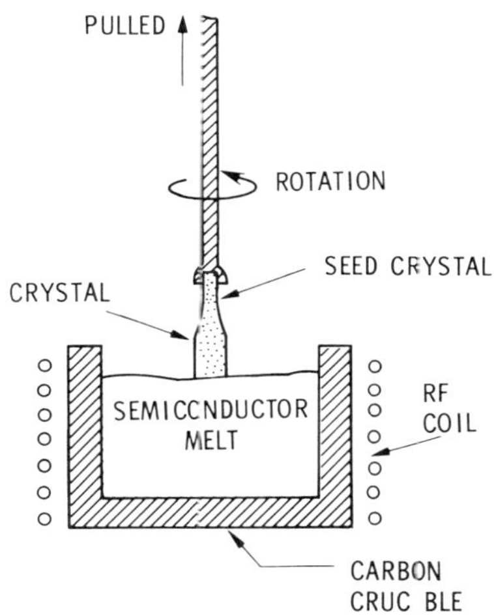

FIGURE 4.2
Basic Czochralski crystal-growing apparatus.

[1]. Other refinement techniques can also be used to purify the semiconductor materials (see [1–3]).

Once the silicon has been zone refined, the next step is to grow large single crystals from it. There are several well-established techniques for growing large single crystals [1–3]. One method that can be used is a simple extension of the floating-zone refinement technique. If a seed crystal of the desired orientation is placed at the end of the rod (Figure 4.1) where the melted zone is initiated, then, under the right conditions, the cooled silicon behind the moving zone will crystallize in the same orientation as the seed crystal. Another common crystal growing process is called the *Czochralski* or *crucible-growing* method. In this process, depicted in Figure 4.2, a seed crystal is dipped into a carbon crucible filled with a molten purified semiconductor. If the seed crystal is slowly withdrawn and rotated continuously, under the right conditions, a rod of single crystal semiconductor with the same orientation as the seed crystal is formed. The semiconductor crystal can be doped by adding a suitable dopant (see Chapter 2) to the melt. If the net melt impurity concentration is changed from donor to acceptor type while the crystal is being pulled, the result will be a "grown" *p-n* junction. This method of forming a *p-n* junction is of historical significance only. In the case of silicon it is necessary to line the carbon crucible with quartz (due to the reactivity of molten silicon). The silicon slowly dissolves the quartz liner, and, as a result, the crystal may contain a significant concentration of oxygen atoms. The oxygen atoms are electrically inactive and do not usually interfere with device fabrication and operation. However, in the case of radiation damage

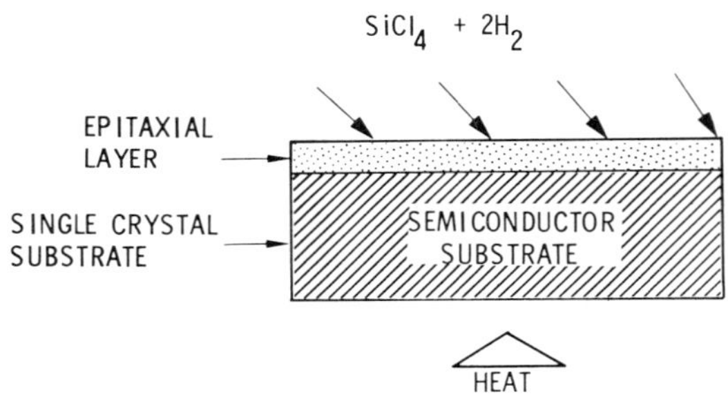

FIGURE 4.3
Epitaxial growth by reduction of silicon tetrachloride on a single-crystal substrate.

to materials, Chapter 5 will show that the oxygen atoms can react with defects to cause stable defect complexes. In this case, the effects of radiation on crucible-grown silicon crystals can be different from the effects of radiation on float-zone grown silicon crystals. (This is because the latter is essentially oxygen-free.)

Another common technique used for growing thin single-crystal layers is the *epitaxial* process. This process is depicted in Figure 4.3. The silicon crystal is grown by the reduction of silicon tetrachloride on a hot silicon single-crystal substrate (usually made of crucible-grown silicon). The reaction that occurs is

$$SiCl_4 + 2H_2 \rightarrow Si + 4HCl \, . \tag{4.1}$$

The resulting layer is a single crystal with the same orientation as the substrate. The epitaxial layer may be doped by the inclusion of gaseous impurities in the chamber while the growth is taking place. Many microwave semiconductor devices are fabricated by starting with a 2-to-20-micron-thick epitaxial layer grown on a heavily doped (for high conductivity) substrate of crucible- or floating-zone-grown single-crystal semiconductor. The substrate thickness is typically a few thousand microns for mechanical strength. Other crystal growing methods are described in references [1–3].

4.3 The Fabrication of *p-n* Junctions

Almost all of the microwave semiconductor devices to be described employ one or more *p-n* junctions. A *p-n* junction is usually formed by starting with a slice sawed from a rod of either floating-zone- or crucible-grown single-crystal semiconductor (which may have an epitaxial layer on it). A region of the slice is then converted to the opposite type (*n* to *p* or *p* to *n*) by changing the local net concentration of dopant atoms. If this is done, a *p-n* junction is formed at the boundary between the *n*- and *p*-regions. There are four basic ways of locally changing the net impurity concentration from one type to the other. These are: (1) the grown-junction method, (2) the alloying method, (3) the diffusion method, and (4) the ion-implantation method.

The Grown Junction. The grown-junction method of forming a *p-n* junction was discussed earlier in this chapter. A grown *p-n* junction can be formed by changing the net impurity concentration in the melt while the crystal is being pulled. This method of forming a *p-n* junction is of historical significance but is no longer used. A modern technique that is a corollary of this technique is the growth of a *p*-type epitaxial layer on an *n*-type substrate or vice versa.

The Alloying Method. The alloying process was once extensively used in the fabrication of transistors and diodes. Today, however, it is only occasionally used in the manufacture of low-cost, low-frequency devices. This technique consists of placing a small pellet on the semiconductor substrate. The pellet contains impurities of the type opposite to the substrate. The melting point of the pellet is considerably lower than that of the single-crystal semiconductor substrate. When the substrate and pellet are heated, the pellet melts. The resulting liquid then begins to dissolve the semiconductor substrate. The temperature is then lowered and the pellet resolidifies. The melted region near the substrate-pellet interface recrystallizes in the same orientation as the substrate and is heavily doped with the impurities from the pellet. In this way, a very abrupt *p-n* junction can be formed. The difficulties with this technique are a lack of control of the area and the depth of the resulting junction. Since both of these quantities must be carefully controlled in microwave semiconductor devices, the alloy technique is not usually used to make these devices.

The Diffusion Method. The diffusion process is the one most frequently employed to make silicon *p-n* junctions in microwave semiconductor devices. The reason for this is that the diffusion process results in excellent dimensional control of the dopant-atom depth distributions. Before a discussion of *p-n* junction fabrication techniques, a brief discussion of diffusion of impurities will be presented. The diffusion of impurity atoms is governed by Fick's law (Eq. 2.40) which is

$$\bar{j} = -D\nabla C \tag{4.2}$$

where ∇ = del operator $= \frac{\partial}{\partial x}\bar{a}_x + \frac{\partial}{\partial y}\bar{a}_y + \frac{\partial}{\partial z}\bar{a}_z$

$\bar{j}$ = impurity-atom current

C = impurity concentration

D = impurity diffusion coefficient (independent of position)

Since the impurity atoms can neither be created or destroyed, they must also obey a continuity equation

$$\nabla \cdot \bar{j} + \frac{\partial C}{\partial t} = 0\,. \tag{4.3}$$

Substituting Eq. (4.2) into Eq. (4.3) and assuming a one-dimensional problem, we may write

$$\frac{\partial C(x, t)}{\partial t} = D\frac{\partial^2 C(x, t)}{\partial x^2}\,. \tag{4.4}$$

If D is independent of concentration and we apply the boundary condition that $C = C_s$ at $x = 0$, then the solution to Eq. (4.4) is

$$C(x, t) = C_s \operatorname{erfc}\left(\frac{x}{2\sqrt{Dt}}\right) \tag{4.5}$$

where erfc = complementary error function

$$\operatorname{erfc}(U_0) = \frac{2}{\sqrt{\pi}}\int_{U_0}^{\infty} e^{-U^2}\,dU\,. \tag{4.6}$$

Figure 4.4. shows a plot of Eq. (4.5). From Figure 4.4 it can be seen that as $U \to 0$ ($t \to \infty$ or $x \to 0$) the impurity concentration approaches the surface concentration and as $U \to \infty$ ($t \to 0$ or $x \to \infty$) the impurity concentration $\to 0$. Equation (4.5) does not hold for all impurity types, since in some cases (arsenic impurities in Si, for example) D is not independent of C. Practically speaking, the diffusion is carried out at elevated temperatures (around 1000°C for Si) because D increases with temperature. The diffusion time varies from several minutes to a few hours. Once the temperature is reduced back to room temperature, the diffusion coefficient becomes vanishingly small and the diffusion process essentially stops. We are left with an impurity profile determined by Eq. (4.5) (with t = constant). It can be seen from Figure 4.5 and Eq. (4.5) that short diffusions result in shallow-abrupt *p-n* junctions while long diffusions (at the same temperature) result in deep-graded *p-n* junctions. A further discussion of diffusion is given in Ref. [3].

There are two basic subclasses of diffusion techniques used for fabricating *p-n* junctions in microwave semiconductor devices. These are the diffused

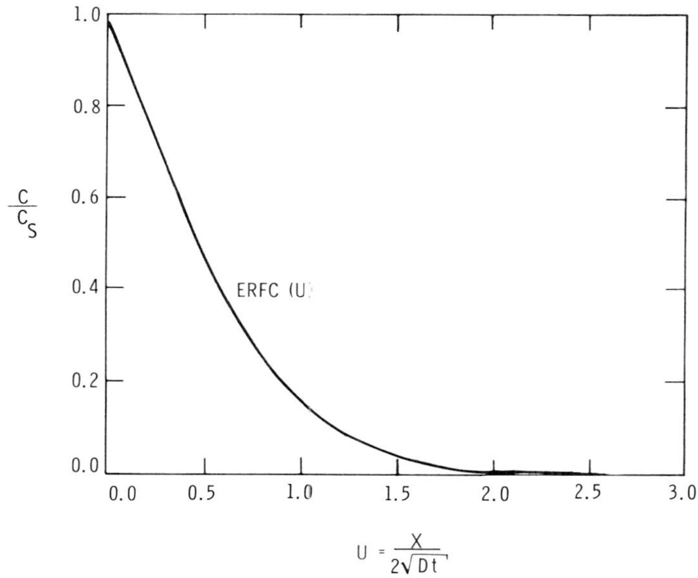

FIGURE 4.4
Normalized impurity concentration or complimentary error function.

mesa-junction technique (used to make IMPATTs and other microwave diodes) and the diffused *planar-junction* technique (used to make microwave transistors).

The diffused mesa fabrication process is depicted in Figure 4.6. In this process a semiconductor substrate (it could be an epitaxial layer on a substrate also) has impurities of the opposite type diffused into one surface (at elevated temperature). The result is a *p-n* junction. The surface is then metalized (see Figure 4.6B). The metalized layer is covered with Kodak Photo Resist (KPR) which is selectively exposed to ultraviolet light. Those regions which are not exposed do not polymerize and are etched away during a resist development cycle. The silicon is now etched from the top and the unprotected areas are etched away (see Figure 4.6D). The result is a diffused mesa diode, which after substate metalization and surface passivation, is ready for bonding into a package.

The second subclass of diffused *p-n* junctions is the planar process [4] depicted in Figure 4.7. The first step in this process consists of thermally

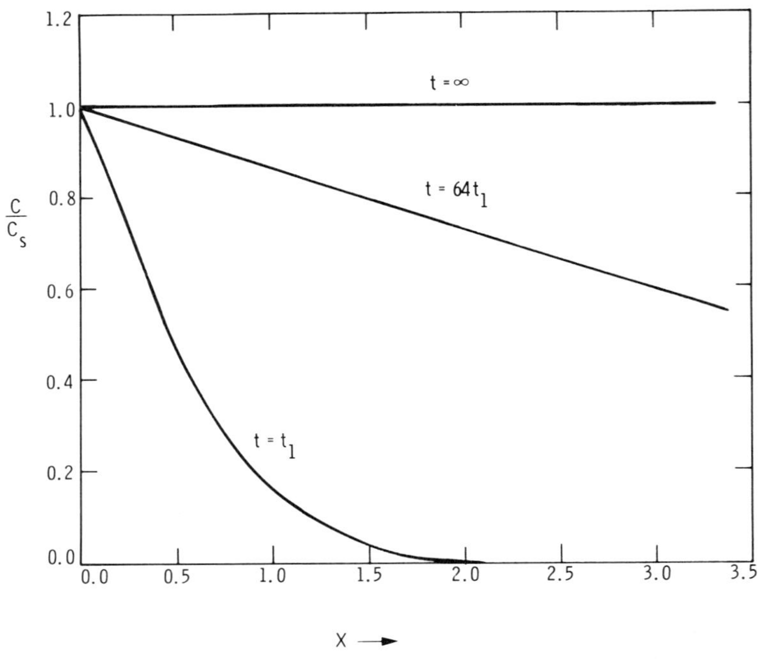

FIGURE 4.5
Impurity concentration versus distance for varying diffusion times.

growing a layer of SiO_2 on the substrate (or on the epitaxial layer on the substrate). This is accomplished by heating the substrate in the presence of pure H_2O or O_2. The SiO_2 layer forms a barrier through which most common dopant impurities do not readily diffuse. Windows are next etched in the SiO_2 (see Figure 4.7B) by the photoresist process described earlier. Diffusion is then carried out through the windows at elevated temperatures using the opposite type of impurity atoms (donors or acceptors) from those in the substrate. This process is depicted in Figure 4.7C. The result is a well-defined planar p-n junction as shown in Figure 4.7D. Surface passivation is not required since the SiO_2 layer protects the surface.

The Ion-Implantation Method. The ion-implantation technique is relatively new and has not yet been fully exploited. In this technique, ions of the dopant atom are accelerated to energies from 50 to 300 keV and used to bombard the semiconductor substrate surface. The result is the formation of a p-n junction near the surface of the substrate. The ions produce damage in the substrate

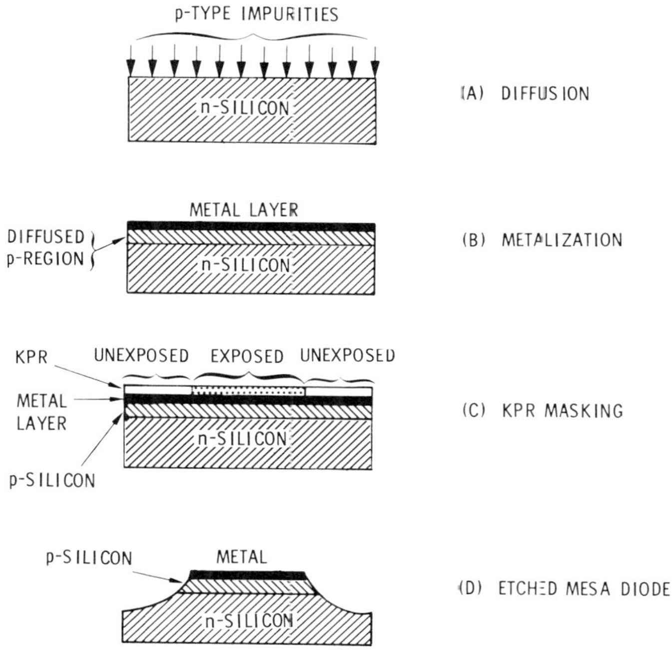

FIGURE 4.6
Diffused mesa-diode fabrication steps.

which is usually removed by a moderate temperature (600°C) annealing cycle. The areas to be implanted can be defined by using a process like the one shown in Figure 4.7 (SiO_2-KPR). The range of the ions is typically 1 micron or less and can be controlled by varying the ion energy. This technique has the advantage of being able to produce shallow abrupt doping profiles. Ion implantation shows promise for making millimeter wave microwave diodes (because of the short penetration range of the ions).

Other methods of forming p-n junctions are described in Ref. [1–3].

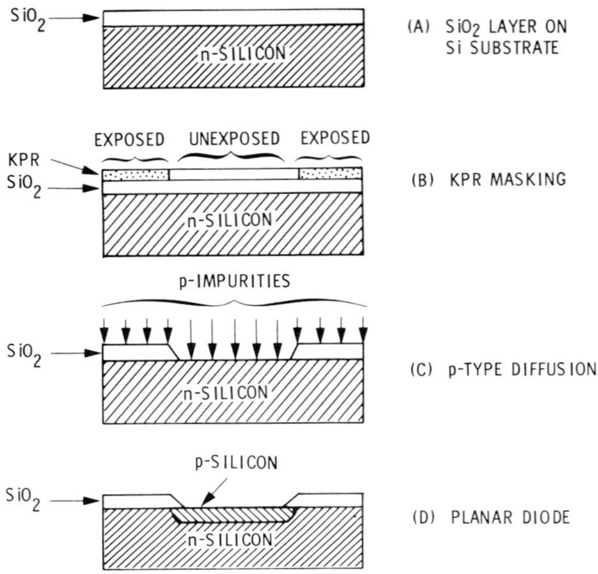

FIGURE 4.7
Diffused planar-diode fabrication steps.

4.4 Measurement of Bulk Semiconductor Properties

The parameters of a finished microwave semiconductor device depend strongly on the characteristic properties of the starting semiconductor material. Such things as carrier concentration, mobility, conductivity, and lifetime (see Chapter 2) must have the correct values before device fabrication can begin if a repeatable device is to be made. Also, the characteristic properties of specific regions of finished devices cannot always be measured, and therefore, in some cases the parameters of the starting semiconductor material are used in the analysis of a finished device. (This is not always correct since the high temperatures involved with diffusion may change some of these parameters, notably the lifetime.) Also, in radiation effects it is usually assumed that radiation-induced changes in bulk material of a given type would affect a region of the same type of material in a finished device in the same way. This has the advantage that the bulk measurements are usually much easier to perform. This method must be applied carefully, however, since depletion regions and other junction effects do not occur in bulk material. There exists a large number of ways of measuring the characteristic properties of bulk semiconductors and only a few of the more common ones will be given here. The reader is directed to Ref. [3] for a much more complete list of available techniques. The bulk measurements to be discussed are (1) conductivity, (2) mobility, and (3) lifetime. From these measurements it is possible to calculate the carrier concentration [through Eq. (2.39)] and the diffusion coefficient [through Eq. (2.47)].

Measurement of Conductivity. Resistivity, ρ, and conductivity, σ, ($\sigma = 1/\rho$) are usually expressed in units of Ω-cm (for ρ). The conductivity depends on both the carrier concentration and the mobility [Eq. (2.39)]. Usually only one type of carrier has to be considered in conductivity measurement because in an extrinsic semiconductor one type of dopant usually dominates [see the discussion following Eq. (2.31)]. The simplest way to measure conductivity would be to take a block of uniform semiconductor L cm long with a cross sectional area A, and measure the resistance between the ends. The result should be

$$R = \frac{L}{\sigma A} \cdot \tag{4.7}$$

If L and A are known, σ may be calculated from Eq. (4.7). This method suffers from the drawback that to measure the resistance a current is passed through the semiconductor block and the resulting voltage drop ($V = I \cdot R$) is measured. The contacts to the ends of the block may rectify, or may inject minority carriers which would void the results. The solution to these problems is to abrade the surfaces where contacts are made (to reduce the lifetime) and to use a four-point measuring system. In the four-point system a known current is passed between two probes and the voltage drop is measured potentiometrically (so that no current flows between the voltage measuring probes) between the other two probes. For a bulk measurement, a van der Pauw sample [6] is usually used. Figure 4.8 depicts a typical van der Pauw sample. For the sample shown in Figure 4.8 the resistivity is given by [6]:

$$\rho = \frac{\pi d}{2 \ln 2} \left[\frac{V_{DC}}{I_{AD}} + \frac{V_{AD}}{I_{BC}} \right] \cdot \tag{4.8}$$

The van der Pauw method is suitable for bulk measurements of specially built samples. However, in some cases it may be desirable to probe various locations of a semiconductor substrate (sliced from a single crystal rod). These slices are typically 2.5 to 5 cm in diameter. Also, a measurement of the resistivity of an epitaxial layer on heavily doped substrate may be desired. These measurements can be achieved with the four-point probe [7,8]. The basic four-point probe setup is shown in Figure 4.9 being applied to an epitaxial layer on a heavily doped (approximated by a conductor) substrate.

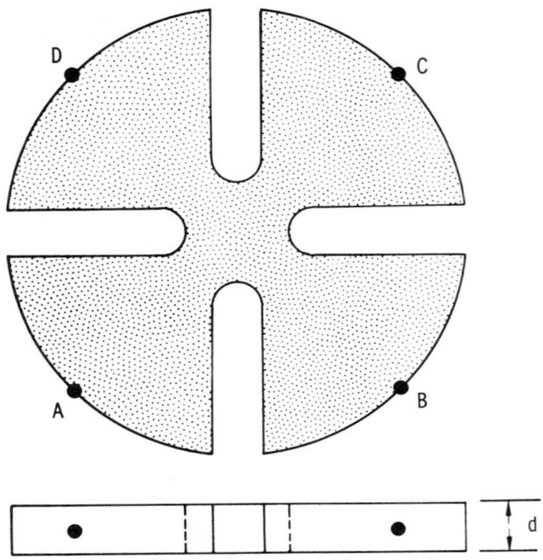

FIGURE 4.8
A typical van der Pauw [6] sample.

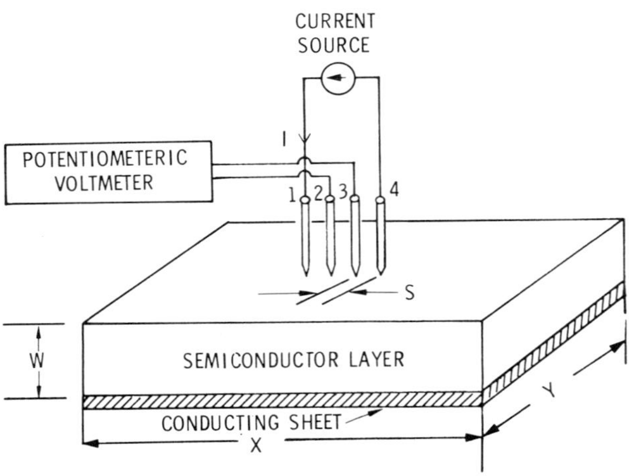

FIGURE 4.9
Four-point probe measurement of resistivity (after Valdes [7]).

For this particular case, Valdes [7] has shown that if $S \ll X,\ Y$, and $W/S \geq 4$ then

$$\rho \cong \frac{V_{23}}{I_{14}} \cdot 2\pi S\,. \tag{4.9}$$

Equation (4.9) is also a general result for a semi-infinite semiconductor substrate of uniform resistivity, ρ. The values of ρ for other cases may be found from Ref. [7,8].

Measurement of Mobility. The concept of carrier mobility, μ, has been discussed previously (see Chapter 2, Section 2.6). At low applied electric

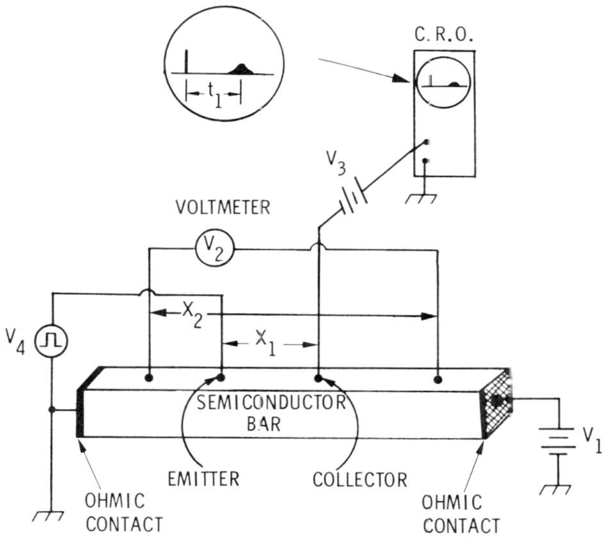

FIGURE 4.10
Haynes-Shockley drift-mobility experiment.

fields the carrier velocity is given by $\bar{v} = \mu\bar{E}$. Mobility has the units cm^2/volt-sec. There are two basic ways of measuring mobility: the *Haynes-Shockley drift-mobility* technique [9] and the *Hall-effect mobility* technique [10]. Figure 4.10 shows a typical experimental setup used to perform the Haynes-Shockley measurement. A uniform bar of semiconductor has a field applied to it by use of the battery, V_1 (see Figure 4.10). The drift field inside the sample is found from a measurement of V_2 and the distance X_2, $E = V_2/X_2$. A pulser V_4, is used to inject minority carriers at an emitter. The collector contact is biased to attract only minority carriers. A spike appears on the C.R.O.

(cathode ray oscilloscope) when the pulser, V_4, is fired. This is due to the voltage divider action of the bar. This spike gives a convenient reference time mark for the time when injection occurs. The pulse of injected minority-carriers drift under the action of the applied field, E, and the peak arrives at the collector at time t_1 seconds after injection. Neglecting recombination, we can easily see that minority-carrier drift velocity is $v = X_1/t_1$, and hence the minority-carrier drift mobility is

$$\mu_D = \frac{X_1 \cdot X_2}{t_1 \cdot V_2} . \tag{4.10}$$

The C.R.O. display (Figure 4.10) also shows a broadening of the injected pulse due to diffusion. Majority-carrier mobility cannot be measured by this experiment.

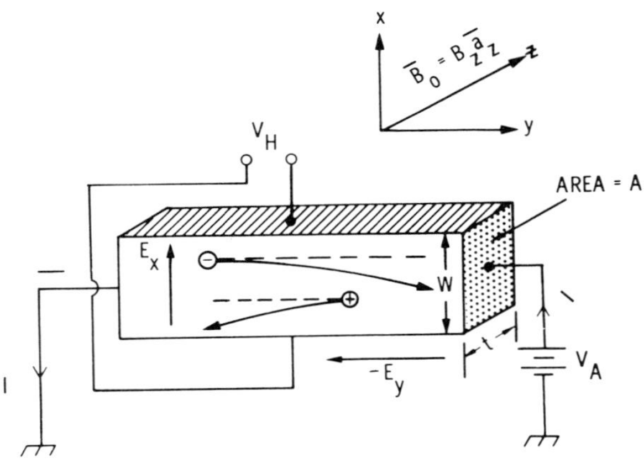

FIGURE 4.11
Depiction of Hall-effect experiment.

The Hall-effect method for measuring mobility is depicted in Figure 4.11. The interaction of moving charged particles with electric and magnetic fields is covered by the well-known Lorentz equation

$$\bar{F} = q(\bar{E} + \bar{v}x\bar{B}) . \tag{4.11}$$

In particular, the drift velocity of the carriers (induced by V_A, Figure 4.11) interacts with a magnetic field in the z direction to cause a deflection of the carriers in the x-direction (see Figure 4.11). For the moment consider just one type of carrier (holes). The deflection of holes will cause an eventual pile-up of positive charge on the bottom edge of the sample (Figure 4.11) and this in turn will set up an electric field E_x which will just balance the $\bar{v}x\bar{B}$ term in Eq. (4.11) (no net current can flow in x direction in the steady state). The Hall voltage is simply

$$V_H = E_x \cdot W = v_y \cdot B_0 \cdot W = \mu_H \cdot E_y \cdot B_0 \cdot W \tag{4.12}$$

solving for μ_H

$$\mu_H = \frac{E_x}{E_y B_0} \cdot \tag{4.13}$$

In general the situation is much more complex than described here. Both electrons and holes, and in some cases more than one kind of hole (i.e., heavy and light holes), must be considered. However, in the case of extrinsic material, an effective Hall mobility for the majority carrier can be readily measured. The Hall mobility and conductivity mobility are in general not equal due to the fact that on a microscopic scale the electrons have a distribution of velocities upon which the magnetic field acts. It can be shown [10] that $\mu_H/\mu_\sigma = 3\pi/8$ if the semiconductor has spherical constant energy surfaces (which is not true for silicon). In general a strict interpretation of the Hall mobility cannot be made for most semiconductors. However, practically speaking, the Hall mobility is usually fairly close to the conductivity mobility (within a factor of 2) even in postirradiated samples. The Hall mobility can be measured with a van der Pauw sample (Figure 4.3). The result is [6]:

$$\mu_H = \frac{d}{B_0 \rho}\left[\frac{V_{CA1}}{I_{BD1}} - \frac{V_{CA0}}{I_{BD0}}\right] \tag{4.14}$$

where the 1 subscript corresponds to the magnetic field turned on and the 0 subscript corresponds to the magnetic field turned off.

The mobilities that have been discussed are the low-field mobilities. At high electric fields the velocity is usually found to saturate (see Figure 2.10). This concept is particularly important to microwave semiconductor devices. Hence, a brief discussion of how characteristics of the type shown in Figure 2.10 are obtained is worthwhile. One of the most accurate methods for obtaining saturated drift velocity measurements is the one proposed by Norris and Gibbons [11], which is depicted in Figure 4.12. The technique consists of injecting electrons and holes at one edge of a reverse biased p^+-i-n^+ diode (i denotes an intrinsic region). The injection is accomplished by bombarding the diode with a 10 keV electron beam (0.5 nanosecond pulse width). The 10 keV energy is low enough so that carriers are created only at the left-hand edge of the diode. However, the 10 keV electrons still have sufficient energy to create a large number of electron-hole pairs for each incident electron. For the setup shown in Figure 4.12, the electrons will immediately recombine in the n^+ contact while the holes will drift across the space-charge region. All the time the holes are drifting, the diode will draw a terminal current:

$$I = \frac{pev_0}{d} \tag{4.15}$$

where p = hole density, v_0 = drift velocity, and d = I region width. This current will produce a voltage drop $I \cdot Z_0$ on the transmission line. When the

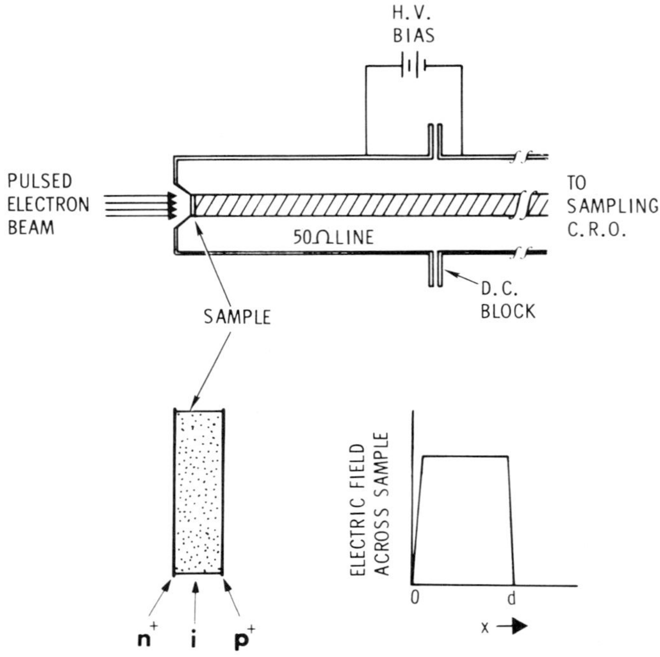

FIGURE 4.12
Apparatus for measurement of saturated drift velocity (after Norris and Gibbons [11]).

holes arrive at the i-p^+ interface the current will ideally drop to zero. If d is accurately known and the voltage pulse ($I \cdot Z_0$) width, T, is accurately measured, the drift velocity can be determined as a function of field, i.e.,

$$E = \frac{\text{High Voltage Bias}}{d}$$

$$v_0 = \frac{d}{T}. \tag{4.16}$$

The technique can be carried out on a repetitive basis so that a sampling C.R.O. can be used to measure T. This allows the measurement of very small values of T which in turn allows the use of narrow diodes and reasonable voltage bias levels.

The Measurement of Bulk Lifetime. In Chapter 2, Section 2.7, the concept of carrier lifetime due to recombination levels was discussed. The value of carrier lifetime has a strong effect on the operation of most microwave semiconductor devices. An accurate method for the measurement of bulk lifetime is shown in Figure 4.13, which depicts a variation of the Stevenson-Keyes experiment [12]. In this experiment the conductivity of a bulk sample is increased through the creation of excess electron-hole pairs by the absorption of an incident pulse of photons whose energy $h\nu > E_g$. The change in conductivity is proportional to the density of excess carriers, and since the density

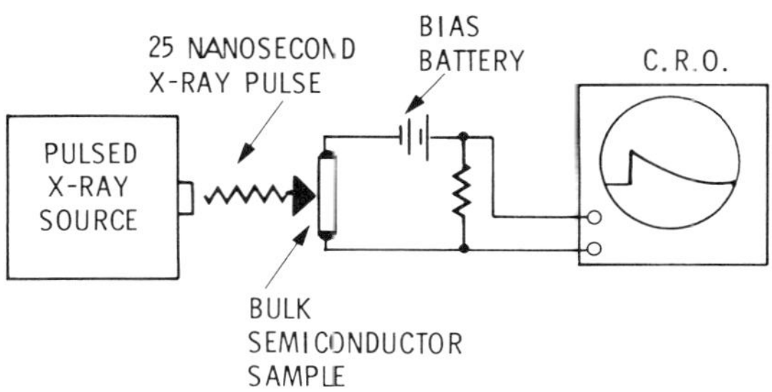

FIGURE 4.13
Setup for measurement of bulk lifetime using an x-ray pulse.

of excess carriers falls off exponentially once the source (radiation pulse) is removed [see Eq. (2.53)], a C.R.O. display like the one shown in Figure 4.13 would be expected. The excess carrier lifetime can be obtained from Eq. (2.53) and the C.R.O. display. Several factors can complicate these measurements. One effect is due to surface recombination. Stevenson-Keyes [12] have shown that the experimentally measured lifetime can be expressed as

$$\frac{1}{\tau_{\text{exp}}} = \frac{1}{\tau_B} + K \tag{4.17}$$

where τ_{exp} = experimentally measured lifetime, τ_B = actual bulk lifetime, and K = a constant dependent on surface recombination effects. K can be made small if a thick sample is used so that surface recombination can be neglected. If a thick sample is used, energetic x-rays make a better photon source than light (as originally proposed [12]). This is because only a small fraction of the x-rays will be absorbed in passing through the sample, resulting in an essentially uniform distribution of excess carriers, even in a thick sample. Haynes and Hornbeck [13] have shown that measurements of this nature can also be complicated by the effect of slow traps in the material. The interpretation of the lifetime measured using this technique is discussed by Moll [14]. The shortest lifetimes that can be measured by this technique must be a factor of 5 or so greater than the fall time of the x-ray pulse, which is typically 20 nsec.

4.5 Measurement of the Characteristic Properties of Semiconductors in Completed Devices

High temperature processing steps can make the characteristic properties of a semiconductor region in a finished device different from the bulk properties of the starting material. Also, in the case of radiation effects it is important to know how the characteristic properties of different regions of a finished device change with radiation. In general, finished devices do not lend themselves well to the types of measurements described in the previous section on bulk measurements. This is because finished devices are usually inhomogeneous. Many techniques have been developed to measure characteristic properties inside finished devices. Several of those techniques pertinent to microwave devices will be described here. Packaged microwave semiconductor devices are designed to have very low parasitic element values, i.e., the series resistance, series inductance, and shunt capacitance are all small. This is required if short time constants, necessary for microwave operation, are to be achieved. The fact that microwave devices have small parasitic values can be exploited in certain measuring techniques, as will be seen shortly. This section will discuss measurements of: (1) series resistance, (2) doping density, (3) lifetime, and (4) mobility in finished devices.

Measurement of Series Resistance. The simplest measurement of resistance that can be made is the one described at the beginning of Section 4.4. The measurement of the resistance of a block of uniform semiconductors of known dimensions allows calculation of the conductivity [see Eq. (4.7)]. This measurement is only meaningful for homogeneous semiconductor devices with ohmic contacts and no *p-n* junctions. Only one class of microwave semiconductor devices fits into this category: these are Gunn and LSA devices biased well below threshold voltage. For these devices the conductivity of the active region can be calculated from Eq. (4.7) even in a postirradiated device.

The preceding method cannot be readily applied to *p-n* junction devices. *p-n* junction devices have equivalent circuits of the type shown in Figure 3.5. If the value of R_s is large (which is usually *not* the case), then at large forward currents the voltage drop across the diode will be $V_f \simeq I \cdot R_s$ (if $V_f \gg V_D$, the diffusion potential). This measurement can be made with varying degrees of success in point-contact and some Schottky-barrier diode mixers and detectors.

If the *p-n* junction is reverse-biased, the only parameters of importance in the equivalent circuit of Figure 3.5 will be R_s and C_d. Measurement of the diode cutoff frequency

$$f_c = \frac{1}{2\pi R_s C_d}, \tag{4.18}$$

or measurement of the resonant quality factor, Q_m, at a frequency f_m ($f_c = Q_m \cdot f_m$, Ref. [15]) can yield the value of R_s if C_d is known.

Measurement of Doping Density. If an abrupt *p-n* junction is one sided (i.e., if one side of the *p-n* junction is much more heavily doped than the other), then a measurement of the *C-V* characteristic in the reverse-bias region can yield the doping density of the lightly doped side. This can be seen from Eq. (3.23) and Figure 6.3 which show that the plot of the slope of the $1/C_d^2$ versus reverse voltage is proportional to $2/(\epsilon_s N_D e)$, assuming $N_A \gg N_D$. This technique must be used carefully when the density of trapping centers becomes appreciable, as is usually the case in irradiated devices. EerNisse [16] has shown that the depletion capacitance is a strong function of measuring frequency in certain ranges. If the measuring frequency is kept very low (a hundred Hz), the effect of traps can be ignored. If the measuring frequency is high (above 1 MHz), C_d is reduced (corresponding to a smaller N_D) due to compensation of the material by the traps.

Measurement of Carrier Lifetime. Because of their small parasitics, microwave semiconductor devices are well suited to lifetime measurement using fast pulses. Two basic techniques will be described here, both of which yield fairly accurate results. The first technique is due to Lederhandler and Giacoletto [17]. This method depends on measuring the voltage decay across a *p-n* junction after a steady forward current is interrrupted. Figure 4.14 shows a typical setup for this experiment. If a p^+-*n* diode is considered, then the only important carriers to be considered at forward bias are the holes injected into the *n*-region. The total hole density at the p^+-*n* boundary is

$$p_n(0) = p_{no} + \Delta p\,. \tag{4.19}$$

Using the arguments similar to those used for Eq. (3.47), we can write

$$p_n(0) = p_{no} \exp\left(\frac{eV}{kT}\right). \tag{4.20}$$

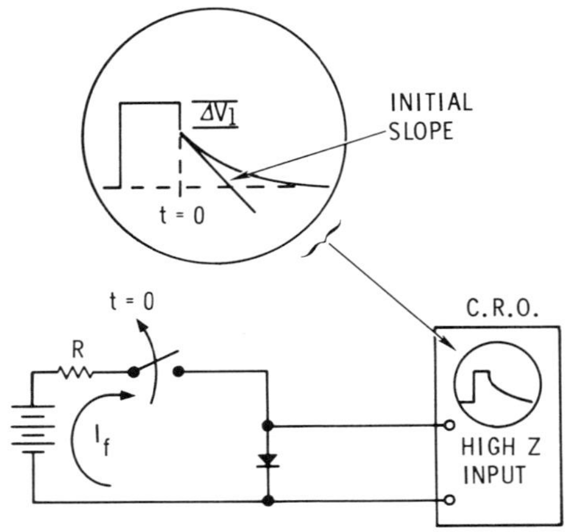

FIGURE 4.14
Open-circuit voltage decay method for lifetime measurement (after Lederhandler and Giacoletto [17]).

Combining Eq. (4.19) and (4.20) and solving for V, we obtain

$$V = \frac{kT}{e} \ln \left(1 + \frac{\Delta p}{p_{n0}}\right) \cdot \tag{4.21}$$

If the supply of injected carriers is suddenly cut off at $t = 0$, then the excess density decays exponentially† and is given by

$$\Delta p = \Delta p_0 e^{-t/\tau} \tag{4.22}$$

where Δp_0 = value at $t = 0$. From Eq. (4.19) and (4.20) we may write

$$\Delta p_0 = p_{n0}\left[\exp\left(\frac{eV_0}{kT}\right) - 1\right] \tag{4.23}$$

where V_0 = voltage at $t = 0$. Inserting Eq. (4.23) and (4.22) into (4.21) and assuming $V_0 \gg kT/e$ and $t \ll \tau$, we can show

$$V(t) = V_0 - \frac{kT}{e\tau} \cdot t\,. \tag{4.24}$$

The initial slope of the voltage decay is just $-kT/e\tau$ or $-0.025/\tau$ at room temperature. Hence, the carrier lifetime, τ, can easily be found from the slope of the decay at $t = 0$ on the C.R.O. display (see Figure 4.14). Figure

† Ref. [17] shows that this is true even in the presence of the resulting diffusion.

4.14 also shows sudden drop in voltage, ΔV_1, at $t = 0$. This drop is just due to the ohmic $I_f \cdot R_s$ drop in the diode and can be used as a measure of R_s. This technique for measurement of lifetime works well in the region 10 μsec $> \tau >$ 50 nsec. At long lifetimes, surface recombination effects dominate the measurement. At short lifetimes, the rise time of most laboratory C.R.O.'s limits the measurement.

The second method for pulse measurement of lifetime in finished devices is attributed to Moll et al. [18]. If a one-sided p^+-n junction is considered, the excess charge stored in the n-region at forward bias is given by Eq. (3.29).

$$Q'_{pn} = ep_{n0}L_p\left[\exp\left(\frac{eV_A}{kT}\right) - 1\right]. \tag{4.25}$$

(The charge stored in the p^+ region can be neglected since the emitter efficiency is near unity.) The resulting hole diffusion current [Eq. (2.42B)] is given by

$$J_p = -eD_p \left.\frac{dp_n(x)}{dx}\right|_{x=0}. \tag{4.26}$$

The distribution $p_n(x)$ is given by Eq. (3.14) (with x replaced by $-x$ if we want to consider positive values of x). Inserting the modified form of Eq. (3.14) into Eq. (4.26) and carrying out the indicated operations, we obtain

$$J_p = \frac{eD_p p_{n0}}{L_p}\left[\exp\left(\frac{eV_A}{kT}\right) - 1\right]. \tag{4.27}$$

Comparing Eq. (4.25) and (4.27) (realizing $L_p = \sqrt{D_p \tau_p}$), we see that

$$J_p = \frac{Q'_{pn}}{\tau_p}. \tag{4.28}$$

The hole-diffusion current represents the total device current for the forward-biased p^+-n junction. Hence, the forward-bias current is simply the ratio of the excess stored minority-charge density divided by the minority lifetime. Since J and Q'_p are defined per unit area, then

$$I_f = \frac{Q_f}{\tau_p} \tag{4.29}$$

where I_f = forward-bias terminal current and Q_f = total stored charge. Equation (4.29) forms the basis for the lifetime measurement method of Moll et al. [18]. If the forward current, I_f, and stored charge, Q_f, can be measured, then τ_p can be calculated. Figure 4.15 shows a typical setup for measuring lifetime in this way. This setup is carefully designed to have good high-frequency response. The operation proceeds as follows: First, a steady current, I_f, is established through the device. Next, a charged transmission line is used to suddenly reverse-bias the diode. The peak reverse current that can flow is limited to $V_{\text{rev}}/50\ \Omega$. This value should be much larger than I_f so that all of the current will be extracted in a time much less than τ (so recombination can be neglected). The charge extracted is

$$Q_R = \int_0^{\infty} I_R \cdot dt\,. \tag{4.30}$$

Q_R may be found from the C.R.O. display as indicated in Figure 4.15. Lindmayer and Wrigley [19] have shown that Q_R must be less than Q_f. For a

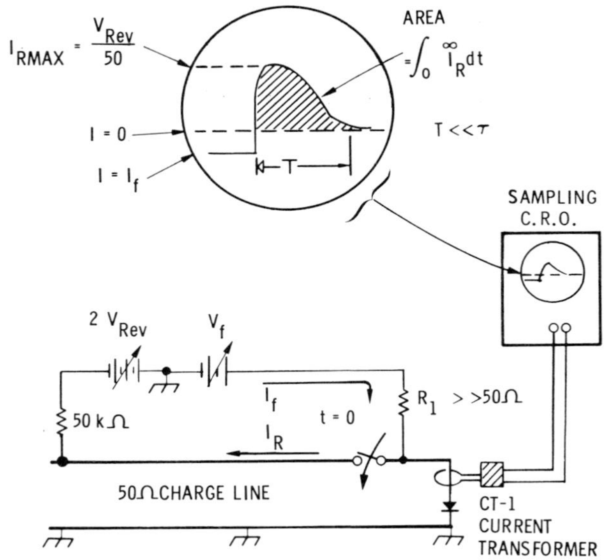

FIGURE 4.15
Charge removal method for lifetime measurement.

wide diode they predict $Q_R = Q_f/2$, and for a narrow diode (width less than L_p) $Q_R = 2Q_f/3$. Hence, the lifetime may be found from

$$\begin{aligned} \tau &= \frac{2Q_R}{I_f} \quad \text{wide diode} \\ \tau &= \frac{3Q_R}{2I_f} \quad \text{narrow diode}\,. \end{aligned} \tag{4.31}$$

Another source of error that must be taken into account is the charging of the depletion capacitance by I_R. This has been discussed by Venkateswaren and Roulston [20], and they show that in some cases this error can be as large as 30%. For carefully designed fixtures and diodes with small parasitics, lifetimes as short as 1 nanosecond can be measured with the apparatus shown in Figure 4.15.

Measurement of Mobility. Only one class of device lends itself well to the measurement of mobility in the finished device and that is Gunn and LSA devices. Jervis and Johnson [21] have shown that the geometrical magnetoresistance (GMR) of a finished Gunn device is essentially equal to the Hall mobility of the starting material. The GMR will be explained with the aid of Figure 4.16. The Gunn device shown in Figure 4.16 has $W/L \gg 1$. A Gunn

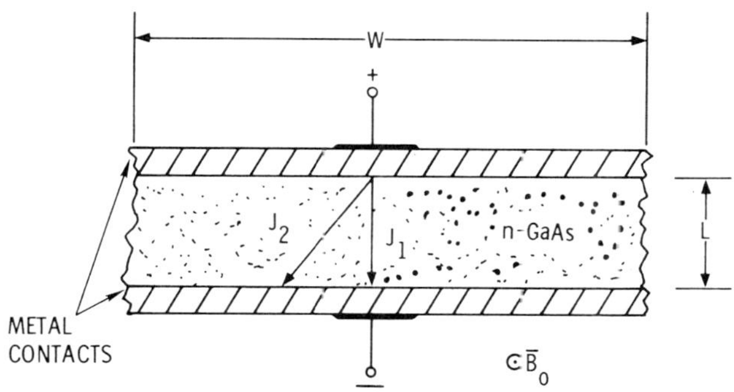

FIGURE 4.16
Explanation of geometrical magneto-resistance effect in a Gunn diode.

(or LSA) device is usually formed by a lightly doped n-type GaAs layer between two very heavily doped GaAs layers (contacts). These heavily doped layers have been modeled as metal contacts in Figure 4.16. When a current flows with the magnetic field, $\bar{B}_0$, turned off, the carriers follow the path J_1 in Figure 4.16. With the magnetic field on (coming out of the page), the carriers deflect and follow the path J_2 in Figure 4.16. Normally this would create a Hall voltage that would tend to straighten out the path of the carriers. However, in the geometry shown in Figure 4.16, the Hall voltage is shorted out by the metal contacts, and the carriers actually must traverse a longer path going from one contact to the other. Jervis and Johnson [21] show that

$$\mu_H \cong \frac{1}{B_0}\left[\frac{\Delta R}{R_0}\right]^{1/2} \tag{4.32}$$

where μ_H = Hall mobility, B_0 = magnetic field strength, R_0 = resistance between contacts when $B_0 = 0$, ΔR = increase in resistance with B_0 turned on. Equation (4.32) was found to be accurate within a few percent as long as the voltage across the sample was kept well below the threshold for oscillation.

REFERENCES

[1] J. P. McKelvey, *Solid State and Semiconductor Physics*, Harper and Row, New York, 1966.
[2] N. B. Hannay, *Semiconductors*, Reinhold Publishing Corp., New York, 1959.
[3] L. P. Hunter, ed., *Handbook of Semiconductor Electronics*, McGraw-Hill, New York, 1970.
[4] Ibid., Chapter 7.
[5] J. W. Mayer et al., *Ion Implantation in Semiconductors*, Academic Press, New York, 1970.
[6] L. J. van der Pauw, "A Method for Measuring Specific Resistivity and Hall Effect of Discs of Arbitrary Shape," *Phillips Res. Rpts.*, **13**, 1 (February 1958).
[7] L. B. Valdes, "Resistivity Measurements on Germanium for Transistors," *Proc. IRE*, **42**, 420 (1954).
[8] F. M. Smits, "Measurement of Sheet Resistivities with the Four-Point Probe," *Bell System Tech. J.*, **37**, 711 (1958).
[9] J. R. Haynes and W. Shockley, "The Mobility and Life of Injected Holes and Electrons in Germanium," *Phys. Rev.*, **81**, 835 (1951).
[10] J. L. Moll, *Physics of Semiconductors*, McGraw-Hill, New York, 1964, Chapter 4.
[11] C. B. Norris and J. F. Gibbons, "Measurement of High Field Drift Velocities in Silicon by a Time-of-Flight Technique," *IEEE PGED*, **ED-14** 38 (1967).
[12] D. T. Stevenson and R. J. Keyes, "Measurement of Carrier Lifetimes in Germanium and Silicon," *J. Appl. Phys.*, **26**, 190 (1955).
[13] J. R. Haynes and J. A. Hornbeck, "Temporary Traps in Silicon and Germanium," *Phys. Rev.*, **90**, 152 (1953); also, Hornbeck and Haynes, *Phys. Rev.*, **97**, 311 (1955).
[14] J. L. Moll, op. cit., p. 105.
[15] N. Houlding, "Measurement of Varactor Quality," *The Microwave Journal*, p. 40 (January 1960).
[16] E. P. EerNisse, "Accurate Capacitance Calculations of PN Junctions Containing Traps," *Appl. Phys. Lett.*, **18**, 183 (1971).
[17] S. R. Lederhandler and L. J. Giacoletto, "Measurement of Minority Carrier Lifetime and Surface Effects in Junction Devices," *Proc. IRE*, **43**, 477 (1955).

[18] J. L. Moll, S. Krakauer, and R. Shen, "P-N Juntion Charge Storage Diodes," *Proc. IRE*, **50**, 43 (1962).

[19] J. Lindmayer and C. Y. Wrigley, *Fundamentals of Semiconductor Devices*, D. Van Nostrand, New York, 1965.

[20] K. Venkateswaran and D. J. Roulston, "Effect of Depletion Layer Capacitance on Lifetime Measurements of p^+-n-n^+ Diodes," *Electronics Lett.*, **6**, 681 (1970).

[21] T. R. Jervis and E. F. Johnson, "Geometrical Magnetoresistance and Hall Mobility in Gunn Effect Devices," *Solid State Elec.*, **13**, 181 (1970).

5

Basic Radiation Damage to Semiconductors

5.1 Introduction

This chapter examines the effects of radiation (energetic electrons, protons, neutrons, and gamma rays) on the electrical properties of semiconductor crystals. A great deal of research has been done on this subject resulting in the publication of several books (Ref. [1–12]) and many papers. There also remains much to be done in terms of understanding the types of defects formed and their kinetics [13]. In view of the large amount of work already published and also due to the fact that the research is not complete, it would be impossible to present a concise, complete treatment in a single chapter of one book. Therefore, the subject will be treated in a manner which will provide some basic insight into the microscopic processes involved in permanent radiation damage as well as present useful data for the prediction of permanent radiation effects.

Section 5.2 discusses the microscopic processes which occur because of the interaction of radiation with matter. This discussion includes the formation of point defects, the reordering of defects, and the formation of defect clusters. Section 5.3 examines the effects of defects on macroscopic semiconductor electrical parameters (carrier concentration, lifetime, and mobility). This is done by means of a simple idealized, energy-level model. The concept of damage constants will also be developed in this section. Section 5.4 will be devoted to a discussion of factors which complicate the prediction of radiation damage in the real world. Experimental data on radiation effects in silicon, gallium arsenide, and germanium will be presented in Section 5.5 along with several useful "rules of thumb" for damage prediction. Section 5.6 will discuss the effect of ionizing radiation on semiconductor materials and devices. Section 5.7 discusses transient annealing effects which must be considered if device operation immediately following a burst of damaging radiation is required.

5.2 Radiation-Induced Defect Formation

There are two basic aspects of the interaction of radiation with matter. A large fraction of the energy of an incident energetic particle goes into electronic processes (excitation and ionization), and this produces a temporary or transient disturbance in the material which disappears shortly after the radiation source is removed. (This will be discussed in Section 5.6.) The remainder of the incident particle energy goes into atomic processes, and produces displacements of atoms within the crystal lattice. The fraction of these displacements which remain after long times (10^3 seconds) at room temperature will be called permanent damage. The discussions in this section will be devoted to the microscopic details of radiation-induced permanent damage in semiconductor crystals. In some low-frequency devices, such as MOSFETs, damage can occur due to charge buildup in the oxide layer. The discussions in this book are devoted to bulk damage due to lattice displacements; therefore, surface effects are ignored. This is in keeping with experimental evidence which indicates that bulk damage dominates microwave semiconductor device degradation mechanisms. We shall discuss each type of radiation (electrons, gamma rays, protons, and neutrons) separately.

Electrons. The interaction of energetic electrons with semiconductor crystals is fairly easy to analyze. Consider an energetic electron bombarding a silicon crystal lattice (see Figure 2.1). If the energetic electron passes sufficiently close to the nucleus (so that coulomb screening is ineffective), then a coulomb scattering event will occur. During this event the electron will transfer some of its energy to the lattice atom. The *maximum* amount of energy that can be transferred is given by [1]

$$T_m = \frac{2(\mathscr{E}_e + 2m_ec^2)}{Mc^2} \cdot \mathscr{E}_\epsilon \tag{5.1A}$$

where $\mathscr{E}_e$, m_e = electron energy, mass, and c = velocity of light, and M = lattice atom mass. For example, Eq. (5.1) predicts that the *maximum* energy that can be transferred to a silicon atom by a 1 MeV electron is 154 eV. Due to the long range interaction obtained with coulomb collisions, one finds that the *average* amount of energy transferred to the lattice is considerably less

[5]. If the energy transferred to the lattice atom is greater than a certain amount, $\mathscr{E}_d$, the lattice atom will be displaced to an interstitial position. ($\mathscr{E}_d$ is called the *displacement threshold energy*, which is between 10 and 25 eV for most semiconductor crystals.) The displaced atom will leave behind a vacancy in the crystal lattice. The resulting vacancy-interstitial pair is called a *Frenkel defect* [4]. These defects are produced uniformly throughout the crystal lattice and for this reason are called *point defects* (in contrast to dislocation lines, etc.). The number of Frenkel defects produced per cm^3 can be calculated if the incident electron energy and fluence (electrons per cm^2) are known. Cahn [14] has derived an expression for the average number of displacements produced by each incident electron as a function of electron energy. It is possible to define a displacement cross section so that

$$N_F = \phi_e \cdot \sigma_d \cdot N_0 \tag{5.2}$$

where N_F = number of Frenkel defects created per cm^3, σ_d = displacement cross section in cm^2, N_0 = number of lattice atoms per cm^3, and Φ_e = electron fluence. Note that σ_d, as used in this book, is an effective cross section used to calculate the *total* number of Frenkel defects due to both primary and secondary displacements. Hence, σ_d has included in it such effects as energy dependence and energy loss through ionization (i.e., only a small fraction of the particle's energy goes into creating displacements). From Eq. (5.2) and the data of Cahn [14] it is possible to calculate σ_d versus electron energy. This result is shown in Figure 5.1 for Si, Ge, and GaAs. The assumed displacement threshold energy used for calculating these curves was 15 eV for all three materials. The curves shown in Figure 5.1 are only valid for thin (<0.01 cm) samples since the absorption of electrons has not been taken into account. From Figure 5.1 and Eq. (5.2) it can be seen that a single 1 MeV electron per cm^2 should produce approximately 4 displacements (Frenkel defects) per cm^3 (assuming $N_0 = 5 \times 10^{22}\ cm^{-3}$). Experimentally, values vary considerably; Wertheim [15] reports a value of 0.26 per cm^3 for silicon, Aukerman and Graft [16] report a value of 0.5 per cm^3 for GaAs for 1 MeV electrons. Cleland and Pacieses [17] report values between 0.1 and 0.4 defects per cm^3 for each incident 1.7 MeV electron per cm^2 in Ge. Other researchers have reported even smaller defect introduction rates. It is important to note that the observed defect introduction rates are always less than would be predicted from Eq. (5.2) and Figure 5.1.

The reason smaller numbers of defects are observed than theory predicts is that not all of the Frenkel defects introduced by the electron bombardment are stable. At finite temperatures, particularly above 100°K, the vacancies and interstitials are mobile in the crystal, and they unite with themselves or impurities in the crystal and form defect complexes in order to obtain a lower energy state (see, for example, [18]). There are several possible reactions that can occur, but only a finite number of vacancies and interstitials are available. Thus, the result is that competing reactions produce several different kinds of defect complexes whose individual densities depend on the energy dependence of the reaction that produces the particular defect type. For this reason the same irradiation performed at different sample temperatures can produce different results. Gregory and Sander [19] have tabulated many of the defect

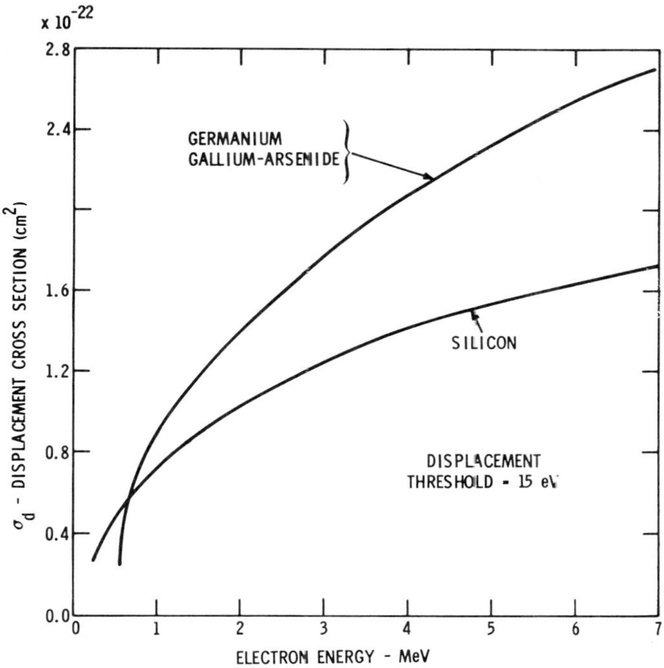

FIGURE 5.1
Displacement cross-section versus electron energy for silicon, germanium, and gallium-arsenide (calculated from Cahn [14]).

species for silicon, and these are given in Table 5.1. One possibility shown in the table is V + I = annihilation (above 76°K), and a certain fraction of the initial defects must suffer this fate.

Not all of the defect complexes are electrically active; consequently they may not contribute to changes in the electrical parameters. From Table 5.1 it can be seen that many of the defect complexes are unstable at room temperature and most anneal out by 600°K. As a result, the electrical properties of electron bombarded silicon can usually be returned to normal by performing a 600°K anneal. However, this is not usually feasible for semiconductor devices mounted in a finished system.

Figure 5.2 shows the typical annealing behavior of electron-damaged p-type silicon. The measurement temperature in Figure 5.2 was 270°K, i.e., the samples were allowed to remain at the annealing temperature for 20 minutes and then the temperature was returned to 270°K for measurement. Such a technique is called *isochronal annealing*. The data shown on Figure 5.2 correspond to an irradiation fluence of approximately 8×10^{14} electrons per cm^2 (energy = 1.7 MeV). The significance of the different sample types is their oxygen content. The relative oxygen content is: crucible-grown > dash > lopex > zone. Figure 5.2 shows that reordering and annealing occur

TABLE 5.1
Defects in Electron or Gamma Irradiated Silicon

Defect Species	Annealing Temperature	Material Type
V^0 (neutral vacancy)	150°K	p-type
$V^=$ (negative vacancy)	(60–80)°K	n-type
I (interstitial)	<4°K	p-type
V-I (vacancy-interstitial pair)	<76°K	n-type
V-V (divacancy)	(500–560)°K	Both
V-V-I (nonreorientable divacancy)	(80–160)°K	n-type
V-O (vacancy-oxygen)	620°K	Both
V–group 5 (vacancy–P, As, Sb)	(400–420)°K	n-type
V–group 3 (vacancy–Al, B, Ga)	~500°K	p-type
I-Al (interstitial-aluminum)	~500°K	p-type
Higher-order complexes	>300°K	Both

Source: Gregory and Sander [19].

at several different temperatures, but most of the damage disappears near 600°K. The meaning of $\Delta p/\Delta\Phi$ will be discussed in Section 5.3.

Gamma Rays. The permanent damage produced by gamma rays is the same type as is produced by electrons, i.e., point defects. In fact, the primary mechanism for damage production is the interaction of energetic electrons (produced by the gamma rays) with the lattice atoms, as described in the previous section. The direct interaction of the gamma rays with the nucleus is usually neglected [4]. The production of energetic electrons by gamma rays can occur by three processes [21]:

1. *The photoelectric effect* is the complete absorption of an incident γ-ray photon by a bound electron, resulting in a free energetic electron

whose energy is equal to that of the absorbed photon minus the electron binding energy.

2. *The Compton effect* is the incomplete absorption of an incident γ-ray photon by a bound electron. The result is an energetic electron and a lower-energy γ-ray photon.

3. *Pair production* occurs when the incident γ-ray photon exceeds 1.02 MeV (twice the rest energy, $m_e c^2$, of an electron). This interaction usually occurs in the nuclear force field near the nucleus. The result is an electron-positron pair.

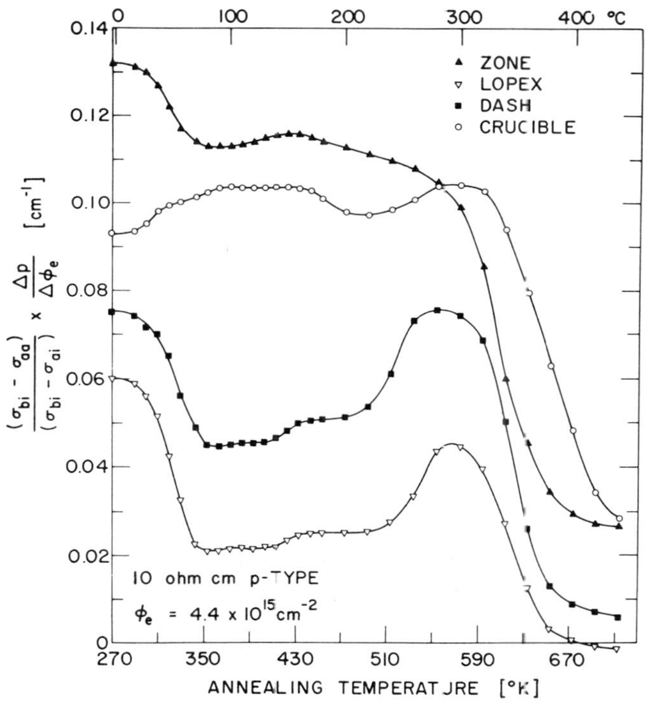

FIGURE 5.2

Isochronal annealing of electron damage between 270° and 690°K in 10 Ω-cm *p*-type silicon samples from four different crystals. The ordinate is the product of the unannealed fraction of the total irradiation-produced conductivity change times $\Delta n/\Delta \phi_e$ from the initial 10% change in unirradiated conductivity. The abbreviations used as σ_{bi} = conductivity before irradiation, σ_{aa} = conductivity after annealing and σ_{ai} = conductivity after irradiation. (After Stein and Gereth [20]).

Evans [21] has calculated the probabilities of occurrence for each of these events.

Evans shows that for silicon the photoelectric effect dominates for photon energies below $h\nu = 0.06$ MeV, that the Compton effect dominates in the range 0.06 MeV $< h\nu <$ 18 MeV, and that pair production dominates above $h\nu = 18$ MeV. For GaAs and Ge the same three energy ranges are $h\nu <$ 0.15 MeV, 0.15 MeV $< h\nu <$ 8 MeV, and $h\nu >$ 8 MeV. (Ge and GaAs have essentially the same average density.)

Gamma ray intensity is usually measured in one of two ways. The first is a fluence, i.e., so many photons of energy, $h\nu$, incident per cm^2, Φ_γ. The second measurement is in terms of absorbed energy dose, rads.

$$1 \text{ rad} \equiv \text{absorption of 100 erg/gram.} \tag{5.3}$$

For a given fluence the absorbed dose can be calculated by the following equation [21]:

$$R(\text{rads}) = \frac{\Phi_\gamma \cdot h\nu \cdot \mu_a}{100 \cdot \rho} \tag{5.4}$$

where μ_a = absorption coefficient, $h\nu$ in ergs, and ρ = crystal density (gram per cm^3). It should be pointed out that Eq. (5.4) is accurate only for thick ($>$0.1 cm) samples. In the case of thinner samples some of the Compton electrons may escape and R will be lower than predicted by Eq. (5.4). In the case of a packaged device one must also consider the effects of energy absorption in adjacent layers (such as heat sinks, etc.). If the semiconductor crystal is adjacent to a layer of high-Z material (such as gold), it may receive a dose somewhat larger than predicted by Eq. (5.4). Eq. (5.4) is usually adequate, however, for the order of magnitude calculation of absorbed dose. From Eq. (5.4) it can be seen that since μ_a/ρ is not necessarily the same for all materials, the material type should be specified, i.e., rads(H_2), rads(Si), etc. However, Evans [21] shows that the ratio μ_a/ρ for most lighter substances is nearly constant. For example, at 1 MeV, μ_a/ρ = 0.028 cm^2/gram for air, 0.030 for water, and 0.027 for aluminum. The same type of scheme can be (and is) used to express the energy deposited by an electron beam in a crystal. In this case, μ_a is replaced by the electron absorption coefficient, ϕ_γ is replaced by ϕ_e, and $h\nu$ is replaced by the energy of the incident electrons.

As was mentioned earlier, the permanent damage produced by gamma rays is just due to energetic electrons produced by the photoelectric, Compton, and pair-production processes. Therefore, once the numbers and energies of these electrons are known, the density of Frenkel defects can be readily calculated. Cahn [14] has calculated the effective displacement cross-section for gamma rays [see Eq. (5.2)] as a function of photon energy. This is plotted in Figure 5.3 for Si, Ge, and GaAs. (Ge and GaAs have essentially the same average atomic mass and hence follow the same curve.) The data in Figure 5.3 is only for thin (0.1 cm) samples since the absorption of γ-rays has not been taken into account. Equation (5.2) (with ϕ_e replaced by ϕ_γ) and Figure 5.3 may be used to calculate the density of Frenkel defects per cm^3 for a given fluence of gamma rays of known energy. If we assume $h\nu = 1.25$ MeV (the average energy of Co^{60} gamma-ray photons), then Eq. (5.2) and Figure 5.3 predict that approximately 0.025 defects per cm^3 should be produced by

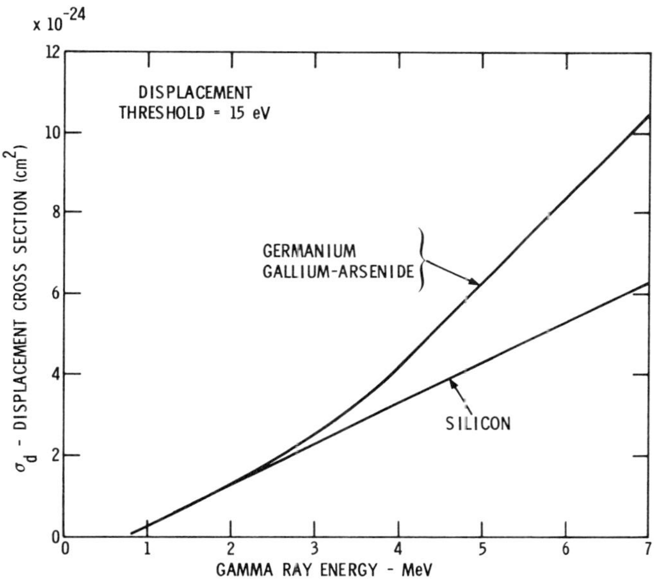

FIGURE 5.3
Displacement cross-section versus gamma-ray energy for silicon, germanium, and gallium-arsenide (calculated from Cahn [14]).

a fluence of 1 photon per cm^2. Experimentally, for Co^{60} radiation, Brehm and Pearson [22] have reported values of $0.0011 \rightarrow 0.0123$ defects per cm^3 per photon in GaAs depending on temperature and material type (n or p); Sonder and Templeton [23] have reported a value of 0.0007 defects per cm^3 per photon for n-type silicon; and Cleland et al. [24] have reported a value of 0.0007 defects per cm^3 per photon for n-type germanium at 76°K. The experimentally measured defect density is less than the predicted value due to defect reordering, as was described previously. Defects formed by gamma-ray irradiation are the same as those formed by energetic electrons, whose possible defect complexes and annealing behavior were discussed at the beginning of this section.

Protons. The bombardment of semiconductor crystals with protons has not been as thoroughly studied as electron and gamma-ray effects. The basic mechanism is the same as in the energetic electron case, i.e., an energetic proton passes sufficiently close to a lattice atom so that coulomb screening is ineffective and a coulomb scattering event occurs. The *maximum* energy that can be transferred to a lattice atom is

$$T_m = \frac{4M \cdot m_p}{(M + m_p)^2} \cdot \mathscr{E}_p , \tag{5.1B}$$

where m_p and $\mathscr{E}_p$ are the proton's mass and energy. For example, a 1 MeV proton can transfer a *maximum* energy of 145 keV to a silicon lattice atom; this is roughly 1000 times the value for the 1 MeV electron. The *average*† energy which is transferred to a lattice atom is [1]:

$$\overline{T} \cong \mathscr{E}_d \ln \left(\frac{T_m}{\mathscr{E}_d} \right) \tag{5.5}$$

where $\mathscr{E}_d$ = displacement threshold (assumed to be 15 eV). Hence, a 1 MeV proton will transfer 138 eV of energy to a silicon lattice atom on the average. This implies that the displaced silicon lattice atom may have enough energy to produce a few more displacements on its own. In this case the initially displaced atom is called the *primary recoil* or *knock-on* atom and the other atoms displaced are called *secondary displacements.*

The problem of calculating the displacement cross-section (including both primary and secondary displacements) as a function of energy has been solved for silicon by Baicker et al. [26]. Figure 5.4 shows the displacement cross section versus proton energy. The displacement threshold assumed by Baicker was 13 eV. The values for Ge and GaAs have been estimated from the Rutherford scattering cross-section for heavy charged particles [1] and by realizing that $\overline{T}$ [Eq. (5.5)] is not a strong function of M [lattice atom mass, see Eq. (5.1)]. The data shown in Figure 5.4 are good only for thin samples (10 μm) since absorption has not been taken into account. Lower-energy protons (below 1 MeV) are not too important in practical radiation-effects applications since the range of these particles is not sufficient to penetrate the encapsulation around most microwave devices. Figure 5.4 and Eq. (5.2) (with Φ_e replaced by Φ_p, protons per cm^2) predict that one 35 MeV proton per cm^2 should create 750 displacements per cm^3 in Si. Bilinski et al. [25] report a measured value of 233 for silicon. Reliable data for Ge and GaAs are not available. However, as will be shown later, lifetime data for proton-irradiated Ge is available. As usual, the reported experimental defect-introduction rate is lower than is predicted theoretically. This is due to defect reordering and

† For nonrelativistic particles only.

annihilation, as discussed earlier. The exact nature of the defect types (see Table 5.1 for examples) are not well cataloged for proton irradiation. However, since the basic damage (Frenkel defects) is the same, it is not unreasonable to assume that the end products (divacancies, V-O, etc.) are probably the same. In some cases these defects may be relatively close together due to defect production by secondary displacements, and this may influence the types of complexes formed through the defect reordering process.

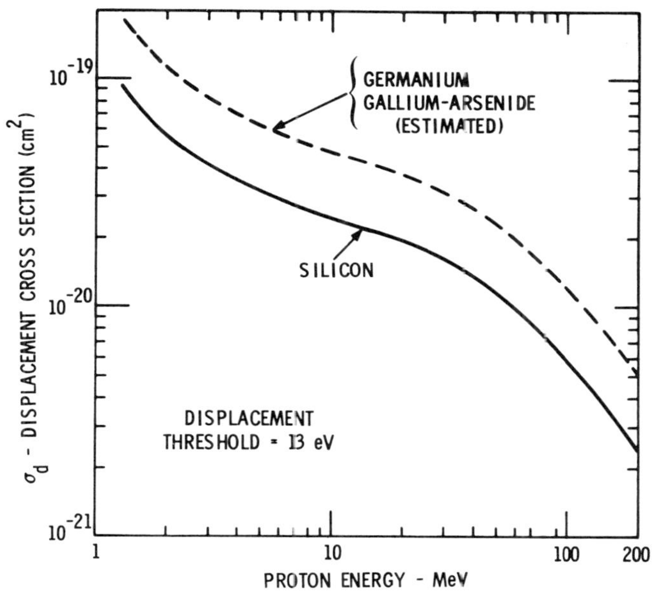

FIGURE 5.4
Displacement cross-section (including primary and secondary displacements) versus proton energy for Si, Ge, and GaAs (calculated from Baicker et al. [26]).

Neutrons. Neutron radiation damage in silicon has been studied extensively, and, although the mechanisms are more complicated than those described earlier, they are fairly well understood. Again, the primary damage mechanism is displacement of lattice atoms to interstitial positions. (The effect of transmutation of the lattice atoms can be ignored [5].) Neutrons are neutral and do not experience coulomb interactions. As a result, the collision cross-section and energy transfer relations are calculated from a hard-sphere model. In general, the cross section for a hard-sphere collision is much smaller than for long-range coulomb interactions, but the *average* energy transferred to the lattice atoms is much higher. The *maximum* energy that can be transferred to a lattice atom by a neutron is given by Eq. (5.1B) (with $\mathscr{E}_p$ and m_p replaced by $\mathscr{E}_n$ and m_n respectively). For example, a 1 MeV neutron can impart a maximum energy of 145 keV to a silicon lattice atom. Since all events ranging from zero to maximum energy transfer are equally likely for a hard-sphere collision, the average energy transferred per collision may be written [1] as

$$\overline{T} = \tfrac{1}{2} T_m \, . \tag{5.6}$$

Hence, the average energy transferred by isotropic elastic scattering of a 1 MeV neutron per collision is 72.5 keV (as opposed to $\overline{T} = 138$ eV for a 1 MeV proton). At higher neutron energies (above 5 MeV) the problem is further complicated by inelastic scattering events [68]. Since the threshold displacement energy is only 15 eV, it can readily be seen that the energetic lattice atom (called the primary recoil or knock-on atom) can displace many more lattice atoms. As a result, regions of localized damage occur in neutron-irradiated semiconductor crystals, and these regions are usually called *defect clusters*. Figure 5.5 depicts the formation of a defect cluster.

In Figure 5.5 an energetic neutron collides with a lattice atom, giving the lattice atom a substantial amount of energy (up to 145 keV). The atom is displaced from its lattice position and moves through the crystal lattice. In the

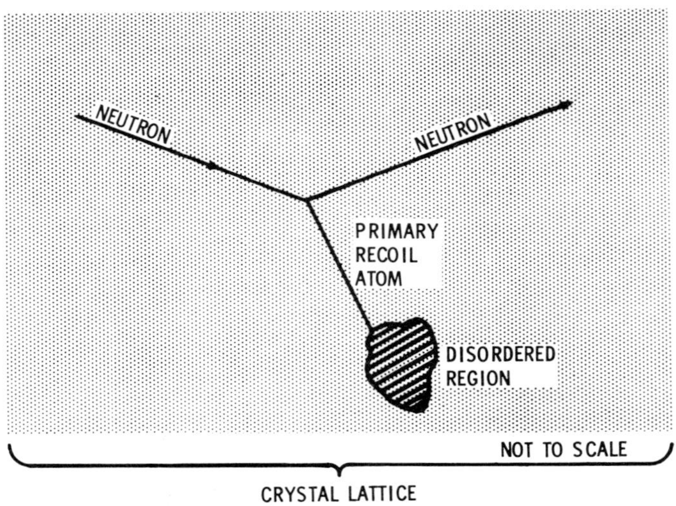

FIGURE 5.5
Formation of a defect cluster in a neutron-irradiated semiconductor crystal (note that the distance between the initial neutron interaction and the defect cluster has been deliberately exaggerated).

initial period this primary recoil atom loses energy primarily by ionization [28] and also produces some displacements. The distance between the initial neutron interaction and the defect cluster has been deliberately exaggerated in Figure 5.5 so that the primary recoil atom could be identified. After it has slowed down below about 10 keV, most of the interactions with the lattice produce displacements; however, energy loss due to ionization is still possible. It is generally conceded that most of the damage is due to displacements produced by the primary recoil atom [19], and all of the damage produced by one neutron interaction is confined to a single, roughly elliptically shaped, defect cluster containing from 200 to 1000 defects. Holmes [34] and Wikner et al. [35] claim that the damage exists in the form of a super cluster which contains several smaller subclusters. As will be shown, experimental evidence seems to favor one large cluster over several smaller subclusters. In Figure 5.5, note also that the energetic neutron which started the process continues on its way, retaining most of its energy (0.93 MeV on the average for a 1 MeV neutron in silicon), and it can create another defect cluster further on in the crystal. However, the collision cross-section for the neutron is so small that the clusters are quite far removed from one another. The cross section for the initial neutron-atom interaction is between 10^{-23} and 10^{-24} cm^2 for Si. If we insert this value into Eq. (5.2), we find that there is less than one interaction per cm^3 for the incident neutron. The major damage comes from the fact that each interaction results in many displacements (200 to 1000 displacements).

Gossick [29] and Cleland and Crawford [30] were among the first to postulate the existence of defect clusters in neutron-irradiated semiconductors. Bertolotti [31] was the first to actually observe these clusters directly. Using a technique called *replication electron microscopy*, he observed defect clusters in Si, Ge, and GaAs bombarded by 14 MeV neutrons. Stein [32] has created a useful sketch of Bertolotti's findings for silicon; this sketch, in Figure 5.6, shows a 500-Å-diameter damaged region surrounded by a space-charge region 2000 Å in diameter. These data are in agreement with calculations made by Gossick [29] and Cleland and Crawford [30]. Clusters similar to Figure 5.6 have been observed in both GaAs and Ge with the exception that no central damaged region has been observed in the GaAs clusters [31]. These data [31] tend to support the concept of a single cluster rather than several smaller subclusters [34,35]. In the model originally proposed by Gossick [29], majority carriers are trapped in the central disordered region creating a depletion layer around the disordered region. The net effect is the creation of a potential barrier for majority carriers and a potential well for minority carriers. This is shown in Figure 5.7 for *n*-type Si. The fact that defects lie in a potential well for minority carriers or on top of a potential hill for majority carriers (see Figure 5.7) greatly complicates the trapping and recombination kinetics of these centers; this has been studied by Gregory [33].

A calculation of the total number of defects produced per cm^3 per neutron per cm^2 is quite involved since calculations must be made, first, of the cross section for the initial neutron-atom collision; next, of the amount of energy which goes into ionization processes; and, finally, of the number of displacements created by the recoil atom as it slows down and finally stops. Smith

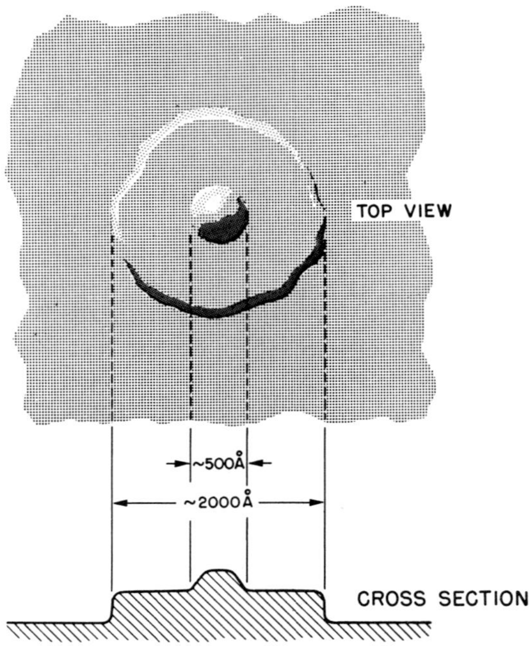

FIGURE 5.6
Damage cluster in neutron-irradiated silicon (after Stein [32] and Bertolotti [31]).

[27] has calculated a damage curve for silicon. Sattler [28] has calculated several discrete points for germanium. Unfortunately, no published data are available for GaAs; however, experimental measurements show that the damage seen in GaAs is the same order of magnitude as that seen in Si and Ge. Figure 5.8 shows the data of Sattler [28] and Smith et al. [27], converted to a total displacement cross-section of the type used to describe the other forms of radiation discussed earlier. If monoenergetic neutrons are used to bombard the semiconductor crystal, then the total number of Frenkel

defects (before annealing and reordering occur) can be calculated from Figure 5.8 and Eq. (5.2) with Φ_e replaced by Φ_n, where Φ_n = neutron fluence (n/cm²).† Quite often a nuclear reactor is used as the neutron source. This results in a neutron spectrum $\Phi_n(\mathscr{E})$, i.e., a neutron fluence versus energy distribution. These neutron energy spectra usually are different for different

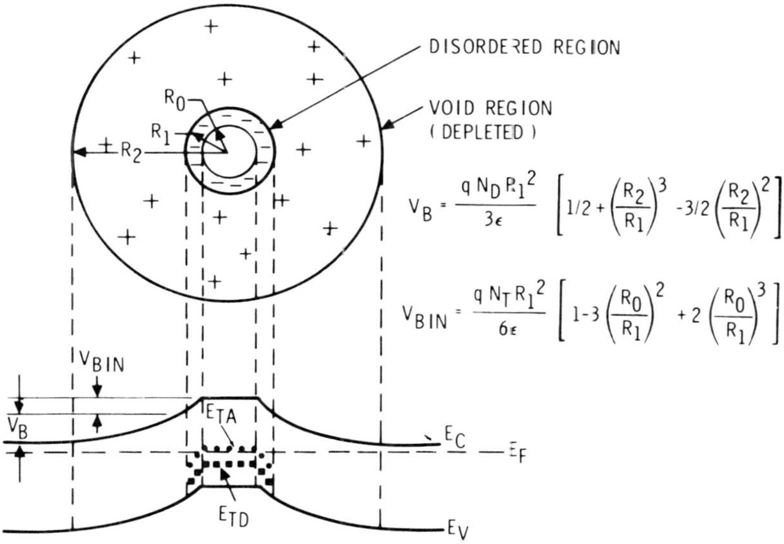

FIGURE 5.7
Schematic model for the neutron-produced cluster in *n*-type silicon. (V_B and V_{BIN} are the external and internal potential barriers, respectively.) (After Gregory [33])

reactor types, and a further discussion may be found in Appendix C. In this case Eq. (5.2) must be put into the form

$$N_F = N_0 \cdot \int_0^{\mathscr{E}_m} \Phi_n(\mathscr{E}) \cdot \sigma_d(\mathscr{E})\, d\mathscr{E} \tag{5.7}$$

where $\mathscr{E}_m$ = maximum expected neutron energy and $\sigma_d(\mathscr{E})$ is given by Figure 5.6. A convenient and useful method of specifying neutron fluence when an energy spectrum is involved has yet to be worked out. For the remainder of this book the following description shall be used:

$$\text{neutron fluence} = \Phi_n \tag{5.8}$$
$$\Phi_n = \text{neutrons/cm}^2,\ \mathscr{E} > 10 \text{ keV (Spectrum } X)$$

† An older form of this notation is nvt = (neutrons/cm³)·(cm/sec)·(sec) = n/cm².

where Φ_n is the total number of neutrons per cm^2 with energy greater than 10 keV that have bombarded the sample. The "Spectrum X" term designates the neutron energy spectrum curve; several of the more common spectra are given in Appendix C. Neutrons with energies less than 10 keV do not produce significant numbers of displacements and hence are not counted.

Figure 5.8 and Eq. (5.2) predict that on the average each 1 MeV neutron per cm^2 in Si should produce 160 Frenkel defects per cm^3 before reordering

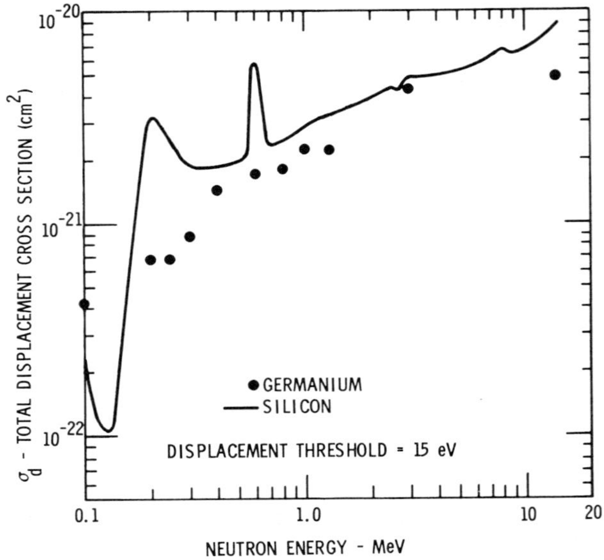

FIGURE 5.8
Displacement cross-section (including primary and secondary displacements) versus neutron energy for Si and Ge (calculated from Smith et al. [27] and Sattler [28]).

occurs. Experimentally, typical values for defect introduction rates† are Ge = 5 [36], Si = 4 [37], GaAs = 3 [38]. The data for Ge were taken using Spectrum C (Appendix C); for Si and GaAs, Spectrum A was used. The number of defects measured is again considerably lower than predicted theoretically; this is partly due to defect reordering and partly due to the effects of the space-charge layer surrounding the cluster discussed earlier.

The initial defects residing inside the cluster are vacancies and interstitials or simple complexes thereof. Cheng and Lori [39] estimate that the defect

† (defects/cm^3)/(neutron/cm^2).

densities inside the clusters are typically 10^{20} per cm^3, and as a result these regions are still crystalline. These defects will reorder at a finite temperature through a competing reaction to form the same types of defects seen in the case of electron irradiation, i.e., V-O, V-V, etc. (see Table 5.1). The key words are *competing reaction* because in the case of the defect cluster, where the defect concentrations are very high and located in a region of modified potential (see Figure 5.7), different reordering reactions will predominate than in the case of point defects. For example, the V-O (vacancy-oxygen) complex is very important in electron-irradiated silicon where point defects predominate, but in neutron-irradiated silicon the V-O complex is not as important since probability dictates that there are only a few oxygen atoms inside the defect cluster. Bass [40] has investigated this and has found that there are still some dissolved impurity effects in neutron-irradiated samples. The primary defect found in the damage clusters (after room-temperature annealing and if the sample is not too heavily doped) is the divacancy [39]. Most of the neutron-induced permanent damage anneals out in the 600°–700°K range [37].

5.3 The Effect of Defects

The previous sections of this chapter were concerned with a microscopic view of the damage produced in semiconductor crystals by energetic radiation. The present section will demonstrate how these microscopic defects affect the macroscopic electrical parameters (carrier concentration, mobility, conductivity, and lifetime) of bulk semiconductor crystals. To do this, a gross oversimplification of the problem will be used in order to analyze the main features of the problem without becoming tangled in a mathematical morass.

For the moment consider only point defects. It can be shown [41] that point defects of the types listed in Table 5.1 (Section 5.2) result in the introduction of allowed energy states within the forbidden gap (Fig. 2.4) of semiconductor crystals. Experimental measurements [18] confirm the existence of these levels in irradiated semiconductors. The effect of energy levels in the forbidden gap has been analyzed in Chapter 2 (Section 2.7), where it was shown that once the characteristics of the radiation-induced levels (energy levels, capture cross-section, and density) are known, it is possible to calculate the effect of radiation on all of the macroscopic electrical parameters. This technique for analyzing radiation damage was first discussed by James and Lark-Horovitz [42].

Assume for the sake of simplicity that radiation-induced defects produce a single energy level in the forbidden energy gap of the semiconductor. Then, if the defect energy level, $\mathscr{E}_T$, and capture cross-sections for electrons and holes, σ_n and σ_p, and density of defects per cm^3, N_T, are known, the problem can be analyzed with the methods given in Chapter 2 (Section 2.7). Assume also that the defect is an acceptor level and that it is located somewhat above midgap $\mathscr{E}_T > \mathscr{E}_i$ in n-type material. Assume that the trap or defect level can have one of two charge states, either zero or negative $(-e)$ corresponding to the trap being empty or occupied by a conduction band electron. The mechanisms that occur in this situation are depicted in Figure 2.12B. The density of traps charged in the negative state (those that have captured electrons from the conduction band) is given by Eq. (2.59) (repeated below for convenience)

$$\frac{N_T^-}{N_T} = \frac{\sigma_n n + \sigma_p p_1}{\sigma_n(n + n_1) + \sigma_p(p + p_1)} \tag{5.9}$$

where all of the symbols are defined by Eq. (2.59). For the assumptions listed above we have $n \gg p$ and $n_1 \gg p_1$, hence we may rewrite Eq. (5.9) as

$$\frac{N_T^-}{N_T} \cong \frac{n}{n + n_1} \cdot \tag{5.10}$$

For a deep trap $\mathscr{E}_T \ll \mathscr{E}_C$, and hence $n_1 \ll n$ and Eq. (5.10) reduces to

$$N_T^- \cong N_T\left(1 - \frac{n_1}{n}\right) \cdot \tag{5.11}$$

Using the definitions of Eq. (2.59) and using Eq. (2.18), we can rewrite Eq. (5.11) as

$$N_T^- = N_T\left\{1 - \exp\left(\frac{\mathscr{E}_T - \mathscr{E}_f}{kT}\right)\right\} \cdot \tag{5.12}$$

Initially, the free electron concentration (in our assumed n-type material) is given by, $n \cong N_D$ [see Eq. (2.27A)]. In the postirradiated case some of the electrons are tied up in traps [Eq. (5.12)], and, therefore, the free electron concentration for this case may be approximated by

$$n = N_D - N_T\left[1 - \exp\left(\frac{\mathscr{E}_T - \mathscr{E}_f}{kT}\right)\right] \cdot \tag{5.13}$$

If the density of traps introduced is proportional to the radiation fluence, then [see Eq. (5.2)]

$$N_T = A \cdot \Phi \tag{5.14}$$

where A = a constant and Φ is the radiation fluence (any type). Substituting Eq. (5.12) into Eq. (5.13), it can be seen that the free majority carrier density is reduced by the radiation fluence. This effect is called *carrier removal* and is usually discussed in terms of a *carrier removal rate* (so many free majority carriers per cm^3 removed by so many neutrons, electrons, etc., per cm^2). Using Eq. (5.14) and (5.13), it can easily be seen that the carrier removal rate for this example is

$$\frac{\Delta n}{\Delta \phi} = -A\left[1 - \exp\left(\frac{\mathscr{E}_T - \mathscr{E}_f}{kT}\right)\right]. \tag{5.15}$$

From Eq. (5.15) and realizing that $\mathscr{E}_T$ is fixed by the type of defect involved, it can be seen that the carrier removal rate is dependent on $\mathscr{E}_f$, the Fermi level. In bulk material the Fermi level depends mainly on the doping density, and therefore a carrier removal rate which is dependent on doping density would be expected. This will be seen in Section 5.5. In the case of a reverse-biased *p-n* junction, Eq. (5.12) does not hold since the Fermi level does not predict the true electron density [43]. This problem can be solved by using the quasi-Fermi-level concept described in Chapter 2, Ref. [11].

The example just described (single-level acceptor trap) will be used now in discussing mobility changes due to radiation damage. The effect of doping atoms on mobility was discussed in Chapter 2 (see Figure 2.8). The mobility in semiconductors is determined by collisions with vibrating lattice atoms and long range coulomb collisions with ionized donor and acceptor atoms. The overall mobility may be written as

$$\frac{1}{\mu_0} = \frac{1}{\mu_I} + \frac{1}{\mu_L} \tag{5.16}$$

where μ_0 = overall mobility, μ_I = ionized impurity contribution, and μ_L = lattice atom scattering contribution. Figure 2.8 shows that at room temperature, μ_L dominates for ionized impurity concentrations below 10^{15} per cm^3] The trapped electrons [Eq. (5.12)] in this example add to the ionized scattering term in Eq. (5.16). The effect can be analyzed by the Brooks-Herring [45] treatment which shows that

$$\mu_I = \frac{2\epsilon^2(kT)^{3/2}}{\sqrt{m_e}e^3N_I}\cdot\frac{1}{\ln(1+b) - b/(1+b)} \tag{5.17}$$

where

$$b = \frac{24\pi\epsilon m_e(kT)^2}{nh^2e^2}, \tag{5.18}$$

and N_I = density of ionized scattering centers. From Eq. (5.17) and (5.18) it can be seen that μ_I depends both on n and N_I; however for n below 10^{18} per cm^3, which is usually the case, one can neglect the variation with n. Hence,

μ_I is then proportional to $1/N_I$. Therefore, using Eq. (5.12), (5.14), (5.16), (5.17), and (5.18), we may write

$$\frac{1}{\mu_f} = \frac{1}{\mu_{ID}} + \frac{1}{\mu_L} + \frac{A\Phi\{1 - \exp[(\mathscr{E}_T - \mathscr{E}_f)/kT]\}}{B}, \tag{5.19}$$

where μ_f = mobility after irradiation, μ_{ID} = mobility due to doping atom scattering, B = a constant (for a given temperature). Hence, it can be seen from Eq. (5.19) that the mobility decreases with increasing radiation fluence. However, since ionized impurity scattering is not important below 10^{15} dopant atoms per cm^3, scattering from trapped electrons will not be important below 10^{15} trapped electrons per cm^3. In discussions of radiation effects on mobility, *mobility degradation* is often defined as

$$\Delta\left(\frac{1}{\mu}\right) = \frac{1}{\mu_f} - \frac{1}{\mu_0}. \tag{5.20}$$

The *mobility degradation rate* is given by

$$\frac{\Delta(1/\mu)}{\Delta\Phi} = \frac{A\{1 - \exp[(\mathscr{E}_T - \mathscr{E}_f)/kT]\}}{B} \tag{5.21}$$

where Φ = radiation fluence per cm^2.

Thus far the discussion has dealt with the changes in majority-carrier concentration and mobility due to an assumed energy level introduced by radiation. Since the carrier concentration and mobility both decrease with radiation (at least in this model), it is obvious that the conductivity $\sigma \cong ne\mu_n$ [Eq. (2.39)] will also decrease with radiation.

Now consider the effect of a radiation-induced single acceptor level on minority-carrier lifetime. The effect of a single acceptor level in an n-type semiconductor was examined in Section 2.7. That section showed that, for the preceding assumption, the low level minority carrier lifetime is given by [see Eq. (2.61B) and (2.62)].

$$\tau_P \cong \frac{1}{N_T v_t \sigma_P} \tag{5.22}$$

where the symbols are all explained by Eq. (2.61B). Note that in Eq. (5.22) it is assumed that $N_T = N_T^-$, i.e., all traps occupied in the negative state. The degradation in minority carrier lifetime is usually expressed as

$$\Delta\left(\frac{1}{\tau}\right) = \frac{1}{\tau_f} - \frac{1}{\tau_0} = K_\tau\Phi \tag{5.23}$$

where τ_f = lifetime after irradiation and τ_0 is the lifetime before irradiation.

K_τ is called the *lifetime damage constant*. The *lifetime degradation rate* is simply

$$\frac{\Delta(1/\tau)}{\Delta\Phi} = K_\tau \,. \tag{5.24}$$

If $\tau_0 \gg \tau_f$, Eq. (5.14), (5.22), and (5.23) can be used to show that

$$K_\tau \cong A v_t \sigma_P \,. \tag{5.25}$$

In some cases the literature discusses a *diffusion-length damage constant* given by

$$\frac{1}{L_f^2} - \frac{1}{L_0^2} = K_L \Phi \tag{5.26}$$

where L_0 = initial diffusion length and L_f = diffusion length after irradiation. From Eq. (2.76) and (2.77) it can be seen that $L = \sqrt{D\tau}$, and hence, $K_\tau = D \cdot K_L$.

The analysis of the effects of a single energy level is very enlightening; however, in the real world, radiation-produced defects introduce a multitude of energy levels in the forbidden gap [7,18]. Mathematical analyses of this problem are difficult at best [47]. They are hampered by a lack of data concerning the exact values of the energy levels and the correct values of the capture cross-sections. One case that can be solved is the divacancy defect in Si (thought to dominate lightly doped, room temperature irradiated samples). Chapter 2, Section 2.7 discussed trapping and recombination effects for a two-level system. The commonly used parameters for the divacancy defect are [18] an acceptor level at $\mathscr{E}_c - 0.54$ eV and a donor level at $\mathscr{E}_v + 0.25$ eV. The capture cross-sections are estimated [46] to be $\sigma_n^0 = 4 \times 10^{-16}$, $\sigma_n^+ = 4 \times 10^{-15}$, $\sigma_p^- = 6 \times 10^{-15}$, $\sigma_p^0 = 6 \times 10^{-16}$ (all in cm^2). The meaning of these symbols is found above Eq. (2.63). The divacancy defect two-level model has been used successfully to predict radiation-induced degradation in IMPATT devices (Chapter 8) and BARITT devices (Chapter 10).

The concept of energy levels in the forbidden gap can still be used when discussing neutron damage clusters, however it is found that the SRH and SL† models do not adequately predict the injection dependence of lifetime. This has been attributed [33] to the presence of the potential barrier around the defect cluster (see Figure 5.7). Hence, the mathematical problem becomes even more complex in this case. The concepts of damage constants and carrier removal rates are, however, still applicable to neutron-irradiated semiconductors.

† Shockley-Read-Hall (SRH) and Shockley-Last (SL), see Chapter 2, Section 2.7.

5.4 Factors Affecting Permanent Damage

It has already been pointed out that the same irradiation performed at a different sample temperature will produce different effects. There are many other factors which also influence the amount and type of damage produced by radiation. Some of these factors can change the results by a factor of 2 or 3, and some of them can change the results by more than an order of magnitude. The reason for bringing up these additional "confusion factors" is to illustrate the importance of designing adequate safety factors to handle these possibilities. For a given radiation fluence (particles per cm^3) the amount of permanent radiation damage is given by

$$\text{Damage} = f(A, B, C, D, E, F, G, H) \tag{5.27}$$

where A, B, C, etc., all are functions which depend on

- A (temperature of sample during irradiation)
- B (energy spectrum of radiation, see Appendix C)
- C (inactive dissolved impurity concentrations O, C, etc.)
- D (p-type or n-type material and doping concentration)
- E (minority-carrier injection ratio during irradiation)
- F (orientation of crystal with respect to radiation direction)
- G (rate of irradiation)
- H (history after irradiation, time temperature, and injection level).

A good illustration of some of these effects can be seen in Figure 5.9 which shows that the amount of permanent damage in gamma-irradiated silicon

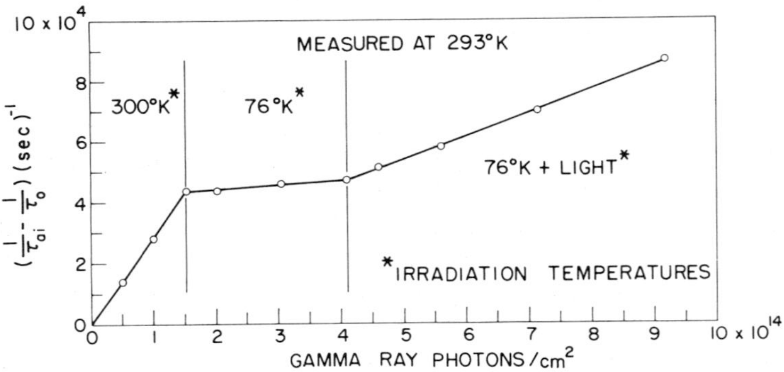

FIGURE 5.9
Effects of irradiation temperature and illumination Co^{60} gammaray-produced minority-carrier-lifetime degradation in vacuum-floating-zone, n-type silicon. τ_0 and τ_{ai} are the initial and after-irradiation lifetimes (after Stein [48]).

(6 Ω-cm, n-type) is a function of irradiation temperature as well as injection during the irradiation (light excites electron-hole pairs in the sample). The data given in the next section will further demonstrate some of these effects. However, in most cases the experimental data is not sufficiently complete so that the damage can be accurately predicted for an arbitrary set of values for the variables in Eq. (5.27). The best that can be done at present is a factor of 2 or 3 in Si. Ge and GaAs have not been studied as thoroughly as yet.

5.5 Experimentally Measured Damage in Si, Ge, and GaAs

This section presents experimentally measured data on radiation effects in bulk semiconductors. The data given here will be used to predict radiation-

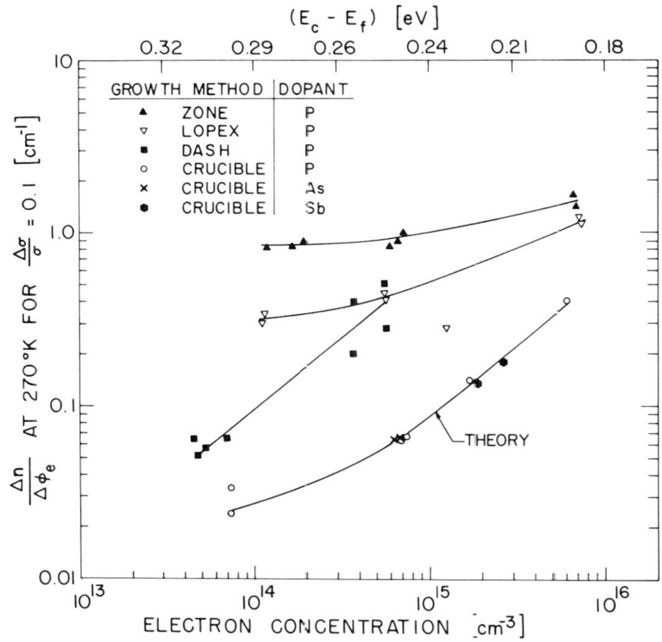

FIGURE 5.10

Electron removal rates $\Delta n/\Delta\phi_e$ for 1.7 MeV electrons at 270°K in n-type silicon versus carrier concentration for crystals differing in growth methods and doping impurities. The values for $\Delta n/\Delta\phi_e$ were calculated from the initial 10% irradiation-produced change in conductivity. The upper abscissa gives positions of the Fermi level at 250°K corresponding to the electron concentration given on the lower abscissa (after Stein and Gereth [20]).

induced failure mechanisms in microwave semiconductor devices later in the book. Experimental data will be presented on the variations of the macroscopic electrical parameters (carrier concentration, mobility, conductivity, and lifetime) with electron, gamma-ray, proton, and neutron irradiation for Si, Ge, and GaAs. The most extensively examined effects are those which occur in neutron-irradiated silicon, while the least measured effect seems to be proton bombardment of GaAs.

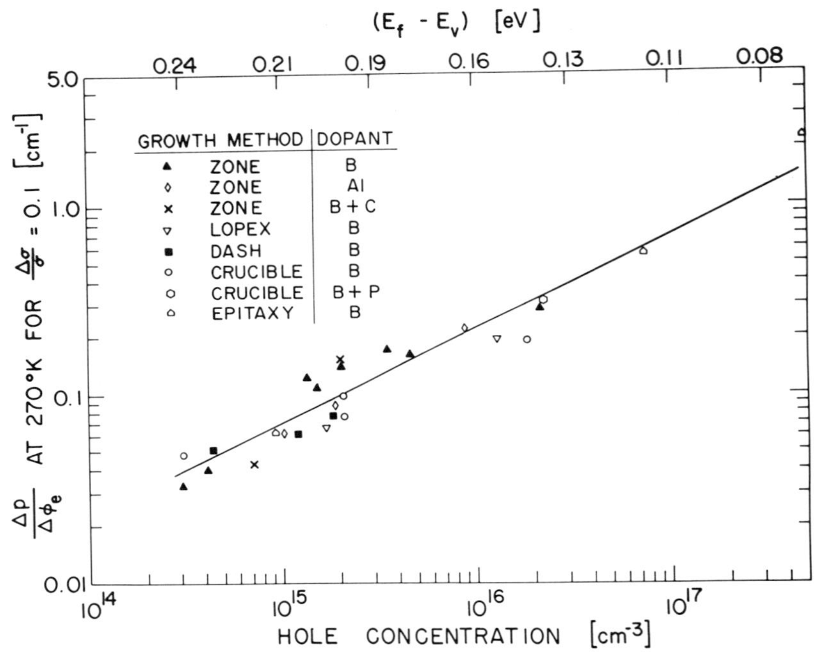

FIGURE 5.11
$\Delta p/\Delta\phi_e$ for 1.7 MeV electrons in *p*-type silicon at 270°K versus hole concentration for crystals differing in growth methods and doping impurities. The values of $\Delta p/\Delta\phi_e$ were calculated from the initial 10% irradiation-produced change in conductivity. The upper abscissa gives positions of the Fermi level at 270°K corresponding to the hole concentration given on the lower abscissa (after Stein and Gereth [20]).

Electrons in Silicon. Typical effects of electron bombardment on the electrical parameters of silicon are shown in Figures 5.10 to 5.15. Figures 5.10 and 5.11 show that the carrier removal rate is a function of the Fermi level (i.e., doping density), and Figure 5.10 is in qualitative agreement with Eq. (5.15) on this point. Figure 5.10 also shows that the carrier removal rate in n-type Si is dependent on dissolved oxygen content (see the text discussion concerning Figure 5.2). Figure 5.12 shows that the carrier removal rate is dependent on the incident electron energy. The curve on Figure 5.12 is qualitatively similar to the displacement cross-section curve shown in Figure 5.1. This is

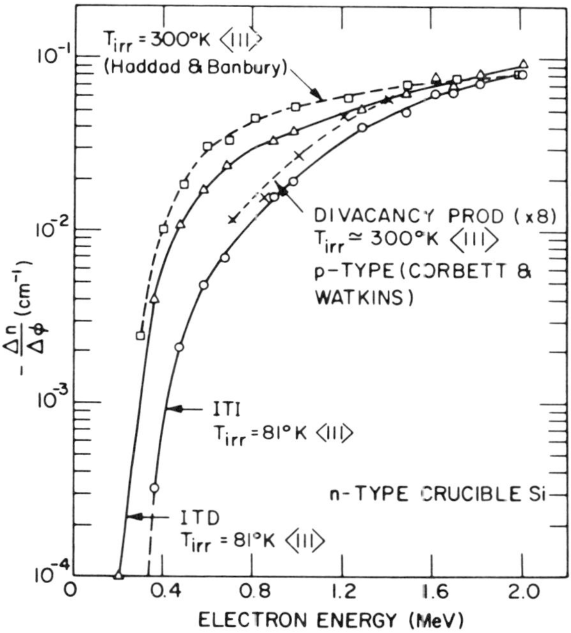

FIGURE 5.12
Carrier removal rate attributable to ITI and ITD defects as a function of electron energy for irradiation at 81°K. Room-temperature irradiation data and divacancy production rates observed in EPR measurements are included in the figure for data comparison (after Stein and Vook [50]).

reasonable since the number of traps introduced by electron bombardment should be proportional to the number of initial defects each electron produces. Figure 5.13 shows the low-temperature carrier removal rate and mobility degradation due to a 10 MeV electron bombardment (plotted versus fluence). It should be noted that the mobility degradation is much larger at 80°K than at room temperature since lattice scattering does not dominate at low temperatures. Figure 5.14 shows that the conductivity of both *p*- and *n*-type silicon decreases with increasing energetic electron fluence. Figure 5.15 shows the predicted and measured values of the diffusion-length damage constant [see Eq. (5.26)] for electron-irradiated silicon. Again, the similarity between the shape of this curve and Figure 5.1 should be noted. The distinction between region I and region II on Figure 5.15 is that the type of defect dominating the damage in *p*-type Si in regions I and II is different [10].

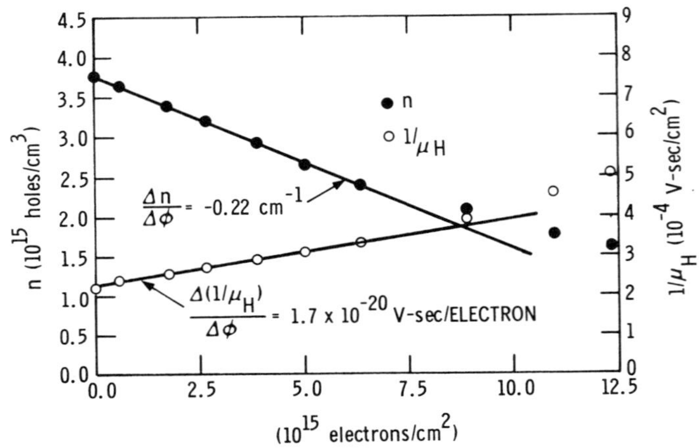

FIGURE 5.13
Carrier concentration and reciprocal Hall-mobility changes with irradiation and measurement at 80°K by 10 MeV electrons for a 10 ohm-cm B-doped silicon sample (after Wikner and Horiye [49]).

Gamma Rays in Silicon. Typical effects of γ-ray bombardment on the electrical properties of silicon are shown in Figures 5.16 to 5.19. Figures 5.16 and 5.17 show that the Hall coefficient ($R \cong 1/pe$ for *p*-type material and $R \cong 1/ne$ for *n*-type material) increases with gamma-ray fluence. These data can be used to calculate a carrier removal rate for both *n*- and *p*-type gamma-irradiated silicon. Figures 5.18 and 5.19 show the Hall mobility as a function of temperature and gamma-ray fluence for both *p*- and *n*-type silicon. The figures show that room temperature mobility is rather insensitive to gamma irradiation while the low-temperature values can be strongly perturbed. The reasons for this have been described previously in the discussion of Figure 5.13. The lifetime damage constant data for gamma-irradiated silicon can be obtained from Figure 5.9 and Eq. (5.23). Hewes [69] has also presented data on this subject.

Protons in Silicon. Published data on the effects of proton bombardment on the electrical properties of silicon are not as complete as for other radiation types. Figure 5.20 shows the diffusion-length damage constant [Eq. (5.26)] for proton-irradiated silicon. Figure 5.20 shows two interesting facts. The first is that the shape of the curves is qualitatively the same as the curve in Figure 5.4

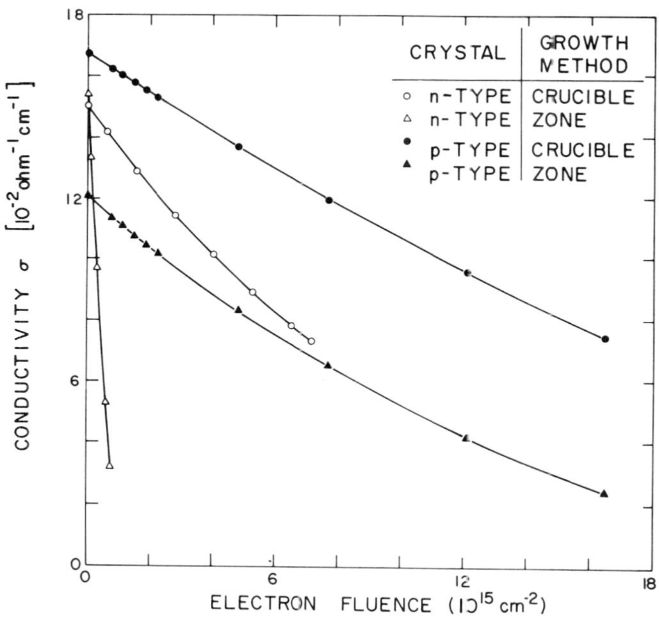

FIGURE 5.14
Conductivity versus electron fluence for *n*- and *p*-type silicon samples at 270°K irradiated with 1.7 MeV electrons (after Stein and Gereth [20]).

for the production of Frenkel defects. The second point to be made about Figure 5.20 is that the diffusion length damage constant is larger for *n*-type material than for *p*-type. (The lifetime effects in a *p*/*n* solar cell are due to the *n*-region.) Unfortunately, adequate data are not available for carrier removal rates and mobility degradation due to energetic protons. However, these parameters can be estimated by assuming that a certain percentage of the Frenkel defects predicted by Figure 5.4 and Eq. (5.2) act as deep traps and hence cause carrier removal and mobility degradation. (See Figure 5.20, p. 119.)

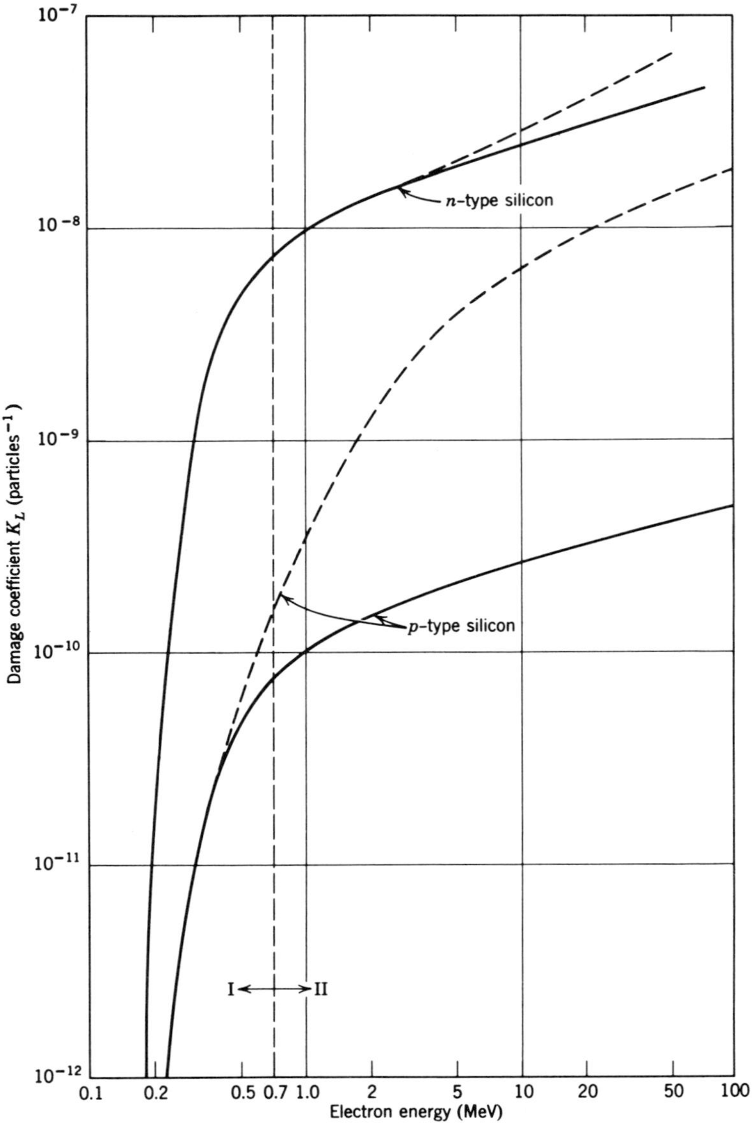

FIGURE 5.15
Diffusion-length radiation damage constant as a function of electron energy for 1 Ω-cm silicon solar cells. The solid line is the predicted value and the dashed line is the typical measured value (after Larin [10]).

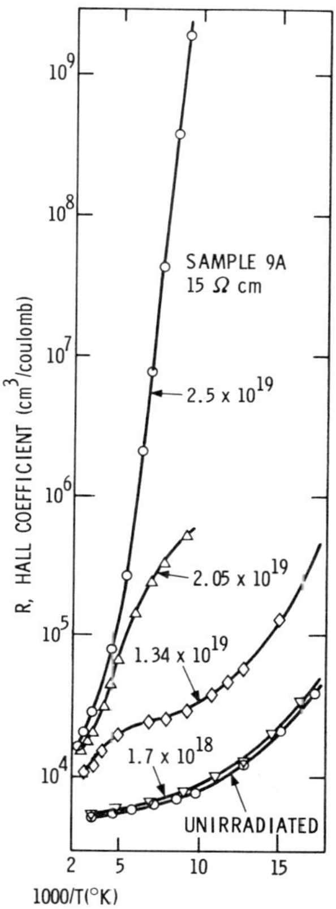

FIGURE 5.16
Hall coefficient for gamma-ray, Co^{60}, irradiated *p*-type (15 Ω-cm) silicon sample versus measuring temperature, total gamma-ray dose (photons per cm^2) as shown (after Sonder and Templeton [51]).

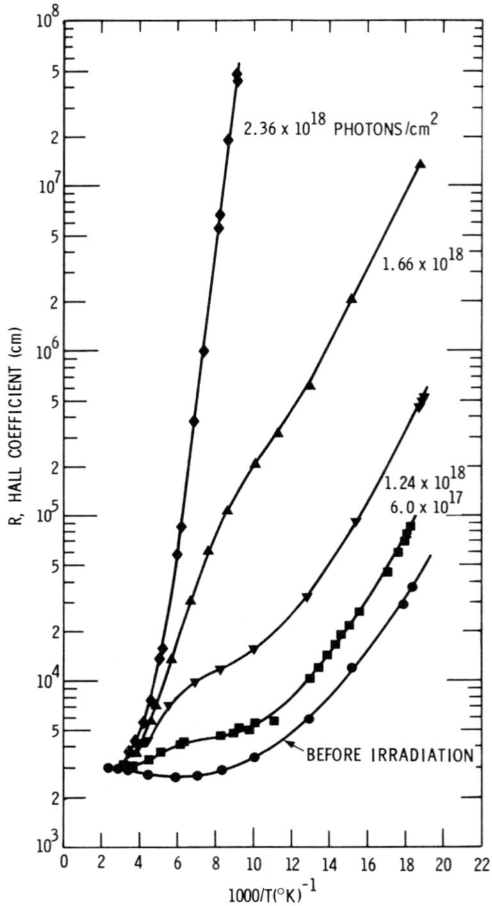

FIGURE 5.17
Hall coefficient as a function of temperature. The different curves were obtained after irradiation with a total Co^{60} gamma-ray dose (photons per cm^2) as shown, *n*-type silicon sample (2 Ω-cm) (after Sonder and Templeton [23]).

Neutrons in Silicon. The effects of neutron bombardment on the electrical parameters of silicon have been extensively studied. Figures 5.21 and 5.22 show that the neutron-induced ($\mathscr{E} > 10$ keV, Spectrum A) carrier removal rate is a function of the Fermi level (i.e., doping density) which is in qualitative agreement with Eq. (5.15). Figure 5.23 shows the effect of neutron irradiation (cutoff energy unknown, Spectrum C) on the Hall mobility of a

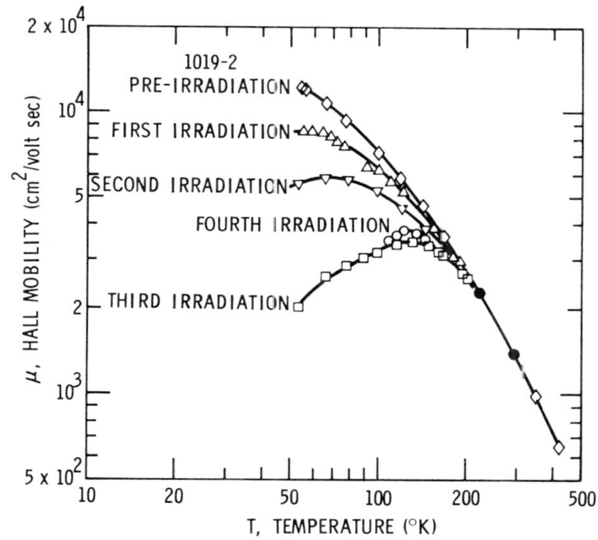

FIGURE 5.18
Hall mobility versus temperature in *n*-type Si (2 Ω-cm) for different gamma-ray fluences. Total irradiation fluence levels: first = 6×10^{17}, second = 1.2×10^{18}, third = 1.7×10^{18}, fourth = 2.4×10^{18}, photons of Co^{60} radiation per cm^2 (after Sonder and Templeton [23]).

moderately heavily doped (0.1 Ω-cm) *n*-type silicon sample. From Figure 5.23 it can be seen that, as usual, the mobility is strongly perturbed at lower temperatures but changes only slightly at room temperature. Figure 5.24 (page 122) shows experimental and empirical values for room-temperature resistivity changes versus neutron fluence ($\mathscr{E} > 10$ keV, Spectrum A). The empirical equations are [53],

$$\rho_p = \rho_{p0} \exp\left[\frac{\Phi}{A}\right], \tag{5.28}$$

$$\rho_n = \rho_{n0} \exp\left[\frac{\Phi}{B}\right], \tag{5.29}$$

where

$$A = 387(p_0)^{0.77} \tag{5.30}$$

$$B = 444(n_0)^{0.77} \,. \tag{5.31}$$

Figure 5.24 is useful for the prediction of resistivity change in neutron-irradiated bulk silicon. Figure 5.25 shows the inverse of the lifetime damage

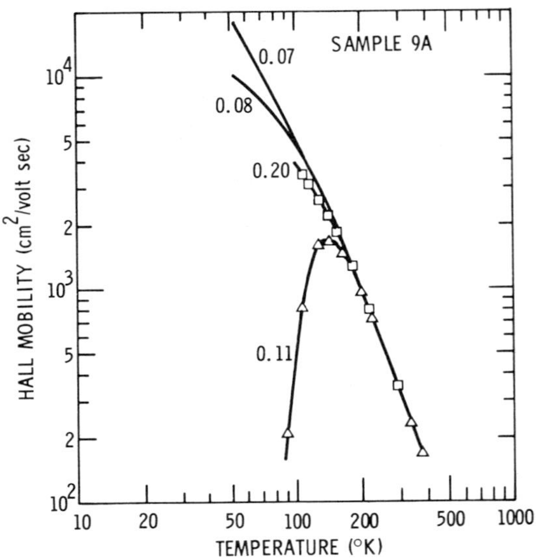

FIGURE 5.19
Hall mobility versus temperature in *p*-type silicon (15 Ω-cm) for different Co^{60} gamma-ray fluences. The numbers next to the curves are the positions of the Fermi level at 100°K. The corresponding total gamma-ray fluences are 0.07 = prerad, 0.08 = 1.3×10^{18}, 0.20 = 1.5×10^{19}, 0.11 = 10^{19} photons per cm^2 (after Sonder and Templeton [51]).

constant [Eq. (5.23)] for neutron irradiated ($\mathscr{E} > 10$ keV, Spectrum A) *n*- and *p*-type silicon as a function of minority-carrier injection ratio. From Figure 5.25 it can be seen that the lifetime damage constant is a slowly varying function of injection ratio.

Electrons in Germanium. Typical effects of electron bombardment on the electrical properties of Ge are shown in Figures 5.26 to 5.28. Figure 5.26

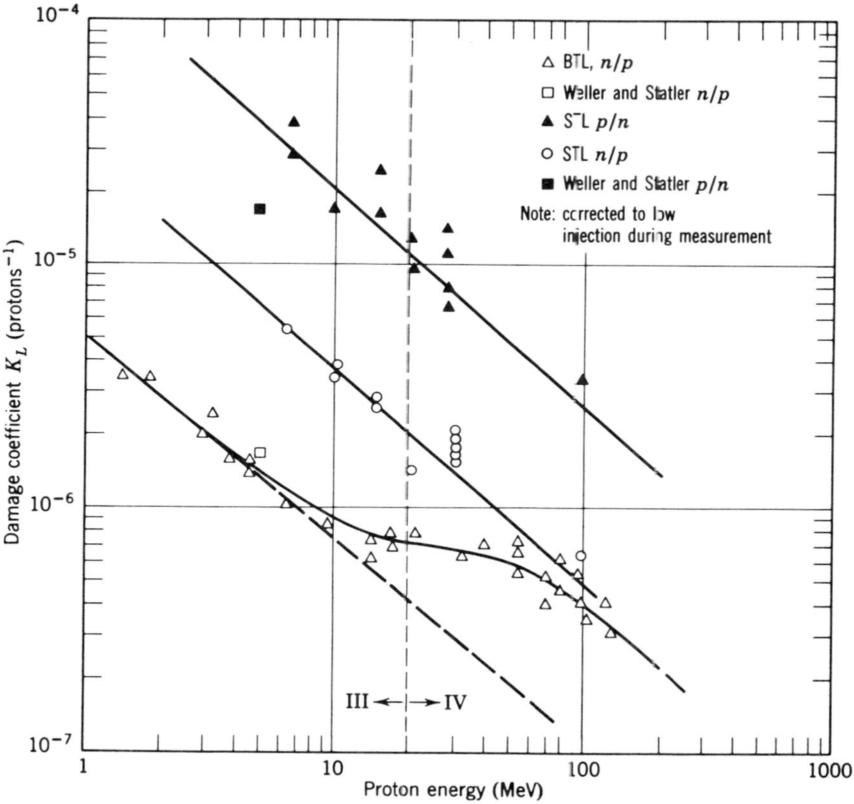

FIGURE 5.20
The diffusion-length radiation damage constant in *n/p* and *p/n* solar cells as a function of proton energy (after Larin [10]).

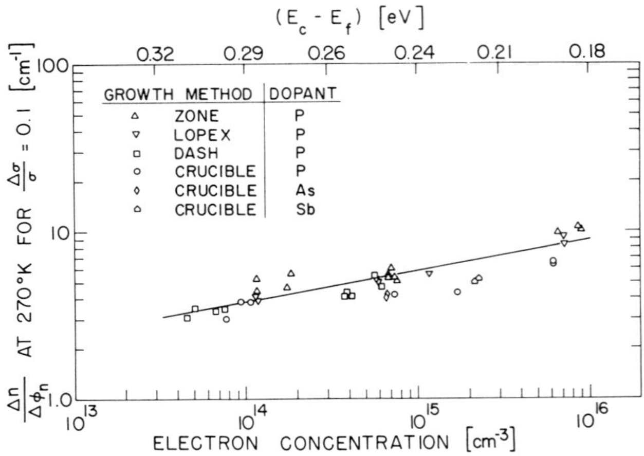

FIGURE 5.21
Neutron-produced removal rates $\Delta n/\Delta \phi_n$ at 270°K in *n*-type silicon versus electron concentration. Coordinates are the same as in Figure 5.10 (after Stein and Gereth [20]).

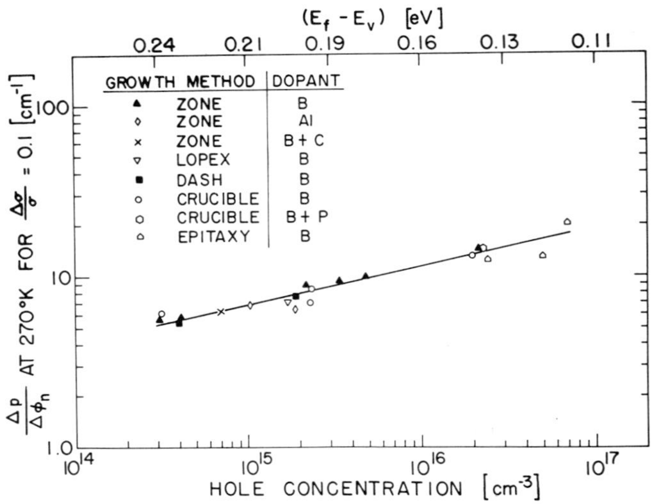

FIGURE 5.22
Neutron-produced hole removal rates $\Delta p/\Delta \phi_n$ at 270°K in *p*-type silicon versus hole concentration. Coordinates are the same as in Figure 5.11 (after Stein and Gereth [20]).

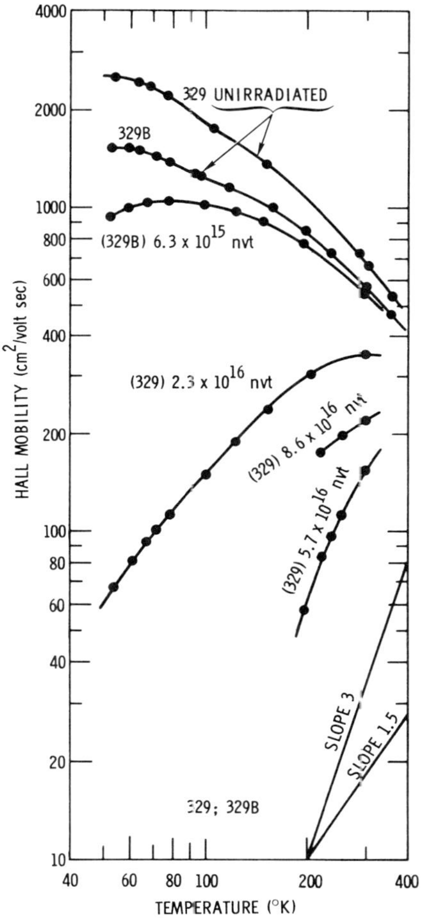

FIGURE 5.23
Hall mobility of *n*-type (0.1 Ω-cm) silicon samples versus temperature and neutron fluence (after Sonder [52]).

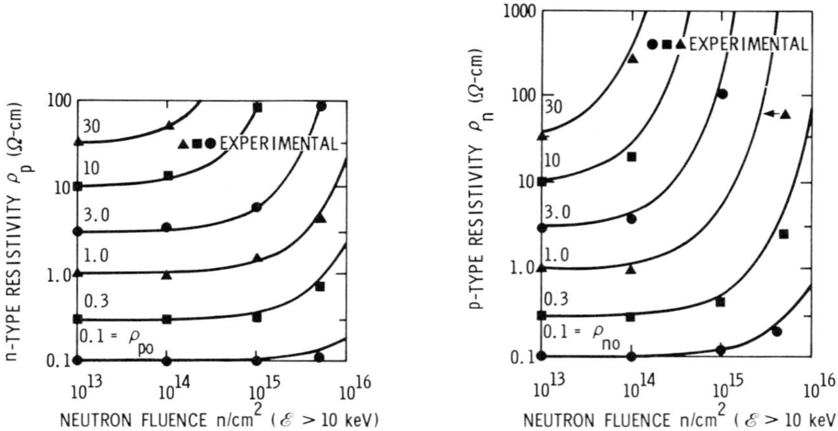

FIGURE 5.24
Exponential resistivity-fluence dependence of uniformly doped silicon irradiated at 270°K (after Buehler [53]).

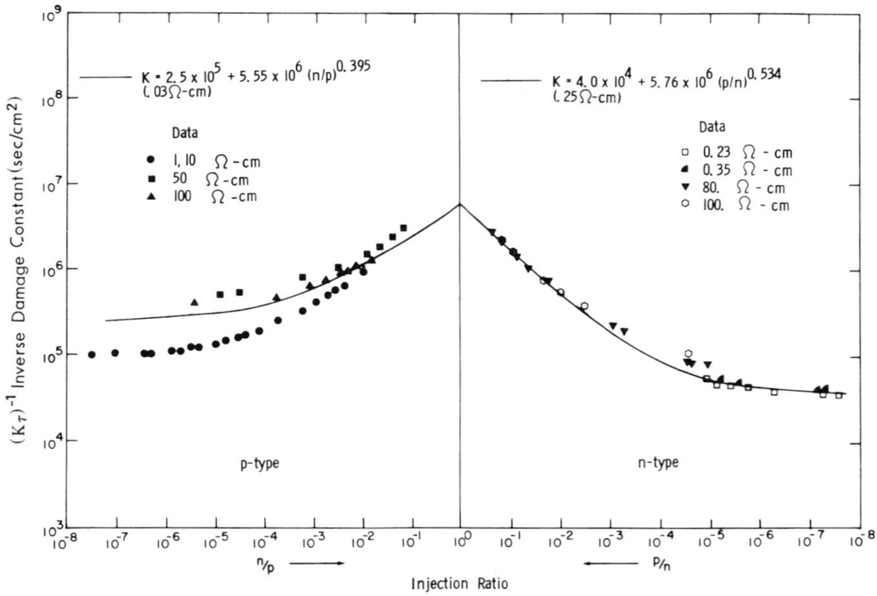

FIGURE 5.25
Inverse lifetime damage constant versus injection ratio for neutron-irradiated silicon (after Gregory and Gwyn [54]).

shows the carrier removal rate versus electron energy for several different temperatures. The shape of the curve is again in qualitative agreement with Figure 5.1, as it should be. Figure 5.27 shows the minority-carrier drift mobility in n-type germanium as a function of electron fluence for 2 MeV electrons. Figure 5.28 shows the variation in $(1/\tau - 1/\tau_0)$ for n- and p-type germanium versus 1 MeV electron fluence. The data on Figure 5.28 is easily converted to the lifetime damage constant via Eq. (5.23).

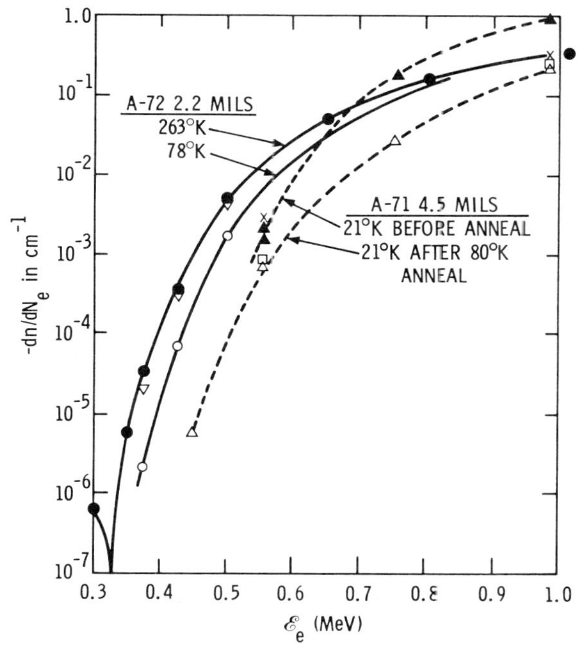

FIGURE 5.26
Carrier removal rate for thin, n-type germanium samples (10 Ω-cm) versus electron energy, at different temperatures (after Brown and Augustyniak [56]).

Gamma Rays in Germanium. Figures 5.29 to 5.31 show the typical effects of gamma rays on the electrical properties of germanium. Figure 5.29 shows the Hall coefficient ($R \simeq 1/ne$) for n-type germanium bombarded with Co^{60} gamma rays. The resistivity passes through a peak as the sample converts from n-type to p-type after 7.4×10^{16} photons per cm^2. Figure 5.30 shows the change in mobility due to gamma irradiation of n-type germanium as a function of temperature. As usual, the mobility is perturbed more at lower

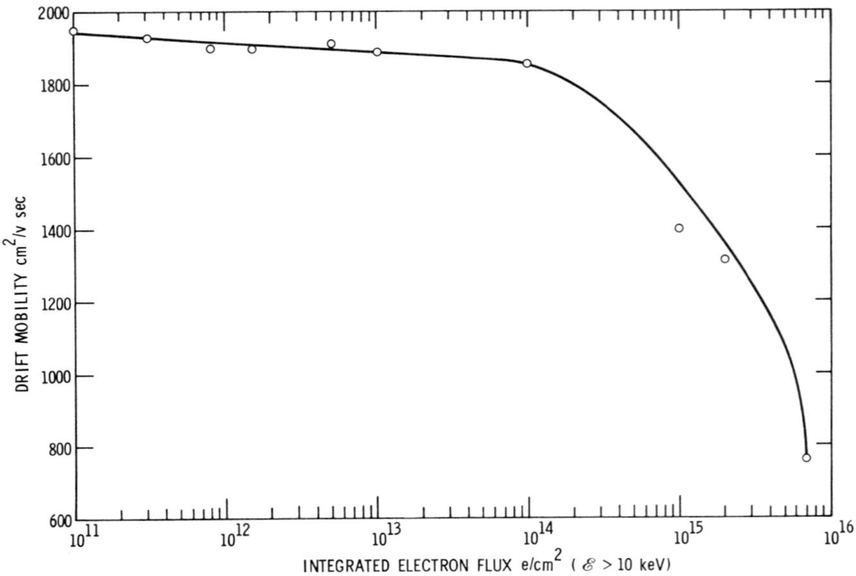

FIGURE 5.27
Drift mobility versus electron fluence for *n*-type germanium (4 Ω-cm) (after Closser [57]).

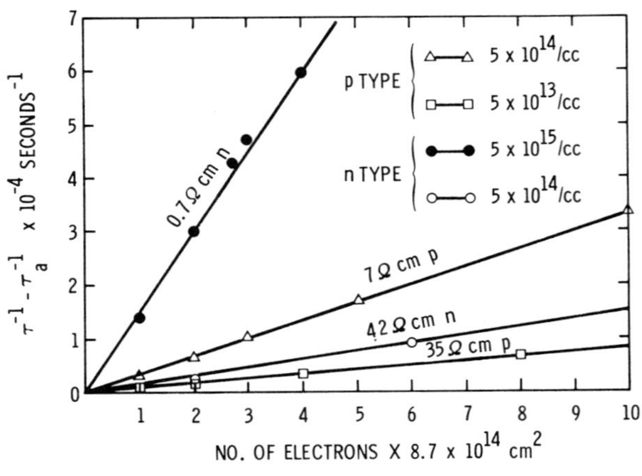

FIGURE 5.28
Plots $1/\tau$ for germanium exposed to 1 MeV electrons (after Loferski and Rappaport [55]).

temperatures due to the lack of lattice scattering at lower temperatures. Figure 5.31 shows the lifetime behavior of a gamma-irradiated n-type germanium sample. The lifetime damage constant may be calculated from Figure 5.31 and Eq. (5.32).

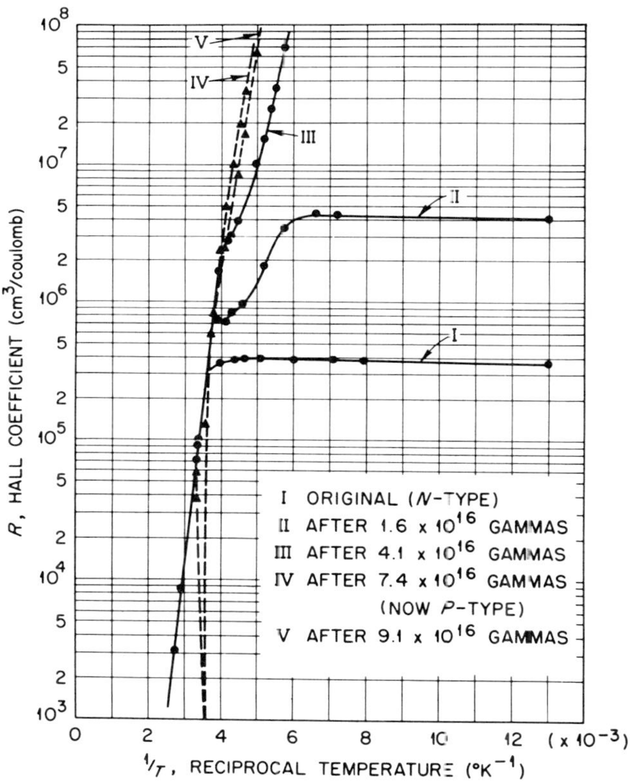

FIGURE 5.29
Hall coefficient versus temperature of gamma-irradiated (Co^{60}) n-type germanium (10 Ω-cm) (after Crawford and Cleland [58]).

Protons in Germanium. Complete data on proton bombardment effects are not available. Figure 5.32 shows the experimentally measured lifetime damage constant for proton-irradiated germanium. The damage constants in Figure 5.32 follow the same general curve as does the displacement cross-section shown in Figure 5.4 (i.e., the number of recombination centers is proportional to the number of initial Frenkel defects). While data on conductivity variation with protons is not readily available, deuteron bombardment data are available [1]. Figure 5.33 shows the effect of energetic deuteron (a proton and a neutron) bombardment on *n*- and *p*-type germanium. Since the charge is the same and the mass is similar, deuteron bombardment should have nearly the same effect as proton bombardment. Figure 5.33 shows that both *p*- and *n*-type germanium become *p*-type with a limiting resistivity of about 10 Ω-cm at high fluence levels.

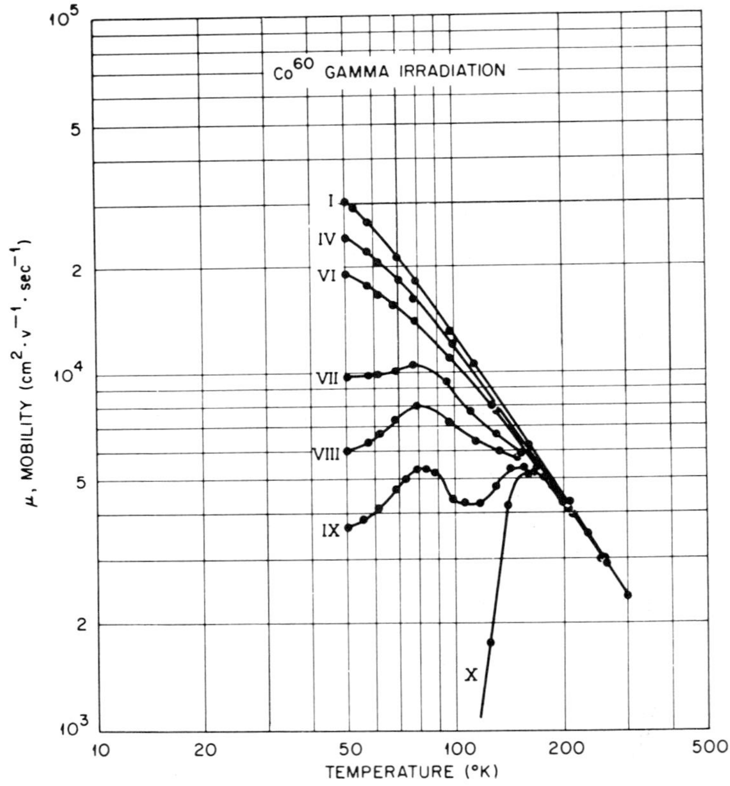

FIGURE 5.30
Electron Hall-mobility versus temperature for *n*-type (10 Ω-cm) Ge after successive exposures to Co^{60} γ-rays. Curve I—before exposure, Curve IV—after 2.3×10^{17} γ-rays/cm², Curve VI—after 5.5×10^{17} γ-rays/cm², Curve VII—after 9.0×10^{17} γ-rays/cm², Curve VIII—after 1.36×10^{18} γ-rays/cm², Curve IX—after 1.73×10^{18} γ-rays/cm², Curve X—after 2.12×10^{18} γ-rays/cm² (after Crawford and Cleland [58]).

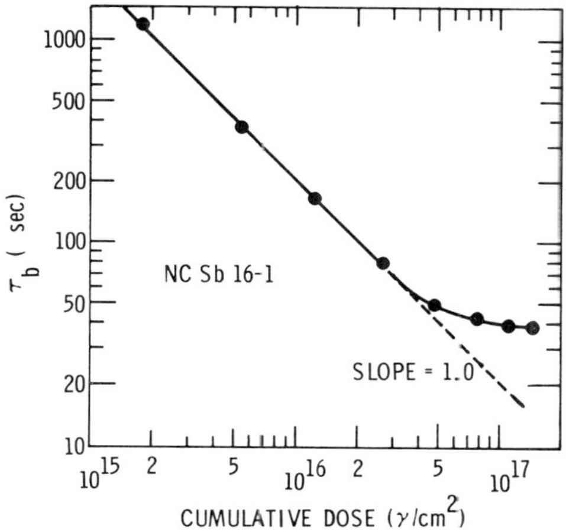

FIGURE 5.31
Lifetime degradation of 1 Ω-cm, *n*-type germanium subjected to Co^{60} gamma-rays (after Streetman [59]).

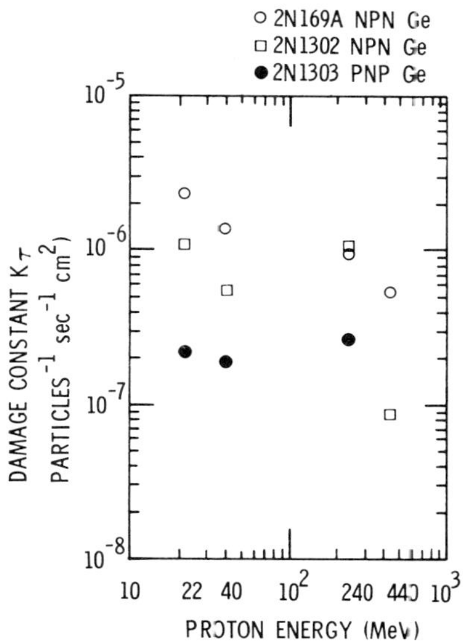

FIGURE 5.32
Lifetime damage constant for proton-irradiated germanium (after Bilinski et al. [25]).

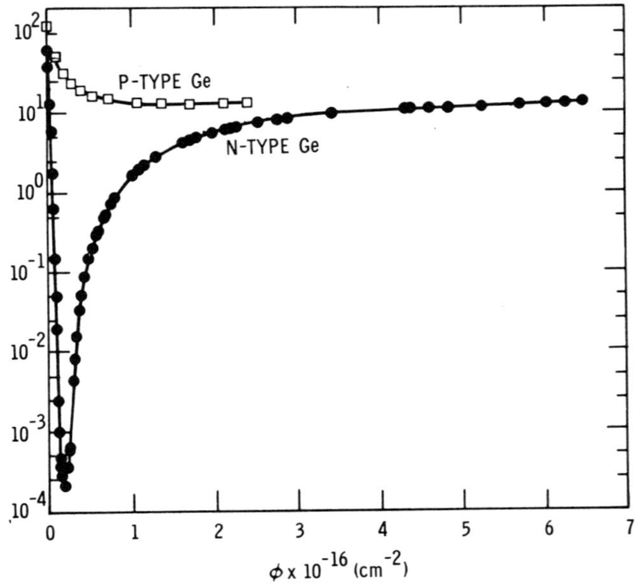

FIGURE 5.33
Conductivity versus fluence curves for an initially *p*-type and an initially *n*-type sample. Irradiation by 9.6 Mev deuterons at 200°K (after Fan and Lark-Horovitz [68]).

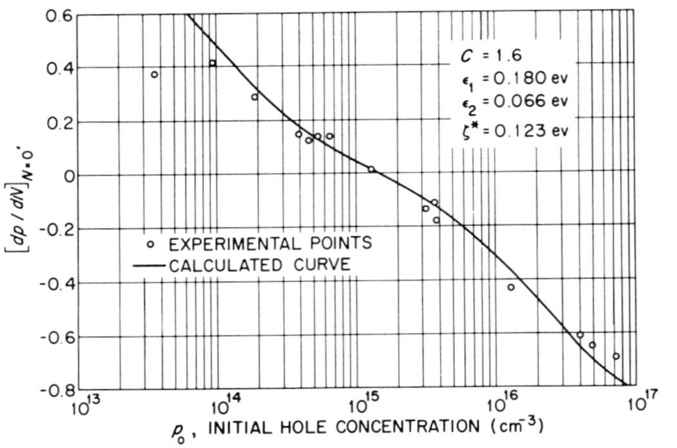

FIGURE 5.34
The initial rate of hole removal as a function of initial hole concentration for *p*-type Ge during neutron bombardment at 195°K (after Cleland et al. [60]).

Neutrons in Germanium. The effect of neutron bombardment in germanium was studied extensively in the 1950s. Some of the more interesting results are shown in Figures 5.34 to 5.36. Figure 5.34 shows the carrier removal rate for *p*-type germanium exposed to fast neutrons ($\mathscr{E} > 1$ MeV, Spectrum C) as a function of initial carrier concentration. What is particularly interesting is that the carrier removal rate is nearly *zero* for initial hole concentrations near 10^{15} per cm^3, *positive* below this level and *negative* above it! This corresponds to the sample trying to approach a limiting resistivity of 1 Ω-cm (see Figure 2.9) which is similar to the behavior seen for deuteron-irradiated Ge (see Figure 5.33). Figure 5.35 shows the variation of the Hall mobility in *p*-type Ge exposed to fast neutron fluence ($\mathscr{E} > 1$ MeV, Spectrum C) versus temperature. From Figure 5.35 it can be seen that, as usual, the mobility is degraded more at lower temperatures due to the dominance of ionized impurity and defect scattering. Figure 5.36 shows lifetime data for neutron-irradiated ($\mathscr{E} > 100$ keV, Spectrum D) *n*- and *p*-type germanium of several different resistivities. Equation (5.23) and Figure 5.36 may be used to calculate a lifetime damage constant for neutron-irradiated Ge.

Electrons in GaAs. The effect of energetic electron bombardment on the electrical properties of GaAs is shown in Figures 5.37 and 5.38. Figure 5.37 shows the low-temperature carrier concentration and mobility degradation and annealing behavior for 2 MeV electron-irradiated *n*-type GaAs. The mobility degradation seen in Figure 5.37 would probably not be seen if the measurement temperature were 300°K as opposed to the 80°K used in Figure 5.37. Figure 5.38 shows experimental data which can be used to calculate the lifetime damage constant for electron-irradiated GaAs. In this experiment [62], a beam of energetic electrons produces electron hole-pairs by ionization in a *p-n* junction device. The resulting short circuit current, J_{SC}, that flows through the diode terminals is related to the incident electron beam current by [62]

$$J_{SC} \simeq J_B\left(\frac{d\mathscr{E}/dx}{\eta}\right)\left[\frac{L_{p0}}{(1 + K_n\tau_{p0}\Phi)^{1/2}}\right] \tag{5.32}$$

where $\dfrac{d\mathscr{E}/dx}{\eta}$ = energy loss term

L_{p0} = initial hole diffusion length

τ_{p0} = initial hole lifetime

K_n = K_τ lifetime damage constant

Φ = electron fluence.

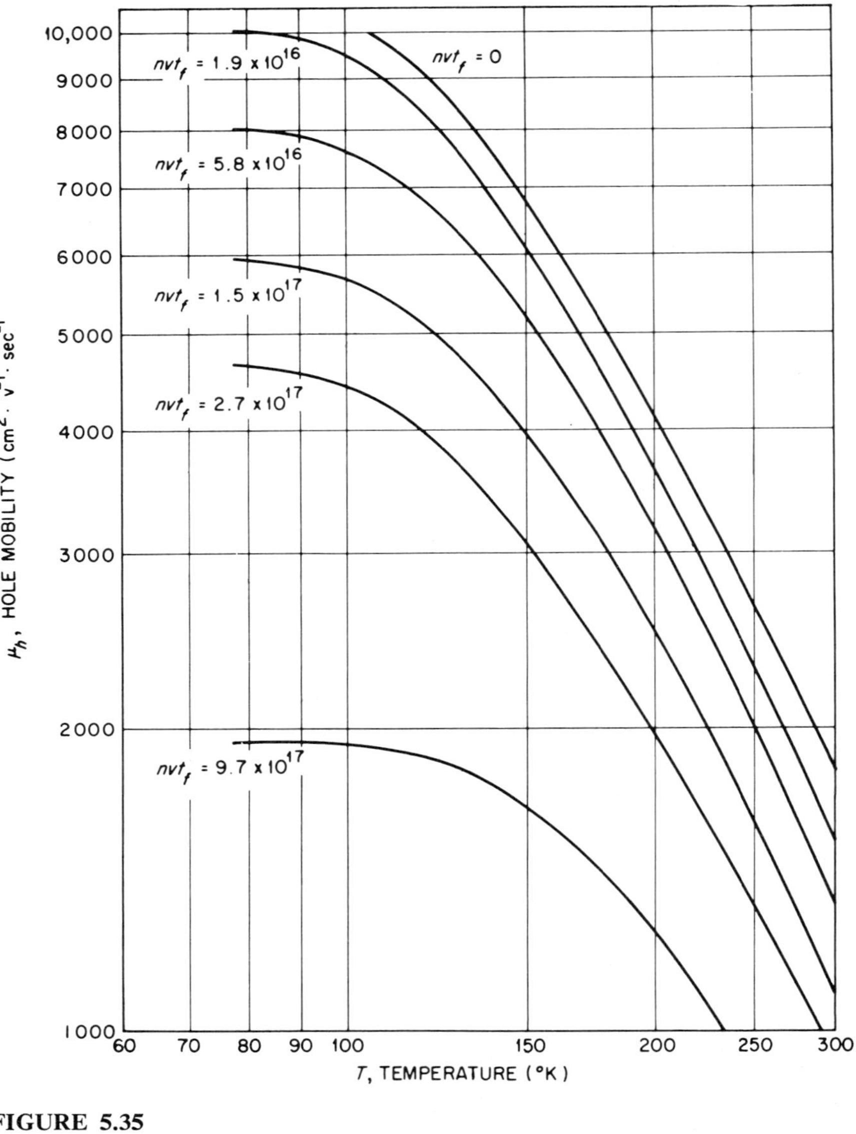

FIGURE 5.35
Hole Hall-mobility versus temperature for *p*-type Ge (2 Ω-cm) after successive fast-neutron exposures (after Cleland et al. [60]).

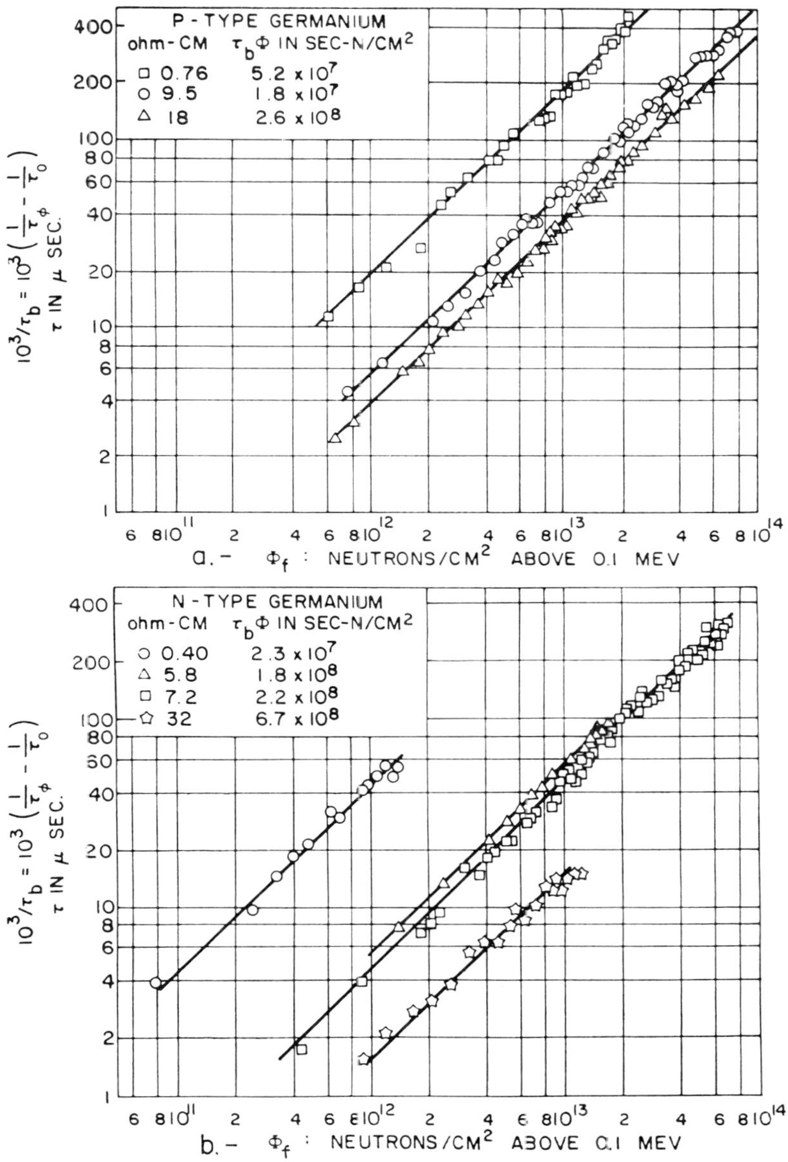

FIGURE 5.36
Change in inverse lifetime for *n*- and *p*-type Ge of several different resistivities exposed to fast-neutron fluence (after Walters [61]).

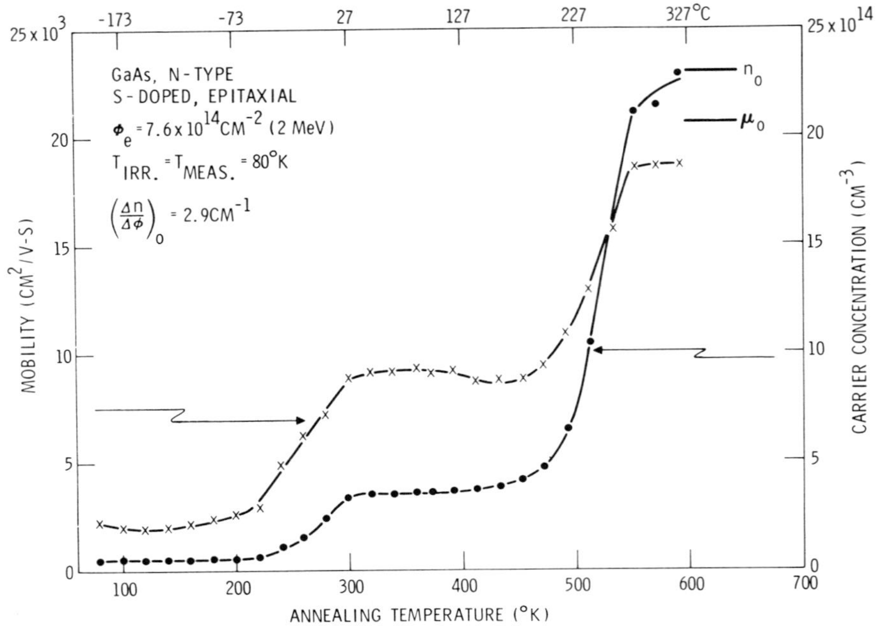

FIGURE 5.37
Low-temperature electron bombardment of epitaxial GaAs, carrier removal and mobility degradation, n_0 and μ_0 are prerad values (after Stein [38]).

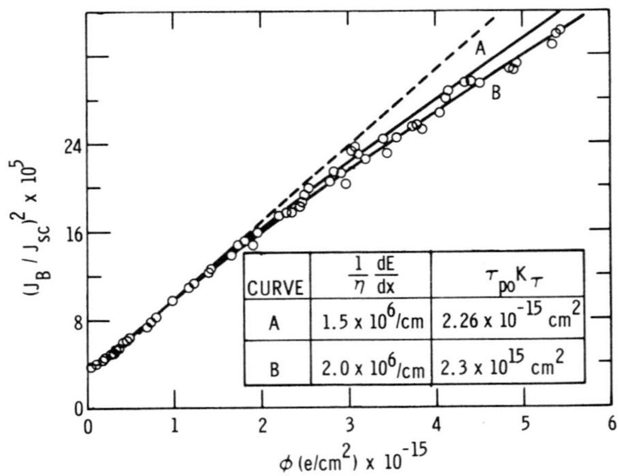

FIGURE 5.38
Variation of the short-circuit-current/beam-current ratio with fluence for a typical GaAs surface barrier diode doped with 2 × $10^{17}/cm^3$ Te (after Aukerman et al. [62]).

As the integrated fluence increases with time, the effective diffusion length decreases [see Eq. (5.32)] and the value of J_{SC} decreases. Using a value of $\tau_{p0} = (10^{-9} N_D)^{-1}$, Ref. [62], we find a lifetime damage constant for 2 MeV electron bombardment of $K_\tau = 5 \times 10^{-7}\ \text{cm}^2\ \text{sec}^{-1}$.

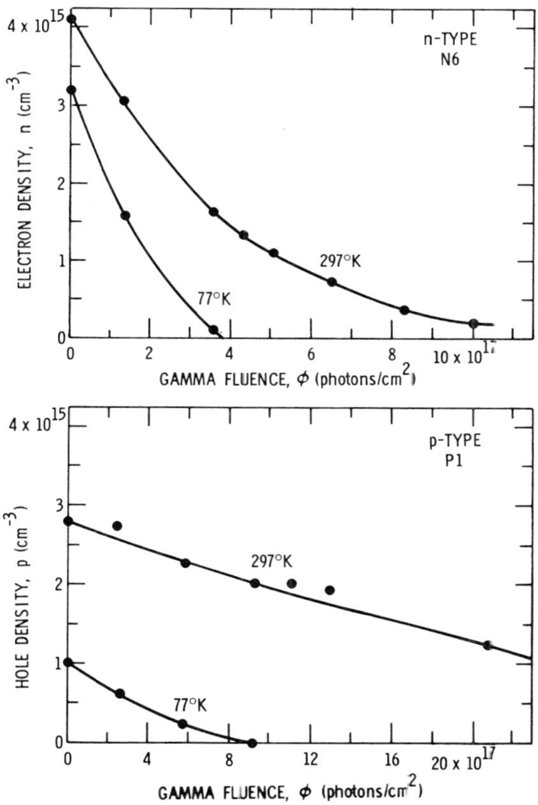

FIGURE 5.39
Carrier removal curves at 77° and 297°K for *n*- (2 Ω-cm) and *p*-type (10 Ω-cm) GaAs samples (after Brehm and Pearson [22]).

Gamma Rays in GaAs. The Co^{60} gamma-ray-induced carrier removal and mobility degradation are shown in Figures 5.39 to 5.41 for *n*- and *p*-type GaAs. Figure 5.39 shows that the carrier removal rate is larger at lower temperatures for gamma-irradiated GaAs and that *n*-type material is somewhat more sensitive to radiation than is *p*-type. Figures 5.40 and 5.41 show the mobility degradation for *p*- (Figure 5.40) and *n*-(Figure 5.41) type GaAs exposed to Co^{60} gamma rays. As usual, the effects are most noticeable at lower temperatures. Adequate lifetime data on electron, proton, and neutron irradiated GaAs are not available; this is probably due to the fact that practical problems in measuring the short lifetimes (typically a few nanoseconds) which occur in this direct semiconductor (see Figure 2.4B) have kept most researchers away from this area.

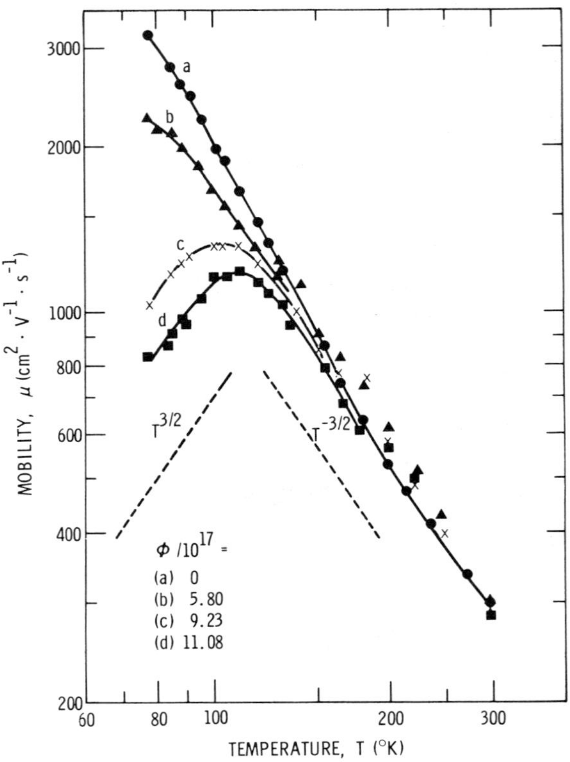

FIGURE 5.40
Hall mobility versus temperature for *p*-type (10 Ω-cm) GaAs sample (after Brehm and Pearson [22]).

Protons in GaAs. Adequate published data for proton-bombardment-induced electrical damage in GaAs is not available. However, it is known that like electron, neutron, and gamma irradiation, proton bombardment eventually changes both *p*- and *n*-GaAs to near-intrinsic GaAs [63].

Neutrons in GaAs. The effects of neutron bombardment on the carrier concentration and mobility are shown on Figures 5.42 and 5.43. Figure 5.42 shows that carrier removal rate in neutron-irradiated ($\mathscr{E} > 10$ keV, Spectrum A) n-type GaAs is a function of Fermi level (or doping concentration) and this is in qualitative agreement with Eq. (5.15). Figure 5.43 shows the mobility

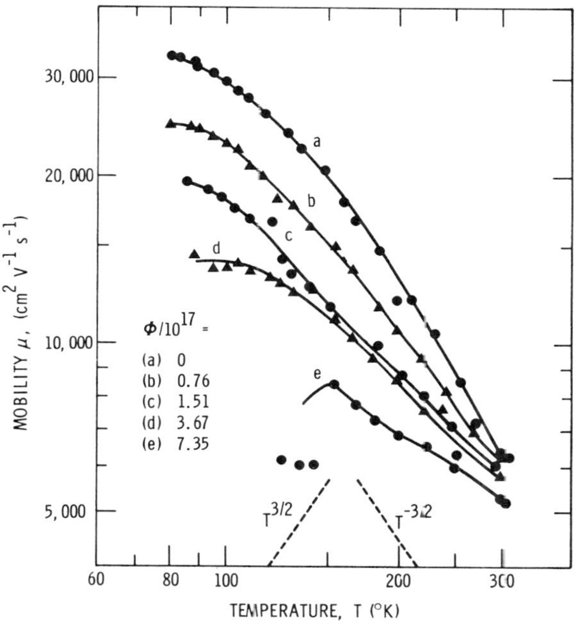

FIGURE 5.41
Hall mobility versus temperature for n-type (1 Ω-cm) GaAs sample (after Brehm and Pearson [22]).

degradation factor versus neutron fluence ($\mathscr{E} > 10$ keV, Spectrum B) for n-type GaAs. The mobility degradation factor is defined [65] as

$$b \cong \frac{\mu_i}{\Phi}\left(\frac{1}{\mu_f} - \frac{1}{\mu_i}\right)\bigg|_{\mu f \to \mu i} \tag{5.33}$$

where μ_i = initial mobility
μ_f = final mobility
Φ = neutron fluence.

Hence, from Figure 5.43 it can be seen that the mobility of lightly doped samples degrades at a faster rate with neutron fluence than does the more heavily doped material.

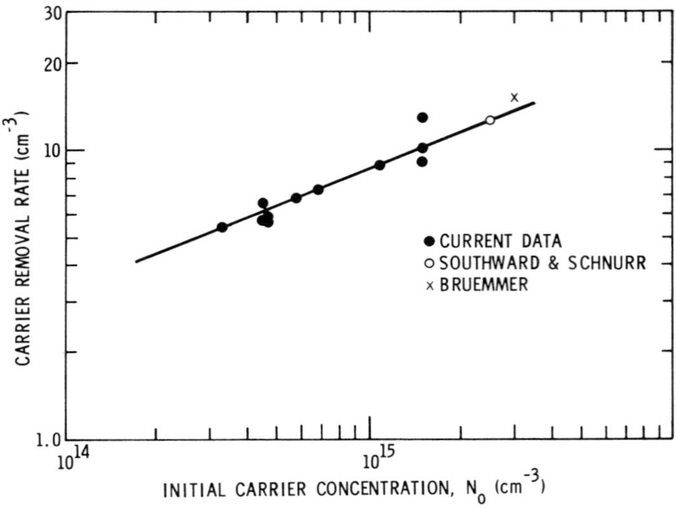

FIGURE 5.42
Carrier removal rate versus doping density for neutron-irradiated GaAs (after Marcus and Bruemmer [64]).

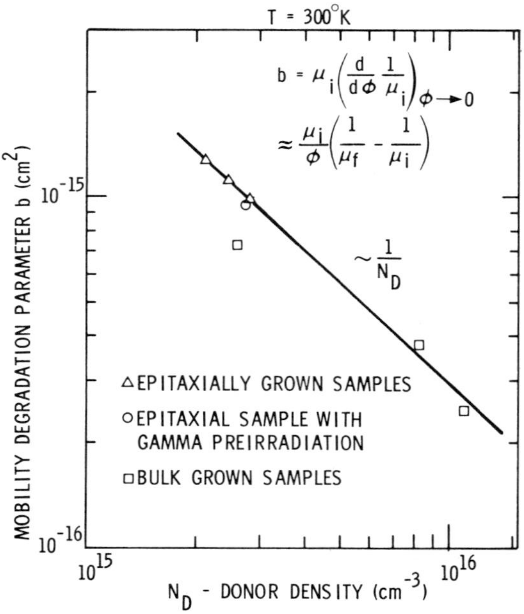

FIGURE 5.43
Mobility degradation parameter for neutron-irradiated *n*-type GaAs of varying resistivities (after Berg and Dropkin [65]).

Summary of Experimentally Observed Effects. At this point a few words spent summarizing the data presented in Figures 5.10 to 5.43 is worthwhile. This is done in Table 5.2. A few generalizations or "rules of thumb" can be made concerning the radiation fluence where the damage becomes important. For example, the conductivity will change appreciably when the density of trapped charges approaches the original doping density [see Eq. (5.13)]. A second useful rule of thumb is that the room temperature mobility will not be appreciably affected until the trapped charge density exceeds 10^{15} per cm^3. (This can be inferred from Figure 2.8.) A third useful rule of thumb is that lifetime damage effects will become important at a fluence where $K_\tau \cdot \Phi \cdot \tau_0 = 1$ [see Eq. 5.23]. This means that longer lifetimes are perturbed at lower fluences than are shorter lifetimes.

TABLE 5.2

Summary of Radiation Effects on Semiconductor Properties

Semiconductor Type	Carrier Concentration	Mobility	Lifetime
Silicon	Becomes intrinsic	Degrades	Degrades
Germanium	Approaches p-type 1 Ω-cm	Degrades	Degrades
Gallium-Arsenide	Becomes intrinsic	Degrades	Degrades

Note: Table represents all types of radiation.

5.6 Ionizing Radiation Effects

The effects of ionizing radiation (other than the creation of permanent displacement-type damage) will be discussed in this section. Energetic radiation, in the form of charged particles or gamma rays, or energetic silicon knock-ons from neutron bombardment, create electron-hole pairs by exciting electrons across the forbidden gap (similar to impact ionization discussed in Chapter 3, Section 3.5). As a result, whenever gamma rays (for example) are incident on a semiconductor, there is an increase in the density of electrons and holes (due to pair generation) which will decay back to the original densities if the ionizing source is removed. Since the bandgap energies of semiconductors are well known (see Appendix A), the amount of energy required to create an electron-hole pair can be readily found. In general, it is found [66] that the *average* energy required to create an electron-hole pair is roughly three times the bandgap of the semiconductor. This is because some of the energy is lost through other mechanisms. Figure 5.44 shows a more exact value for pair creation for several different semiconductor types, irradiated with protons (alphas), electrons, and x-ray or gamma-ray photons. Once the average energy for pair formation is known, the calculation of the number of carriers produced is very simple if the ionizing radiation is expressed in units of rads [see Eq. (5.3) and (5.4)]. Since the rad corresponds to

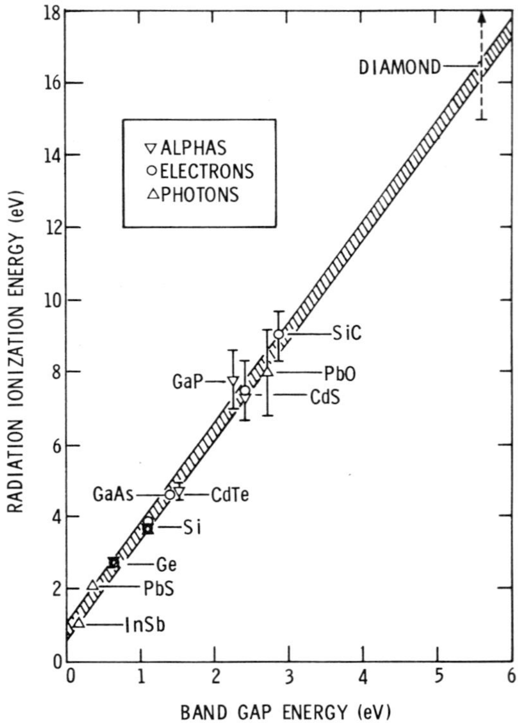

FIGURE 5.44
Average energy required to create an electrical-hole pair in several different semiconductors (after Klein [66]).

100 erg per gram, then rads $\cdot\, \rho$ (density grams per cm^3) is equal to rads per cm^3, i.e., if $\mathscr{E}_p$ = pair formation energy (in ergs)

$$\Delta n = \Delta p = \frac{100 \cdot \rho \cdot (\text{dose in rads})}{\mathscr{E}_p} \text{ excess carriers per cm}^3 . \quad (5.34)$$

The results for Si, Ge, and GaAs are (Si) 4×10^{13}, (Ge) 1.2×10^{14} and (GaAs) 7.25×10^{13} pairs per cm^3 per rad.

Unless the radiation dose is received in a time short compared to the lifetime, the excess density predicted by Eq. (5.34) will be too large. A more general approach is to utilize the incident dose rate (rads per sec) to calculate a generation rate, g, for excess holes and electrons given by

$$g_p = g_n = g = \frac{100 \cdot \rho \cdot (\text{rads per sec})}{\mathscr{E}_p} . \tag{5.35}$$

This generation rate can then be substituted into Eq. (2.69) and (2.70) (the continuity equations) and the exact density as a function of time can be solved. This effect can be used to measure lifetime in bulk semiconductors (see Figure 4.13 and its discussion; see also Figure 2.13 and its discussion).

In the case of a *p-n* junction device, it can be shown [10,67] that the absorption of steady state ionizing radiation produces a terminal current (called the photocurrent) given by

$$J = g \cdot e \cdot (L_p + L_n + W_T) \tag{5.36}$$

where L_p = hole diffusion length

L_n = electron diffusion length

W_T = total depletion width $W_A + W_D$ [see Eq. (3.6) and (3.7)].

Equation (5.36) simply states that all pairs created in the depletion region or within a diffusion length of the depletion region are collected.

In the case of a transient, or pulsed, radiation dose the diode photocurrent [Eq. (5.36)] takes on a more complex form. It can be shown [67] that the short-circuit current resulting from a rectangular pulse of ionizing radiation of duration t_0 is given by

$$\begin{aligned} J(t) &= g \cdot e \cdot \left[W_T + L_n \operatorname{erf}\left(\sqrt{\frac{t}{\tau_n}}\right) + L_p \operatorname{erf}\left(\sqrt{\frac{t}{\tau_p}}\right) \right] U(t) \\ &\quad - g \cdot e \cdot \left[W_T + L_n \operatorname{erf}\left(\sqrt{\frac{t - t_0}{\tau_n}}\right) + L_p \operatorname{erf}\left(\sqrt{\frac{t - t_0}{\tau_p}}\right) \right] U(t - t_0) \end{aligned} \tag{5.37}$$

where τ_n, τ_p = electron, hole lifetime

$U(t)$ = unit step turned on at $t = 0$

$U(t - t_0)$ = unit step turned on at $t = t_0$

erf = error function.

From Eq. (5.37) it can be seen that the diode current response to a pulse of ionizing radiation consists of a prompt component and a delayed component. The delayed component results from the fact that it takes time for the carriers to diffuse in from outside the depletion region. In the case of a diode, the photocurrent is not too bothersome provided that the total power dissipated (voltage × current) does not damage the device by heating. In the case of a transistor, there is further cause for concern in that the photocurrent is amplified by the transistor [10,67], and, unless current limiting is provided by the external circuit, the transistor may burn out.

5.7 Transient Annealing Effects

Sections 5.2 and 5.4 pointed out that the primary defects produced by gamma, electron, and neutron irradiations reorder into defect complexes.

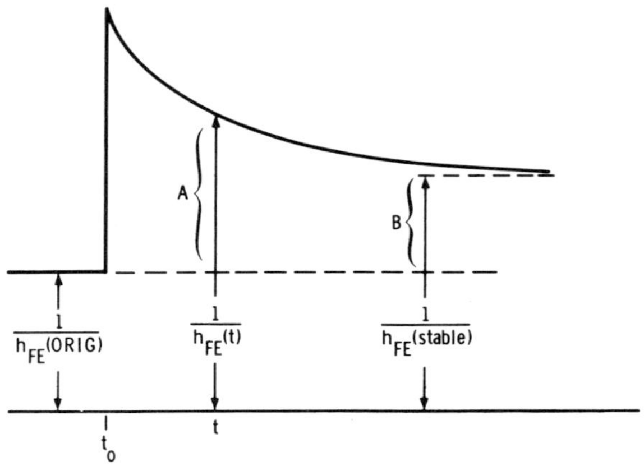

FIGURE 5.45
Plot of $(h_{FE})^{-1}$ versus time after exposure to a burst of fast neutrons (after Gregory and Sander [19]).

This reordering takes a finite time to complete and as a result the amount of damage seen will depend on the amount of time which has elapsed since the irradiation occurred. At room temperature most of the reordering in silicon is complete and the damage has reached its stable value in approximately 1000 seconds after exposure. The data presented in Section 5.5 for Si, Ge, and GaAs is for times sufficiently long after the irradiation so that the damage may be considered stable. Some electronic systems have a requirement that they must function very shortly after exposure to a pulse of damaging radiation. In this case the predictions of the damage curves in Section 5.5 are not accurate since the reordering of the damage is not yet complete at these times.

Gregory and Sander [19] have recently reviewed this problem for neutron-irradiated silicon. They define a time-varying annealing factor, $AF(t)$ given by

$$AF(t) = \frac{\dfrac{1}{h_{FE}(t)} - \dfrac{1}{h_{FE}(\text{ORIG})}}{\dfrac{1}{h_{FE}(\text{stable})} - \dfrac{1}{h_{FE}(\text{ORIG})}} = \frac{A}{B} = \frac{N_T(t)}{N_T(\text{stable})} = \frac{\Phi^*}{\Phi} \tag{5.38}$$

where h_{FE} = transistor current gain (if one is considering a transistor), N_T = radiation induced trap density, Φ = actual neutron fluence and Φ^* = effective (higher) neutron fluence at a time t. Some of the quantities in Eq. (5.38) vary with time after the irradiation. Figure 5.45 shows the typical behavior of $(h_{FE})^{-1}$ versus time following a burst of neutrons from a pulsed reactor. The definitions of the annealing factor may be extended to devices other than transistors by using the $N_T(t)/N_T(\text{stable})$ ratio in Eq. (5.38). Gregory and Sander [19] have discussed the variations in $AF(t)$ with temperature and electron density. In one particular case they report an increase in the AF of an *n-p-n* transistor by a factor of 2.5 when the temperature is reduced from 348°K to 213°K. More importantly they find a strong dependence of the AF on electron density in the damaged region of interest. The AF for neutron-irradiated silicon has been empirically predicted in general terms by Gregory and Sander [19]. Their predictions may be presented in the form of the nomograph shown in Figure 5.46 [70]. This figure relates the damage at a given time after irradiation to the damage seen 1000 seconds after the irradiation. To predict the $AF(t)$ for a silicon semiconductor device, it is necessary only to know the electron density in the critical (most easily damaged) region of the device. In most microwave semiconductor devices this density will vary considerably with the applied bias. As an example, let us consider a *p-i-n* diode (see Chapter 7) which is only slightly forward biased. The electron concentration in the *i*-region will be on the order of $10^{14}/\text{cm}^3$. At 10 msec after the irradiation burst, an inspection of Figure 5.46 predicts an $AF \cong 1.9$. From Eq. (5.38) we find $\Phi^* = 1.9 \cdot \Phi$, which implies that at 10 msec after the irradiation burst the device will behave as if it had been irradiated at nearly twice the actual fluence. If operation at this time is a system requirement, the radiation tolerance limit of the system must be based on the effective fluence $\Phi^* = AF \cdot \Phi$ rather than the actual fluence. Since transient measurements of device parameters are very difficult to perform, all of the data presented throughout this book will be given in terms of the stable radiation damage. In the cases where the transient annealing may be important, refer to Figure 5.46 to calculate the AF. The junction temperature of many microwave devices is near 423°K (150°C). It should be noted that at this elevated temperature the AF will be appreciably smaller than that predicted by Figure 5.46.

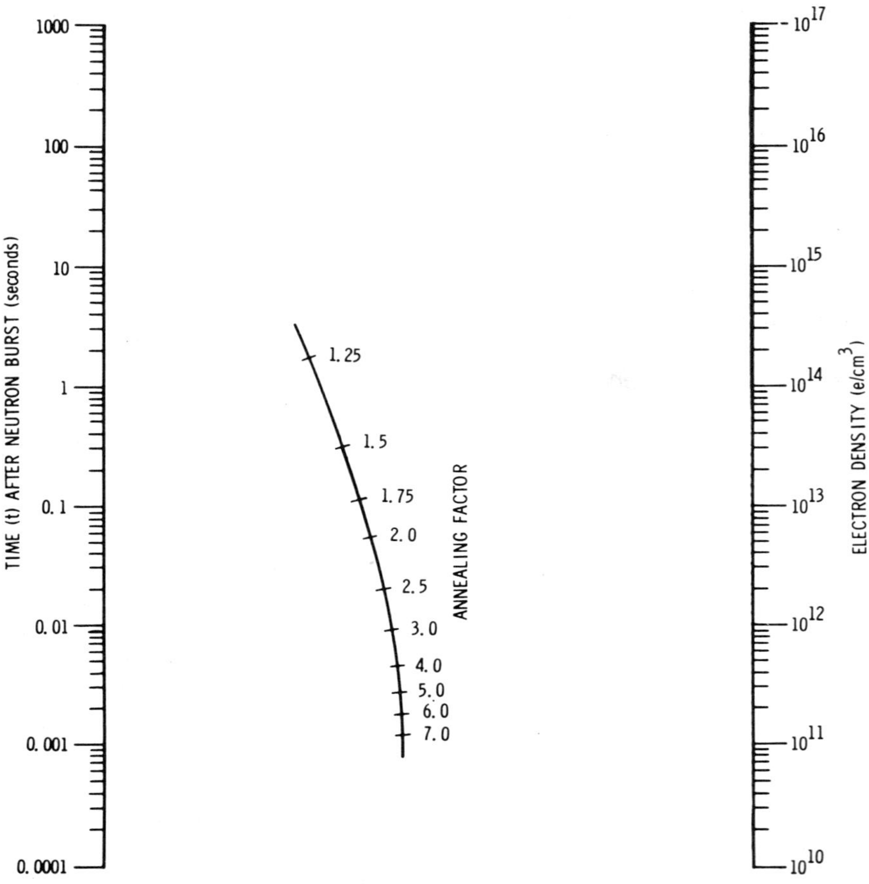

FIGURE 5.46
Nomograph for prediction of annealing factor in neutron-irradiated silicon at room temperature. The annealing factor is given by the intersection of the annealing factor curve with a straight line drawn between the electron density in the device's active region and the time after the irradiation burst. (After Sander and Gregory [70]).

REFERENCES

[1] G. J. Dienes and G. H. Vineyard, *Radiation Effects in Solids*, Interscience Publishers, New York, 1957.

[2] D. S. Billington and J. H. Crawford, Jr., *Radiation Damage in Solids*, Princeton University Press, Princeton, N.J., 1961; 2nd printing, 1965.

[3] *Radiation Damage in Solids, Vol. III*, Proceedings of Symposium on Radiation Damage in Solids and Reactor Materials held in Venice, May 1962 (International Atomic Energy Agency, Vienna, 1963).

[4] V. S. Vavilov, *Effects of Radiation on Semiconductors*, Moscow State Press, 1963 (translated version published by Consultants Bureau, New York, 1965).

[5] J. F. Kircher and R. E. Bowman, *Effects of Radiation on Materials and Components*, Reinhold Publishing Co., New York, 1964.

[6] P. Baruch, ed., *Radiation Damage in Semiconductors*, Proceedings of the 7th International Conference on the Physics of Semiconductors held at Royaumont, France, 1964 (Academic Press, 1964).

[7] J. W. Corbett, *Electron Radiation Damage in Semiconductors and Metals*, Academic Press, New York, 1966.

[8] A. Bishay, ed., *Interaction of Radiation with Solids*, Proceedings of the Cairo Solid-State Conference held in Cairo, N.Y., 1966 (Plenum Press, New York, 1967).

[9] F. L. Vook, ed., *Radiation Effects in Semiconductors*, Proceedings of 1967 Santa Fe Conference on Radiation Effects in Semiconductors (Plenum Press, New York, 1968).

[10] F. Larin, *Radiation Effects in Semiconductor Devices*, John Wiley & Sons, New York 1968.

[11] R. Hasiguti, ed., *Lattice Defects in Semiconductors*, Pennsylvania State University Press, University Park, Pa., 1968.

[12] J. W. Corbett and G. D. Watkins, eds., *Radiation Effects in Semiconductors*, Gordon and Breach, New York, 1971.

[13] A. M. Stoneham, "The Theory of Defects in Irradiated Semiconductors," Ref. [12], p. 7.

[14] J. H. Cahn, "Irradiation Damage in Germanium and Silicon due to Electrons and Gamma Rays," *J. Appl. Phys.*, **30**, 1310 (1959).

[15] G. K. Wertheim and D. N. E. Buchanan, "Electron Bombardment of Oxygen-Free Silicon," *J. Appl. Phys.*, **30**, 1232 (1959).

[16] L. W. Aukerman and R. D. Graft, "Annealing of Electron-Irradiated GaAs," *Phys. Rev.*, **127**, 1576 (1962).

[17] J. W. Cleland and W. S. Paciesas, "Impurity Dependence of Defect Introduction and Annealing in Electron-Irradiated *n*-Type Germanium," *IEEE Trans. Nuc. Sci.*, **NS-16**, 6, 19 (1969).

[18] G. D. Watkins, "EPR and Optical Absorption Studies in Irradiated Semiconductors," Ref. [9], p. 67.

[19] B. L. Gregory and H. H. Sander, "Transient Annealing of Defects in Irradiated Devices," *Proc. IEEE*, **58**, 1328 (1970).

[20] H. J. Stein and R. Gereth, "Introduction Rates of Electrically Active Defects in *n*- and *p*-Type Silicon by Electron and Neutron Irradiation," *J. Appl. Phys.*, **39**, 2890 (1968).

[21] R. D. Evans, *The Atomic Nucleus*, McGraw-Hill, New York, 1955.

[22] G. E. Brehm and G. L. Pearson, "Gamma-Radiation Damage in Epitaxial Gallium Arsenide," *J. Appl. Phys.*, **43**, 568 (1972).

[23] E. Sonder and L. C. Templeton, "Gamma Irradiation of Silicon I. Levels in *n*-Type Material Containing Oxygen," *J. Appl. Phys.*, **31**, 1279 (1960).

[24] J. W. Cleland, et al., "Effects of Gamma Radiation on Germanium," *Phys. Rev.*, **102**, 722 (1956).

[25] J. R. Bilinski et al., "Proton-Neutron Damage Equivalence in Semiconductors," *IEEE Trans. Nuc. Sci.*, **NS-10**, 6, 71 (1963).

[26] J. A. Baicker, et al., "Proton-Induced Lattice Displacements in Silicon," *App. Phys. Lett.*, **2**, 104 (1963).

[27] E. C. Smith et al., "Theoretical and Experimental Determination of Neutron Energy Deposition in Silicon," *IEEE Trans. Nuc. Sci.*, **NS-13**, 6, 11 (1966).

[28] A. R. Sattler, "Partition of Energy of Silicon and Germanium Recoils and Application to Neutron-Induced Effects," Ref. [9], p. 243.

[29] B. R. Gossick, "Disordered Regions in Semiconductors Bombarded by Fast Neutron," *J. Appl. Phys.*, **30**, 1214 (1959).

[30] J. H. Crawford and J. W. Cleland, "Nature of Bombardment Damage and Energy Levels in Semiconductors," *J. Appl. Phys.*, **30**, 1204 (1959).

[31] M. Bertolotti, "Experimental Observation of Damage Clusters in Semiconductors," Ref. [9], p. 311.

[32] H. J. Stein, "Electrical Properties of Neutron-Irradiated Silicon at 76°K: Hall Effect and Electrical Conductivity," *IEEE Trans. Nuc. Sci.*, **NS-15**, 6, 69 (1968).

[33] B. L. Gregory, "Minority Carrier Trapping by Defect Clusters," *App. Phys. Lett.*, **16**, 67 (1970).

[34] R. R. Holmes, "Carrier Removal in Neutron-Irradiated Silicon," *IEEE Trans. Nuc. Sci.*, **NS-17**, 6, 137 (1970).

[35] F. G. Wikner, et al., *Transient Radiation Effects, Final Report*, Contract number, DA-49-186-AMC-64(X)-1, February 1967.

[36] J. W. Cleland, et al., "Fast-Neutron Bombardment of *n*-Type Ge," *Phys. Rev.*, **98**, 1742 (1955).

[37] H. J. Stein, "Electrical Studies of Neutron-Irradiated *n*-Type Silicon: Defect Structure and Annealing," *Phys. Rev.*, **163**, 801 (1967).

[38] H. J. Stein, "Electrical Studies of Low Temperature Neutron and Electron-Irradiated Epitaxial *n*-Type GaAs," *J. Appl. Phys.*, **40**, 5300 (1969).

[39] L. J. Cheng and J. Lori, "Characteristics of Neutron Damage in Silicon," *Phys. Rev.*, **171**, 856 (1968).

[40] R. F. Bass, "Influence of Impurities on Carrier Removal and Annealing in Neutron-Irradiated Silicon," *IEEE Trans. Nuc. Sci.*, **NS-14**, 6, 78 (1967).

[41] H. Y. Fan, "Theories of Point Defects in Valence Semiconductors," Ref. [9], p. 17.

[42] H. M. James and K. Lark-Horvitz, "Localized Electronic States in Bombarded Semiconductors," *Z. Phys. Chem.*, **198**, 107 (1951).

[43] M. G. Buehler, "The Space Charge Region of Neutron-Irradiated Silicon p^+n Junctions," *IEEE Trans. Nuc. Sci.*, **NS-17**, 6, 341 (1970).

[44] P. P. Debye and E. M. Conwell, "Electrical Properties of *n*-Type Germanium," *Phys. Rev.*, **93**, 693 (1954).

[45] Discussed in Ref. [44].

[46] E. P. EerNisse and R. J. Chaffin, "Carrier Trapping and Recombination in Avalanche Diodes," *IEEE Trans. ED*, **ED-17**, 520 (1970).

[47] J. S. Blakemore, *Semiconductor Statistics*, Pergamon Press, New York, 1962.

[48] H. J. Stein, "Comparison of Neutron and Gamma-Ray Damage in *n*-Type Silicon," *J. Appl. Phys.*, **37** (1966).

[49] E. G. Wikner and H. Horiye, "Study of *n*- and *p*-Type Silicon Exposed to Highly Energetic Radiation," *IEEE Trans. Nuc. Sci.*, **NS-13**, 6, 18 (1966).

[50] H. J. Stein and F. L. Vook, "Characteristics of Electron-Induced Defects in *n*-Type Silicon," Ref. [9], p. 115.

[51] E. Sonder and L. C. Templeton, "Gamma Radiation in Silicon III, Levels in *p*-Type Material," *J. Appl. Phys.*, **36**, 1811 (1965).

[52] E. Sonder, "Magnetic and Electrical Properties of Reactor-Irradiated Silicon," *J. Appl. Phys.*, **30**, 1186 (1959).

[53] M. G. Beuhler, "Design Curves for Predicting Resistivity Changes in Silicon," *IEEE Proc.*, **56**, 1741 (1968).

[54] B. L. Gregory and C. W. Gwyn, "Application of Neutron-Damage Models to Semiconductor Device Studies," *IEEE Trans. Nuc. Sci.*, **NS-17**, 325 (1970).

[55] J. J. Loferski and P. Rappaport, "Electron Bombardment-Induced Recombination Centers in Germanium," *J. Appl. Phys.*, **30**, 1181 (1959).

[56] W. L. Brown and W. M. Augustyniak, "Energy, Orientation and Temperature Dependence of Defect Formation in Electron Irradiation of *n*-Type Germanium," *J. Appl. Phys.*, **30**, 1300 (1959).

[57] W. H. Closser, "Drift Mobility in Neutron Irradiated *n*-Type Ge," *J. Appl. Phys.*, **31**, 1693 (1960).

[58] J. H. Crawford and J. W. Cleland, "Nature of Bombardment Damage and Energy Levels in Semiconductors," *J. Appl. Phys.*, **30**, 1204 (1959).

[59] B. G. Streetman, "Recombination and Trapping in Co^{60} Gamma-Irradiated *n*-Type Germanium," *J. Appl. Phys.*, **37**, 3145 (1966).

[60] J. W. Cleland et al., "Fast Neutron Bombardment of *p*-Type Germanium," *Phys. Rev.*, **99**, 1170 (1955).

[61] A. E. Walters, "Minority Carrier Lifetime in Neutron-Bombarded Germanium," in *Proceedings of the Second Conference on Nuclear Effects on Semiconductor Devices, Materials and Circuits*, Copyright Cowan Publishing Co., New York, 1959, p. 35.

[62] L. W. Aukerman et al., "Diffusion Lengths of Electrons and Holes in GaAs," *J. Appl. Phys.*, **38**, 685 (1967).

[63] A. G. Foyt et al., "Isolation of Junction Devices in GaAs using Proton Bombardment," *Sol. State Elec.*, **12**, 209 (1969).

[64] G. H. Marcus and H. P. Bruemer, "Radiation Damage in GaAs Gunn Diodes," *IEEE Trans. Nuc. Sci.*, **NS-17**, 6, 230 (1970).

[65] N. Berg and H. Dropkin, "Neutron Displacement Effects in Epitaxial Gunn Diodes," *IEEE Trans. Nuc. Sci.*, **NS-17**, 6, 233 (1970).

[66] C. A. Klein, "Bandgap Dependence and Related Features of Radiation Ionization Energies in Semiconductors," *J. Appl. Phys.*, **39**, 2029 (1968).

[67] J. L. Wirth and S. C. Rogers, *The Transient Response of Transistors and Diodes to Ionizing Radiation*, Sandia Corporation Report SC-R-64-194, (1964).

[68] H. Y. Fan and K. Lark-Horovitz, "Fast Particle Irradiation of Germanium Semiconductors," *Report of the Bristol Conf. on Defects in Crystalline Solids*, The Physical Society London, 1955, pp. 232–45. Copyright by the Institute of Physics.

[69] R. A. Hewes, "Recombination Lifetimes in Gamma-Irradiated Silicon," *J. Appl. Phys.*, **39**, 4106 (1968).

[70] H. H. Sander and B. L. Gregory, "Circuit Applications of Transient Annealing," *IEEE Trans. Nuc. Sci.*, **NS-18**, 6, 250 (1971).

6

Schottky-Barrier Diode Mixers and Detectors

6.1 Introduction

Metal semiconductor diodes are surely the oldest of the component types being described in this book. In various forms they date back as early as 1874; a good historical review of this subject is given by Henisch [1] in his book. The diodes to be discussed in this chapter are called *Schottky-barrier diodes* after W. Schottky whose early work [2] formed the basis for most modern theories. The basic difference between the Schottky-barrier diodes and the *p-n* junction diodes discussed in Chapter 3 is that the Schottky-barrier diode is a majority carrier device and hence does not exhibit any minority carrier storage effects such as diffusion capacitance. In the Schottky-barrier device the forward-bias current consists of energetic electrons which are injected over a potential barrier into the metal from an *n*-type semiconductor. The earliest microwave devices of this type were the *point-contact* diodes developed for radar mixers during World War II. A point-contact device is usually fabricated by pressing a finely pointed tungsten wire onto a piece of *p*-type silicon [3]. These devices were used for the majority of the microwave mixer and detector applications until the late 1960s. The silicon point-contact mixer is thought to be a Schottky-barrier device. In the late 1960s photomasking techniques were sufficiently refined so that evaporated Schottky-barrier diodes with small capacitances could be made. For these devices, a thin layer of metal in the shape of a small dot was evaporated onto a clean silicon surface. Devices made in this way will be called Schottky-barrier devices. The Schottky-barrier device has better V-I and noise characteristics than does the point-contact diode, and it is also known to be more mechanically rugged and reproducible than the point-contact diode.

Military and space microwave systems utilize Schottky-barrier mixer and detector diodes. These systems may be exposed to radiation environments; hence, it is important for the designer to know and understand radiation effects in these devices so he can minimize undesirable permanent and transient effects.

Section 2 of this chapter examines basic metal-semiconductor theory; in Section 3 the V-I characteristic is derived. Section 4 considers several different device geometries. Radiation effects on dc and low-frequency parameters are covered in Section 5. Sections 6 and 7 cover detectors and mixers, respectively, starting with basic concepts and finishing with permanent neutron-radiation damage effects. The final section of this chapter, Section 8, considers the effect of transient ionizing radiation on Schottky-barrier devices.

6.2 Ideal Barrier Theory

The metal-semiconductor Schottky-barrier effect is best understood using the band model of solids developed in Chapter 2. Figure 6.1A shows a metal and an *n*-type semiconductor not in physical contact. Φ_m is the work function of the metal, which is the energy required for an electron at the Fermi energy to escape into a vacuum. (Φ_m is a few electron volts.) Φ_s is the work function of the semiconductor, and χ_s (the difference between the conduction band edge and vacuum free energy level) is called the electron affinity of the semiconductor. When the metal and semiconductor are brought into intimate contact (as in Figure 6.1B), thermal equilibrium requires that the Fermi levels in the metal and semiconductor align (as shown). When the metal and the semiconductor are first brought into contact there will be a net flow of electrons from the conduction band of the semiconductor into the metal. This flow will cause a pile-up of negative charges on the surface of the metal and produce a positive space-charge or depletion region in the semiconductor due to the unneutralized, ionized donor atoms left behind. The combination of the negative charge on the metal and positive space charge in the semiconductor sets up a retarding field for electrons trying to go from the semiconductor to the metal. It is known [4] that this process will reach equilibrium when the electrostatic potential difference between the metal and the semiconductor reaches the diffusion potential of the metal-semiconductor junction.

$$V_{DO} = \frac{1}{e}(\Phi_m - \Phi_s) \tag{6.1}$$

This is the height of the potential barrier at zero bias which impedes the flow of electrons from the *semiconductor into the metal.* The potential barrier (for any bias) retarding the flow electrons from the *metal to the semiconductor* is

$$V_{BN} = \frac{1}{e}(\Phi_m - \chi_s) . \tag{6.2}$$

If it is assumed that the semiconductor is uniformly doped with a donor density, N_D, the depletion width W (Figure 6.1B) can be solved for by using Poisson's equation in a way similar to that done in Chapter 3.

$$\frac{d^2V}{dx^2} = \frac{-\rho(x)}{\epsilon_s} \tag{6.3}$$

where

$$\rho(x) = +eN_D \qquad 0 \le x \le W$$

$$\rho(x) = 0 \qquad x > W .$$

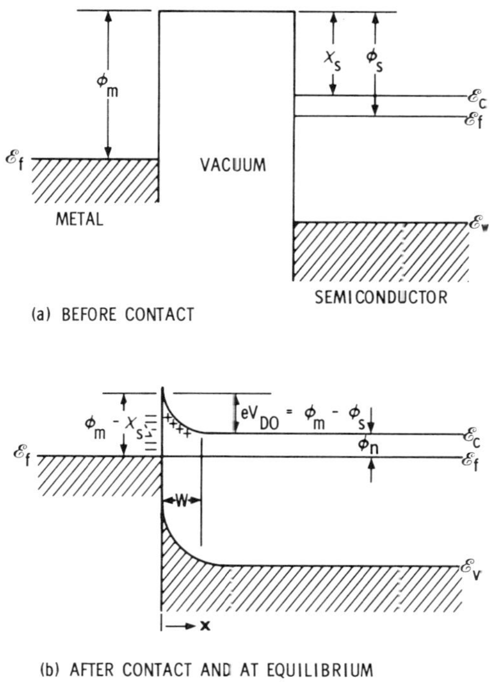

FIGURE 6.1
Metal-semiconductor junction.

The boundary conditions to be satisfied are $dV/dx = 0$, at $x = W$ and $V = -V_{BN}$ at $x = 0$. The solution to Eq. (6.3) may be shown to be†

$$V(x) = \frac{eN_D}{\epsilon_s}\left(Wx - \frac{x^2}{2}\right) - V_{BN}\,. \tag{6.4}$$

The difference of potential between $x = 0$ and $x = W$ is simply V_{DO} (see Figure 6.1B), hence

$$V_{DO} = V(x)|_{x=W} - V(x)|_{x=0}\,. \tag{6.5}$$

Performing the operations indicated in Eq. (6.5) and solving for W, we get the width of the zero-bias depletion layer as

$$W = \left[\frac{2\epsilon_s V_{DO}}{eN_D}\right]^{1/2}. \tag{6.6}$$

If an external bias is applied to the metal-semiconductor junction, the result will be a shift in relative Fermi levels of the metal and semiconductor by an amount equal to the applied bias. This will have the effect of changing W. From Figure 6.2 it can be seen that the applied voltage, V_A, adds to or subtracts from the diffusion voltage, V_{DO}, depending on whether there is reverse or forward bias applied. Forward bias will be defined as the case when the metal is positive with respect to the semiconductor. The new value of depletion width (as a function of applied bias) may be solved for exactly as before, using Eq. (6.5) replacing V_{DO} by $V_{DO} - V_A$ (V_A is positive for forward bias). The result is

$$W(V_A) = \left[\frac{2\epsilon_s(V_{DO} - V_A)}{eN_D}\right]^{1/2}. \tag{6.7}$$

From Eq. (6.7) it can be seen that the depletion region widens for reverse bias and narrows for forward bias. This behavior is similar to that of an ordinary *p-n* junction (see Chapter 3). If a small ac voltage were applied to the junction, the depletion region would periodically narrow and widen [see Eq. (6.7)]; as it narrows and widens the amount of charge stored in the depletion region ($Q = AeN_DW$, where A = device area) will also vary. Hence, the depletion capacitance may be calculated from

$$C_d = \frac{\partial Q}{\partial V} = -\frac{\partial Q}{\partial V_A} = A\left[\frac{\epsilon_s eN_D}{2(V_{DO} - V_A)}\right]^{1/2} = \frac{\epsilon_s A}{W}\ \text{farad}\,. \tag{6.8}$$

This result is identical to that obtained for the one-sided abrupt junction in Chapter 3. By rearranging Eq. (6.8), we see

$$\frac{1}{C^2} = \frac{2(V_{DO} - V_A)}{\epsilon_s N_D e A^2} \tag{6.9}$$

† This neglects the contribution of mobile charges in the space-charge region which is equivalent to increasing the barrier height by kT/e (see Goodman [5]).

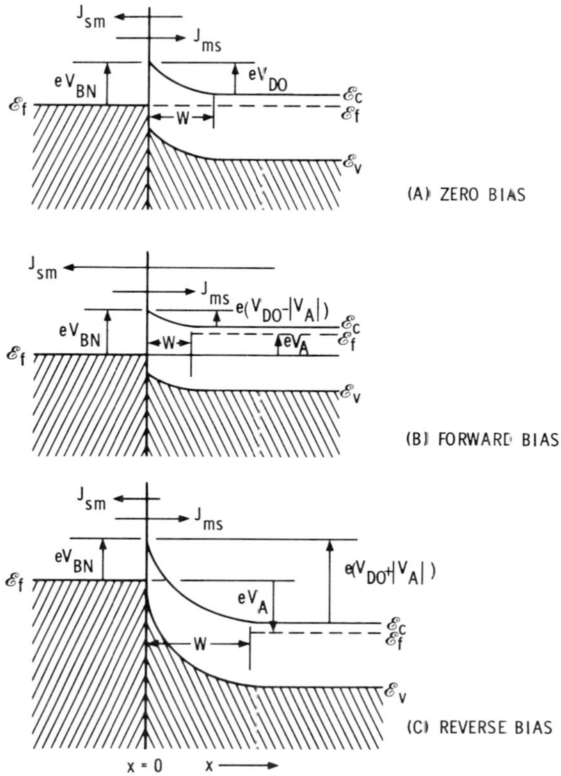

FIGURE 6.2
Schottky-barrier diode with bias applied.

and also

$$\frac{d}{dV_A}\left[\frac{1}{C^2}\right] = \frac{-2}{\epsilon_s N_D e A^2} \,. \tag{6.10}$$

Hence, if $1/C^2$ is plotted versus V_A, the result is a straight line the slope of which gives the value of N_D, while the intercept where $1/C^2 \to 0$ (obtained by extrapolation) may be used to find the diffusion voltage and hence the barrier height.† This procedure is demonstrated in Figure 6.3.

† The diffusion voltage differs from the barrier height because of the location of the Fermi level, i.e., $\Phi_{BN} = eV_{DO} + \Phi_n$, see Figure 6.1.

6.3 Current-Voltage Characteristics

There are two basic theories [7] used to describe current flow in a Schottky-barrier diode. The diode (or thermionic) theory assumes that the barrier is narrower than the mean free path of the electrons in the semiconductor, and therefore collisions inside the barrier may be neglected. The diffusion theory assumes that the barrier is thick compared to the mean free path, and thus the motion through the barrier region is controlled by diffusion processes. Of the two theories, the thermionic theory is the most widely accepted and is the only one that will be treated here.

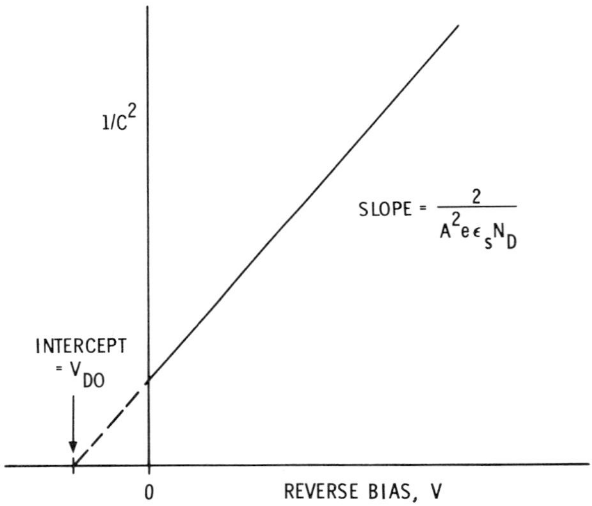

FIGURE 6.3
Capacitance-voltage characteristic of Schottky-barrier diode.

The basic goal in the development of the thermionic theory is to determine the number of electrons in the semiconductor which are sufficiently energetic to surmount the potential barrier, $V_{DO} - V_A$, see Figure 6.2. These energetic or "hot" carriers are then injected into the metal over the potential barrier. Schottky diodes are sometimes called "hot carrier" diodes for this reason. Once the hot electrons are injected into the metal they very quickly give up their excess energy through collisions (in approximately 10^{-13} sec, [9]) and return to the Fermi energy. This means that if the polarity reversal time of an applied sinusoidal voltage takes longer than approximately 10^{-12} sec, the injected electrons cannot be returned (the barrier Φ_{BN} holds them back, see

Figures 6.1 and 6.2). Hence, this device could theoretically rectify at frequencies up to 10^{12} Hz. This limit cannot practically be achieved because of the shunt depletion capacitance associated with the junction. The forward current may be found [10,11] by using the Maxwell-Boltzmann statistics to describe the electron distribution in the semiconductor. The number of carriers per unit volume with velocity between v and $v + dv$ in the $-x$ direction is

$$dn_x = n\left(\frac{m_{de}^*}{2\pi kT}\right)^{1/2} \cdot \exp\left[\frac{-m_e^* v_x^2}{2kT}\right] dv_x . \tag{6.11}$$

The number of carriers per second impinging on the barrier with energy sufficient to surmount it is

$$\int_{e(V_{DO}-V_A)}^{\infty} \left(\frac{dn_x}{dE}\right) dE = \tfrac{1}{2} n\bar{v} \exp\left[\frac{e(V_A - V_{DO})}{kT}\right] \tag{6.12}$$

where

$$\bar{v} = \left(\frac{2kT}{\pi m_{de}^*}\right)^{1/2} .$$

The current density flowing from the semiconductor to the metal will just be e times this or

$$J_{sm} = \tfrac{1}{2} ne\bar{v} \exp\left[\frac{e(V_A - V_{DO})}{kT}\right] . \tag{6.13}$$

At zero bias and thermal equilibrium ($V_A = 0$), the current from the metal to the semiconductor must be equal to the current from the semiconductor to the metal, i.e., $|J_{sm}| = |J_{ms}|$, see Figure 6.2A. It has been stated previously that the barrier height Φ_{BN} does not vary appreciably with bias, hence it will be assumed that J_{ms} = constant = $\tfrac{1}{2} ne\bar{v} \exp(-eV_{DO}/kT)$. Considering both J_{sm} and J_{ms} we may express the total electron current as

$$J = J_{sm} + J_{ms} = \tfrac{1}{2} ne\bar{v} \exp\left(\frac{-eV_{DO}}{kT}\right)\left[\exp\left(\frac{eV_A}{kT}\right) - 1\right] . \tag{6.14}$$

Watson [12] shows that Eq. (6.14) may be reduced to

$$J = J_s\left[\exp\left(\frac{eV_A}{kT}\right) - 1\right] \tag{6.15}$$

where $J_s = A_R T^2 \exp(-\Phi_{BN}/kT)$ and A_R = effective Richardson constant, and T = temperature. The V-I characteristic given by Eq. (6.15) for the Schottky-barrier diode is very similar to the one derived in Chapter 3 for the ideal p-n junction, i.e., the current rises exponentially in the forward direction and quickly reaches a saturation value in the reverse direction.

The theory presented here for Schottky-barrier diodes is the simplest one available. In practice there are a great many complicating effects which should be included in the theory in order to improve the model. The treatment of these effects in detail is beyond the scope of this book but can be found in a very excellent text by Sze [13]. Some of these complicating effects are:

1. *Image force barrier lowering*. As an electron approaches the metal it induces an equal and opposite image charge on the metal, which attracts the electron and helps it over the barrier.
2. *Quantum mechanical tunneling*. There is a finite probability that a carrier with insufficient energy to surmount the barrier can tunnel through it.
3. *Diffusion theory*. The neglect of collisions within the barrier is not always possible.
4. *Minority carrier injection*. At high forward currents the effect of hole motion cannot be ignored.
5. *Interfacial layer*. In many Schottky-barrier devices, intimate contact between the semiconductor and metal is not achieved, instead a thin oxide interfacial layer is found between them.
6. *Surface states*. The surface of the semiconductor is usually not neutral due to the presence of surface states; hence, a depletion region can exist even before the metal and semiconductor are brought into contact.

The last point (number 6) will be expanded upon briefly since it may be important in the radiation effects case. Mead [14] shows that surface states, caused by the termination of the lattice, have a localized density of states peaked at an energy level $E_0 = E_c - \frac{2}{3}E_g$ for most materials. These are acceptor-type states and will be filled with electrons if they lie below the Fermi energy. In the limiting case of a large density of surface states, the surface states and not the work function of metal determines the barrier height [14].

6.4 Diode Construction and Equivalent Circuits

The Schottky-barrier diode has several advantages over a conventional *p-n* junction diode. The fact that it is a majority-carrier device and exhibits no charge storage can be seen quite clearly from the preceding section. Another advantage this device has over a *p-n* junction is the absence of the *p*-layer. This implies that devices with lower series parasitic resistance can be made.

The first Schottky-barrier microwave devices were in the form of point-contact diodes. The point-contact diodes were usually fabricated [15] by pressing a small diameter, pointed tungsten wire onto a piece of *p*-type silicon followed by an "adjustment"† to improve the V-I characteristic. The p^+-type, 0.01 Ω-cm [19], substrate is usually processed by out-diffusion and subsequent polishing so that it has a thin lightly doped layer at the surface. This was found to give better reverse characteristics and a lower depletion capacitance while maintaining a low series resistance. The narrow lightly doped region has a wider depletion region for a given reverse bias, and this explains both the reduced capacitance and increased breakdown voltage. Figure 6.4 shows the cross section of a typical point contact device. The diameter of the point can vary from 2 to 75 μm [17]. A good review of present-day point-contact devices is given by Burrus [18].

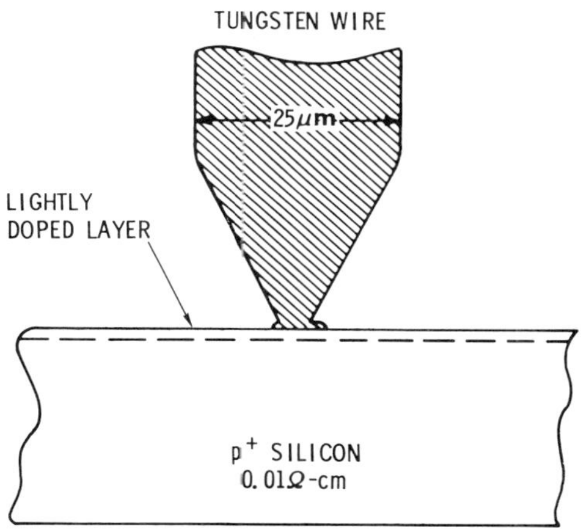

FIGURE 6.4
Cross-sectional view of point-contact diode.

In recent years the evaporated metal Schottky-barrier diode has replaced the point-contact device in most microwave mixer and detector applications. This is because these devices have greater mechanical strength, superior V-I characteristics, and lower noise. Evaporated Schottky-barrier devices are fabricated by depositing a metal (usually Pt or Au) onto a vacuum-cleaned silicon surface. There is no need to "adjust" the device made in this way; therefore they tend to be much more reproducible than point-contact devices. The devices are usually fabricated using a thin lightly doped n-type epitaxial layer grown on an n^+ silicon substrate. The thickness of the epitaxial material is typically [19] 1 μm with a resistivity of 1 Ω-cm. The function of this lightly

† This usually consists of tapping the unit with a small hammer or passing a semi-destructive current pulse through the device (see [16]).

doped layer is the same as for the point contact, i.e., it widens the depletion region which reduces the capacitance and increases the breakdown voltage. Several typical geometries for evaporated Schottky-barrier devices are shown in Figure 6.5. The diameter of the contacts shown in Figure 6.5 are typically 25 μm [19]. The guard-ring structure [20] shown in Figure 6.5C uses a diffused p^+ guard ring to eliminate unwanted edge breakdown and channeling effects. Devices of the type shown in Figure 6.5C have extremely low reverse leakage currents.

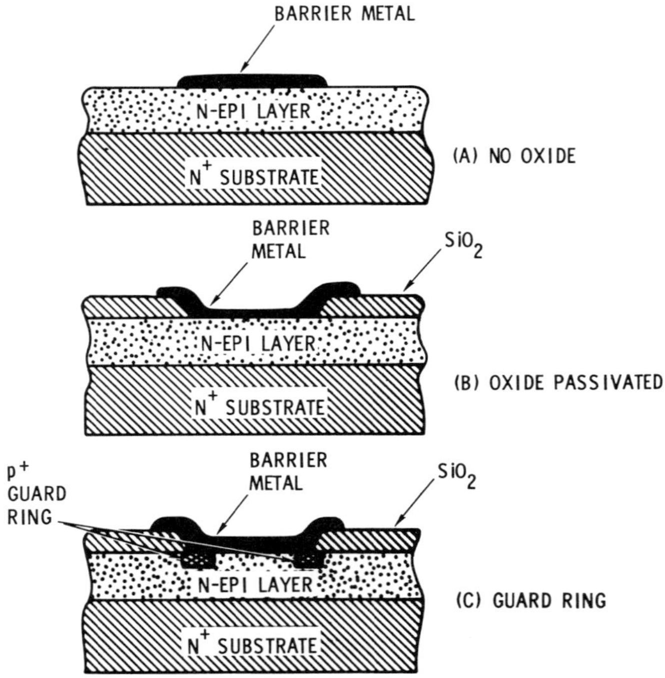

FIGURE 6.5
Schottky-barrier diode construction.

The equivalent circuit of an unpackaged Schottky-barrier diode is shown in Figure 6.6A. In Figure 6.6A, R_s represents the series resistance due to the bulk material and the contacts. R_s can be expressed as

$$R_s \cong \frac{d_1 - W}{\sigma_1 A} + \frac{d_2 - d_1}{\sigma_2 A} + \frac{1}{4\sigma_2 r} \tag{6.16}$$

where W = depletion width, d_1 = epitaxial region thickness, $d_2 - d_1$ =

substrate thickness, σ_1, σ_2 = conductivity of undepleted epitaxial and substrate respectively, A = area of the device, and r = radius of the barrier contact. The first term on the R.H.S. of Eq. (6.16) is the resistance of the undepleted epitaxial region, the second term is the substrate resistance, and the third term is the spreading resistance. R_s is usually 10 ohms or less for a good microwave device. C_J on Figure 6.6A represents the depletion capaci-

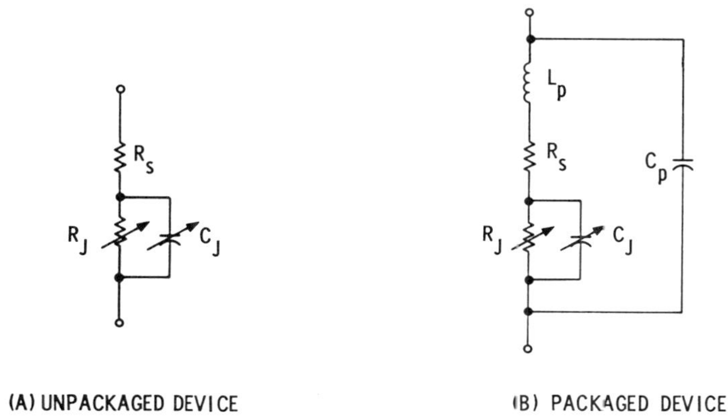

FIGURE 6.6
Equivalent circuits of Schottky-barrier diode.

tance and is given by Eq. (6.8). R_J is the dynamic resistance of the junction, $R_J = 1/G_J$, where $G_J = dI/dV$. From Eq. (6.15)

$$G_J = \frac{d}{dV}\left\{J_s A\left[\exp\left(\frac{eV}{kT}\right) - 1\right]\right\}$$

for $V \gg e/kT$ we get

$$R_J \cong \frac{kT}{eI} \quad \text{or}^\dagger \quad \frac{26}{I} \text{ ohms}. \tag{6.17}$$

The completed devices, either point-contact or Schottky-barrier are packaged so that they can be safely handled. The package adds parasitic elements to the equivalent circuit as shown in Figure 6.6B. These additional circuit elements can be quite important at microwave frequencies.

It was mentioned earlier that the Schottky-barrier diode has a superior V-I characteristic when compared to a point-contact diode. Figure 6.7 shows typical V-I characteristics of both types of devices.

† I in milliamperes, at room temperature $kT/e \cong 0.026 V$.

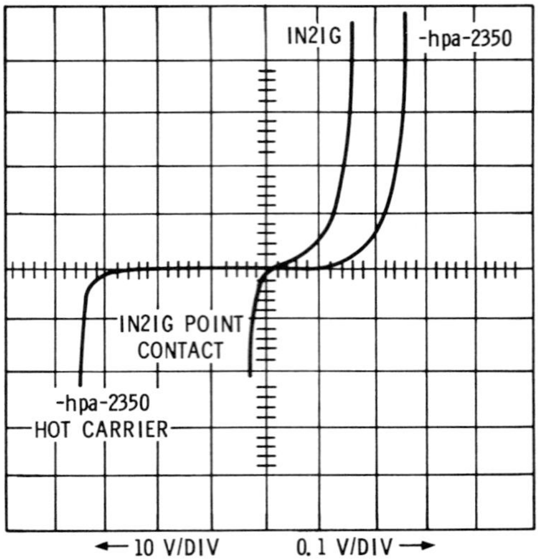

FIGURE 6.7
V-I characteristic of point-contact and Schottky-barrier diodes (after Kaposhilin [21]).

6.5 DC and Low-Frequency Radiation Effects

Schottky-barrier devices find applications at dc and low frequencies as well as at microwave frequencies. For this reason several authors (see [22–29]) have investigated the effects of radiation on only the dc and low-frequency properties of these devices. These papers will be reviewed here since some of the effects seen at dc and low frequencies will also be seen at microwave frequencies.

In 1949, Johnson and Lark-Horovitz [22] found that the series resistance, R_s on Figure 6.6B, of point-contact diodes biased well into the forward region (where the ohmic loss can be measured easily) varied with neutron irradiation fluence. In particular, they found that for diodes made with *p*-type germanium R_s decreased with fluence and for devices made of *n*-type germanium R_s first increased, then decreased (as the germanium was converted from *n*-type to *p*-type with neutron fluence, see Chapter 5). For devices made from silicon R_s was found to increase in both *n*- and *p*-type devices. All of these observations can be explained through Eq. (6.16) and the known variations in the conductivity of silicon and germanium with neutron fluence as set forth in Chapter 5.

In 1966, Wilson [23] irradiated some gold-silicon and some platinum-silicon Schottky-barrier diodes with γ-rays from a Co^{60} source and also with fast neutrons from a reactor. Wilson measured the barrier height using the method outlined in Section 6.2 and reported that no change in barrier height occurred for γ-irradiations up to 10^8 rads, or for fast-neutron fluences up to 10^{14} n/cm^2. Wilson felt that, since the surface state density was very high to begin with, it would take a great deal of radiation (greater than 10^8 rads) to increase the surface state density significantly. Wilson also found that in some irradiated devices the reverse leakage increased by four orders of magnitude. This was attributed to edge breakdown effects caused by trapped positive charge in the oxide passivation layer (see Figure 6.5B).

In 1968, Schnurr and Southward [24,25] studied radiation effects on silicon and gallium-arsenide Schottky-barrier diodes. They exposed the diodes to a flash x-ray source and could detect no permanent damage. They also irradiated the devices with fast neutrons from a reactor, and the results they obtained for their gallium-arsenide device are shown in Figure 6.8. They also

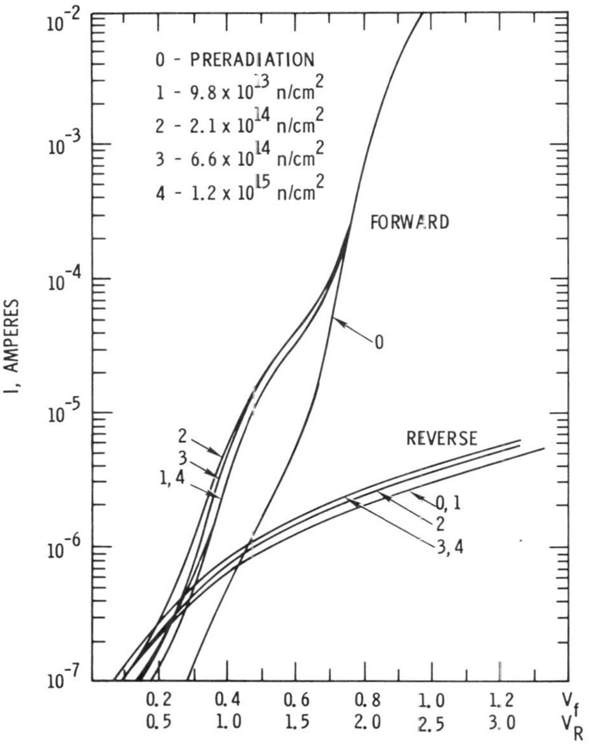

FIGURE 6.8
Effect of neutron radiation on GaAs Schottky-barrier diodes (after Schnurr and Southward [25]).

reported seeing similar effects in silicon. No explanation for the observed effects was given. The authors also reported being unable to detect any transient annealing effects in these devices.

Also in 1968, Lee and Sze [26] reported on the effects of neutron and gamma irradiation in silicon Schottky-barrier diodes with a platinum-silicide barrier. They pointed out that, while the density of surface states is high ($10^{14}/cm^2$) in unirradiated devices, radiation-induced defect production

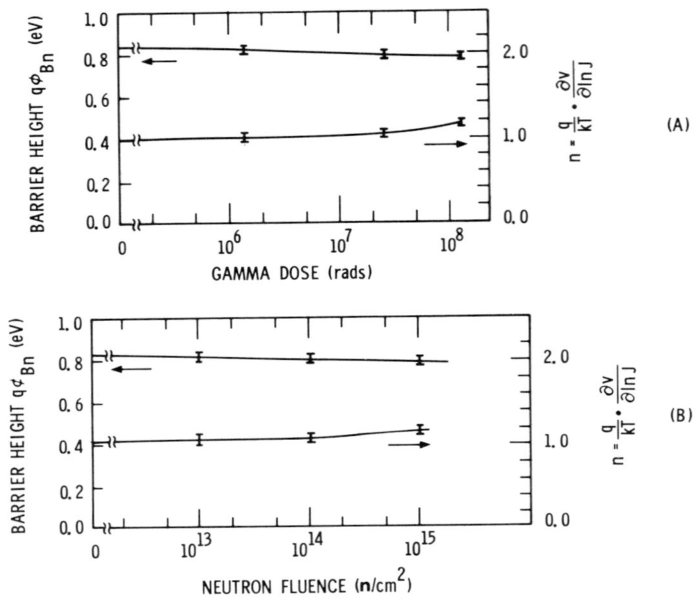

FIGURE 6.9
Barrier height and "n" factor as a function of radiation (after Lee and Sze [26]).

rates may be much greater for the surface than for the bulk, since the surface may serve as a sink for vacancies or interstitials. Therefore, careful attention was given to barrier height measurements as a function of radiation in this study. Figure 6.9 shows the results obtained by Lee and Sze for barrier height and "n" factor† for both neutron- and gamma-radiation damage. From Figure 6.9 it can be seen that these quantities do not change appreciably at γ doses of 10^8 rads or fast-neutron fluences of 10^{15} n/cm^2. Lee and

† The "n" factor comes from the fact that experimentally Eq. (6.15) is observed to follow the form $J = J_s[\exp(eV_A/nkT) - 1]$ where n varies from 1 to 10 but is close to 1 for good devices.

Sze [26] also showed that good stability of the reverse characteristics of irradiated devices could be obtained by using the guard-ring structure (Figure 6.5C). Figure 6.10 shows the results obtained for Co^{60} irradiation of Schottky-barrier diodes with and without a guard ring. From Figure 6.10 it can be seen that guard-ring structure gives the device parameters greatly improved stability in a radiation environment. They attributed this improvement to the fact that the guard ring eliminates the edge breakdown problems reported by earlier workers.

In 1969, Yu and Snow et al. [27,28] evaluated the effect of low-energy electron (15 to 20 keV) and neutron irradiations at room temperature on

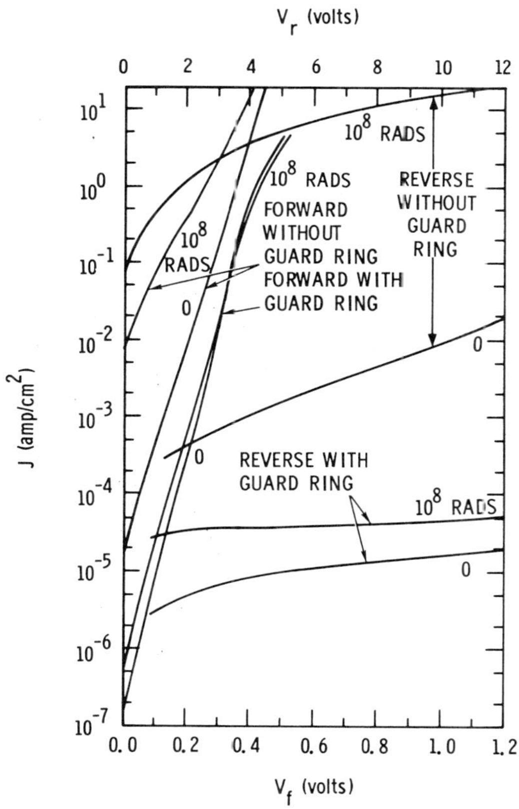

FIGURE 6.10
Comparison of radiation effects in Schottky-barrier diodes with and without guard rings (gamma radiation). (After Lee and Sze [26]).

silicon Schottky-barrier diodes made with aluminum and platinum-silicide barriers. They reported the effects of electron irradiation as increasing the surface recombination velocity and increasing the buildup of trapped charge in the oxide layer. They found a large increase in reverse leakage current with electron irradiation if a guard-ring structure was not used. Some of their

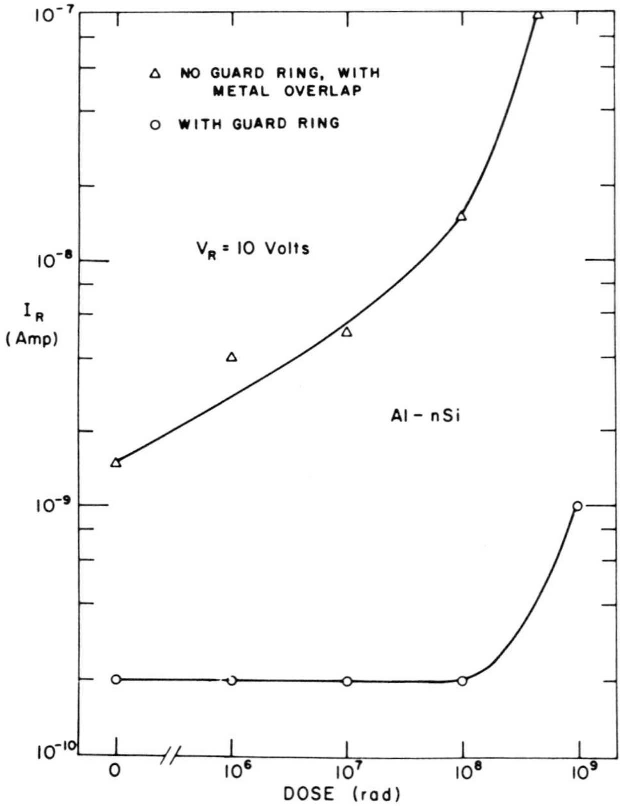

FIGURE 6.11
Comparison of radiation effects in Schottky-barrier diodes with and without a guard ring for low-energy electron radiation (after Yu and Snow [27]).

results are shown in Figure 6.11. From this data it can be seen that the reverse leakage characteristics essentially do not change for doses of up to 10^8 rads of low-energy electron irradiation if a guard ring is used. The authors also concluded that the barrier height was unaffected for electron doses up to 10^8 rads. Yu and Snow reported seeing excess current at low forward voltage and an increase in series resistance, R_s, after exposure to 10^{15} fast-neutrons/cm^2.

Aukerman [29] has studied the effect of low-energy (0.2 to 0.4 MeV) proton irradiations done at room temperature on lightly doped n-type (800 Ω-cm) gold-silicon Schottky-barrier devices. The predominant effect seen was the

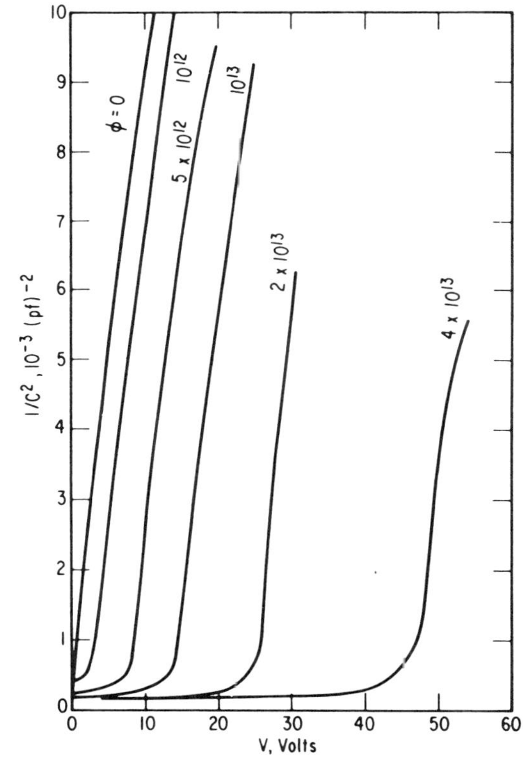

FIGURE 6.12
$1/C^2$ versus voltage for several proton fluences. Proton fluences per cm^2 (after Aukerman [29]).

introduction of a deep donor defect (see Chapter 5) at the energy $E_c - 0.47$ eV. He attributed this to the Si-P3 defect (see [30]). The introduction rate was found to be 1.3×10^{-2} net donor defects per proton. Aukerman found that the charge state of the defect was dependent on the position of the Fermi level with respect to the defect level, and this caused some interesting changes in C-V characteristics of the device. This effect, shown in Figure 6.12, can be explained as follows: The donor defects are neutral as long as they are occupied by an electron. If the depletion region is widened by an increase in the reverse bias, the defects in the depletion region lose their electrons and become positively charged. This increases the effective net donor density, and from Poisson's equation it can be seen that this would tend to reduce the depletion width. The end result is that capacitance (and hence $1/C^2$) remains virtually constant until the depletion region widens past the disordered region (the range of the protons was 4.5 μ).

Chaffin [31] has made measurements of reverse leakage current and forward voltage drop for the three types of Schottky-barrier structures shown in Figure 6.5. The devices used for the study were silicon Schottky-barrier devices supplied by Hewlett-Packard Associates. The devices were exposed first to γ-radiation (to 10^6 rad) then to neutron irradiation to 10^{16} n/cm^2 ($E > 10$ keV)† in the Sandia ACPR reactor. The results of the study are tabulated in Table 6.1. The values in Table 6.1 are ten sample averages. From

TABLE 6.1

	Leakage Current at −3 V			Forward Voltage at 1 mA		
Diode Type	Prerad	10^6 rad-γ	10^{16} n/cm^2	Prerad	10^6 rad-γ	10^{16} n/cm^2
Figure 6.5A H-P 5082-2300	14.9 nA	19.0 nA	17.3 nA	330 mV	326 mV	346 mV
Figure 6.5B H-P 5082-2835	350 nA	478 nA	313 nA	309 mV	303 mV	305 mV
Figure 6.5C H-P 5082-2800	3.9 nA	6.0 nA	8.7 nA	367 mV	359 mV	4000 mV

Table 6.1 it can be seen that even after 10^{16} n/cm^2 the reverse leakage current for the device without a guard ring (5082–2835) has not increased appreciably from its prerad value. This result seems to be in opposition to those obtained by the other authors [22–29]. The difference may lie in the fact that the device tested here may have a thicker, or thinner, oxide layer, or different impurities may be present in the oxides. The point to be made by Table 6.1 is that it may not be necessary to have a guard ring to prevent high reverse leakage currents; the answer very probably depends on the specific device being used. The increase in forward drop (due to an increase in R_s) seen in Table 6.1 is in

† The devices also received approximately 10^7 rads of gamma radiation during the exposure in the reactor.

agreement with the results of the other authors. The reason for the differences between the three diode types in the magnitude of the increase in forward drop with radiation is that the doping concentrations and thickness of the epitaxial layers are different for the three device types. The 5082-2800 has a very thick epitaxial region which is fairly lightly doped and shows the largest increase in forward drop at 10^{16} n/cm^2. The 5082-2835, on the other hand, has a very thin epitaxial layer which is more heavily doped and shows practically no increase in forward drop.

At this point a brief summary of the findings of all the above works may be helpful. The barrier height remains essentially constant up to fairly high radiation doses. At fairly high neutron fluences (greater than 10^{15} n/cm^2) one can expect the series resistance to increase (for silicon devices) due to carrier removal effects. This can be predicted through Eq. (6.16) and information given in Chapter 5. Trapping effects in the depletion region may produce an offset in the C-V characteristic at fairly high radiation doses. A guard ring is seen to reduce edge leakage and breakdown effects which can occur as the result of radiation, although not all devices seem to have this problem. However, in light of the edge breakdown problem it would be advisable to use devices of the type shown in Figure 6.5A or 6.5C if a testing program could not be carried out for the device to be used.

6.6 Microwave Detectors

A microwave video detector is the simplest type of microwave receiver available. It consists of a single Schottky-barrier, or point-contact, microwave diode.† The circuit has a microwave input port (usually matched to 50Ω) and a video output port. The function performed by the diode is simple rectification. Several authors (see [32–35]) have discussed the basic parameters of such circuits. A well-established measure of the threshold sensitivity of a detector is called its *tangential sensitivity* (TSS). The tangential sensitivity of a detector is defined‡ as the amount of signal required to raise the baseline level of a received pulsed signal to the peak noise amplitude when the signal is absent. Figure 6.13 shows a pulsed signal at the tangential sensitivity (TSS) level. Tangential sensitivities of 10^{-9} W (-60 dBm) have been achieved by Schottky-barrier detector diodes [37]. The achievable TSS depends on such factors as diode impedance, video frequency, microwave frequency, noise

† Other types of detectors include backward diodes, tunnel diodes, and thermoelectric detectors. These other devices are not nearly as widespread in their use as the Schottky-barrier is, and they will not be discussed here.

‡ This measurement is somewhat subjective in nature and dates from the early radar days coming from the "A"-scope presentation of radar signals. Cowley and Sorenson [36] have proposed an alternative measure of sensitivity.

properties of the diode, and the video amplifier used. The current sensitivity, β, of a diode is defined as

$$\beta = \frac{\Delta i}{P}\ \mu\text{A}/\mu\text{W} \tag{6.18}$$

where Δi is the rectified current that flows into a short circuit for a given microwave input power P. The limiting value of β can be calculated [32,33] by performing a Taylor series expansion of the diodes V-I characteristic,

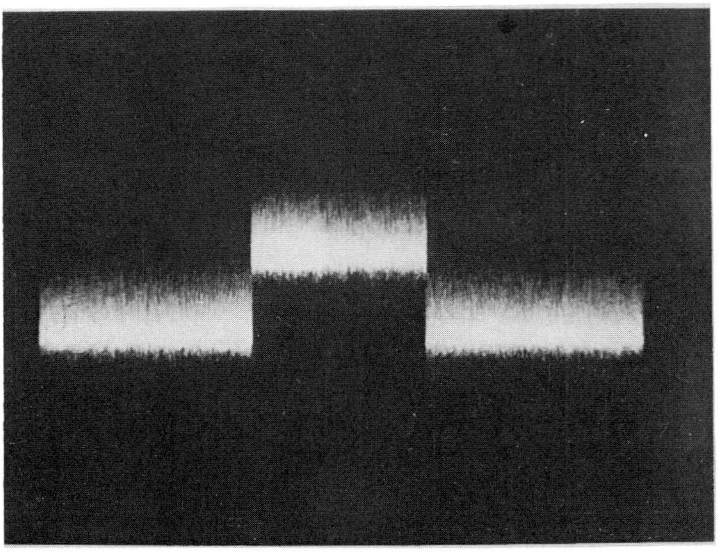

FIGURE 6.13
Tangential sensitivity level of a pulsed signal in noise.

$J = J_s[\exp(eV_A/kT) - 1]$. The result for low amplitude signals is $\beta = e/2kT \cong 19\ \mu\text{A}/\mu\text{W}$ at $T = 300°\text{K}$. In practice, not all of the power applied to the terminals of the diode appears across the junction. If we assume that the package inductance and capacitance are tuned out by a matching network, then we are still left with the diode equivalent circuit shown in Figure 6.6A. Cowley and Sorenson [36] show that the resulting value of current sensitivity in this case is not simply $e/2kT$ but rather

$$\beta = \frac{e}{2kT}\frac{1}{(1 + R_s/R_J)[1 + (f/f_c)^2]}\ \mu\text{A}/\mu\text{W} \tag{6.19}$$

where f = frequency of operation and f_c is the diode's cutoff frequency:

$$f_c = \frac{(1 + R_S/R_J)^{1/2}}{2\pi C_J(R_s R_J)^{1/2}} \cdot \tag{6.20}$$

To derive the tangential sensitivity it is necessary to consider the noise current inputs to the video amplifier from all sources, such as resistor noise ($i_n = \sqrt{4kTB/R}$) and diode excess noise. Ideally the diode would contribute a noise current just due to its junction resistance of $i_n = \sqrt{4kTB/R_J}$; however, in practice it is known [9] that the diode exhibits an excess noise temperature ratio (noise temperature effectively $= t_d \cdot T$) and so the noise current is actually $i_n = \sqrt{4kTBt_d/R_J}$. The excess noise temperature ratio is a function of both diode current and frequency. It is usually expressed as [33]

$$t_d = t_w + \frac{K_n I_d}{f} \tag{6.21}$$

where t_w = white noise temperature ratio (typically near unity), I_d = diode current, and $K_n \cong 1.8$ Hz/μA for a good Schottky-barrier diode. Typical plots of t_d for a point-contact and a Schottky-barrier diode are shown in Figure 6.14, and it can be seen that noise temperature ratio increases with current and decreases with frequency. Figure 6.14 also shows that Schottky-barrier devices contribute significantly less noise at the lower frequencies than

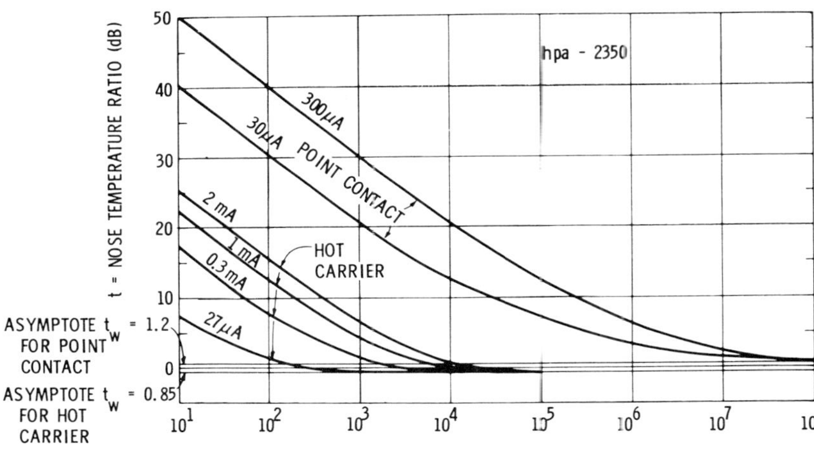

FIGURE 6.14
Comparison of noise temperature ratio of point-contact and Schottky-barrier diodes (after Cowely and Sorenson [36]).

do the point-contact devices. Cowley and Sorenson [36] have derived the tangential sensitivity (TSS) using the parameters discussed above, and the result is

$$\mathrm{TSS} = \frac{2\cdot 5}{\sqrt{B}}\frac{2kT}{e}\left[\frac{4kTt_w}{R_J}\right]^{1/2}\left[1 + \frac{R_s}{R_J}\right]^{1/2}\left[1 + \left(\frac{f}{f_c}\right)^2\right]\left[1 + \frac{K_n I}{f_v}\right] \quad (6.22)$$

where B = bandwidth of the detector and f_v = frequency of the video modulation. Equation (6.22) expresses the TSS in terms of known quantities and will be used to discuss variations in the TSS with radiation. Before discussing radiation effects it should be pointed out that the Schottky-barrier detector usually operates with a forward bias of a few μA. This is because R_J decreases with bias [see Eq. (6.17)], and its impedance becomes easier to match to the impedance of a 50 Ω circuit. However, Figure 6.14 and Eq. (6.21) show that the noise temperature ratio increases with bias, and so the bias is usually limited to a few μA with resulting values of R_J near several kΩ [see Eq. (6.17)]. The previous dc and low-frequency radiation studies indicated that several important changes occurred in the diode model with radiation. Most authors saw an increase in R_s due to carrier removal. The effect of R_s can be seen quite clearly for the 5082-2800 device in Table 6.1 after 10^{15} n/cm^2. (The forward drop at 1 mA increases from 0.36 V to 4.0 V postrad.) The effect of R_s may be included through Eq. (6.22), from which we can see that terms on the order of R_s/R_J are important. Since R_J is on the order of a few thousand ohms and R_s is initially a few ohms a large increase in R_s can be tolerated without the R_s/R_J terms becoming significant. R_s also enters Eq. (6.22) through the f_c term [Eq. (6.20)]. In general, for $R_J \gg R_s$ the value of f_c is proportional to $(R_s)^{-1/2}$ and some degradation in TSS can be expected to occur if the operating frequency is close to f_c (i.e., $f \simeq f_c$). For operation appreciably below the cutoff frequency, no change in TSS due to increases in R_s should be seen (as long as $R_s \ll R_J$). Another possible radiation-induced change in the TSS could occur if the noise properties of the diode given by Eq. (6.21) should degrade with irradiation. Since the $1/f$ contribution to noise is usually associated with surface states [34], it is conceivable that if a sufficient change in surface state density were to occur with radiation the noise performance would degrade. However, Wilson ([23], see Section 6.5) has pointed out that the surface state density of an unirradiated device is already very high, and it would take a great deal of radiation to change it. Therefore, this is not expected to be a problem. The edge breakdown effects reported by several authors in the previous section could lead to serious problems due to the inherent noisiness of avalanche breakdown. One might conclude from this discussion that microwave detectors made with Schottky-barrier devices that do not suffer edge breakdown problems should be quite radiation tolerant.

An experiment was performed to measure radiation-induced changes in the

TSS of detectors made with the three types of Schottky-barrier diodes shown in Figure 6.5. The devices used were described in Section 6.5 and are the same devices as listed in Table 6.1, i.e., the Hewlett-Packard 5082-2300, 2835, and 2800. From Table 6.1 it can be seen that these devices show very little increase in reverse leakage after irradiation and hence are probably free from edge breakdown effects. The devices (ten of each type) were tested in the broadband detector circuit shown in Figure 6.15. The video frequency, f_v, was 1 kHz and the bandwidth was 20 Hz. The untuned circuit does not allow the ultimate TSS to be achieved because of the presence of a mismatch between the

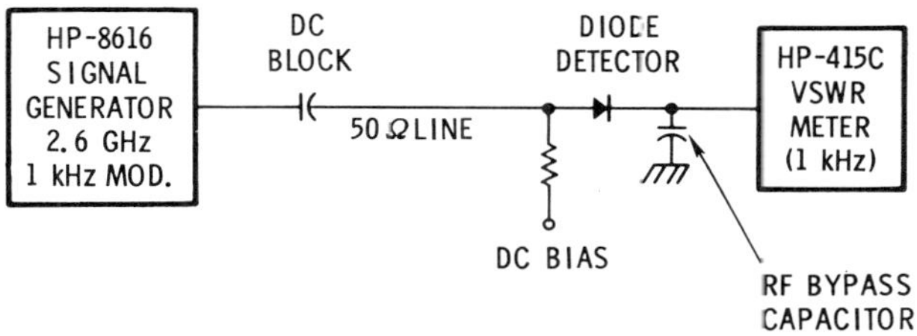

FIGURE 6.15
Untuned detector test setup for measuring tangential sensitivity.

generator and the load. Absence of a tuner to match the diode was found to decrease the sensitivity by 3 dB typically. Results of the test for both gamma and neutron irradiation are shown in Figure 6.16. The points shown on the figure are 10 unit averages; it can be seen that even after 10^{16} n/cm^2 the tangential sensitivity of the detector is essentially unchanged. This holds for all three types of diodes tested (corresponding to the three different geometries shown in Figure 6.5). It is interesting to note that the 5082-2800 diode shows no change in TSS after 10^{16} n/cm^2 even though its R_s increased grossly (see Table 6.1) at that level. This is to be expected from the previous discussion.

6.7 Microwave Mixers

A microwave mixer is a device which utilizes a nonlinear element (such as a Schottky-barrier diode) to achieve conversion of an input rf signal to another (usually lower) frequency. In a typical microwave receiver the incoming microwave signal is converted to a lower frequency (usually in the 20 to 100 MHz range) called the intermediate frequency. The signal is then amplified at the intermediate frequency before being processed. The reasons

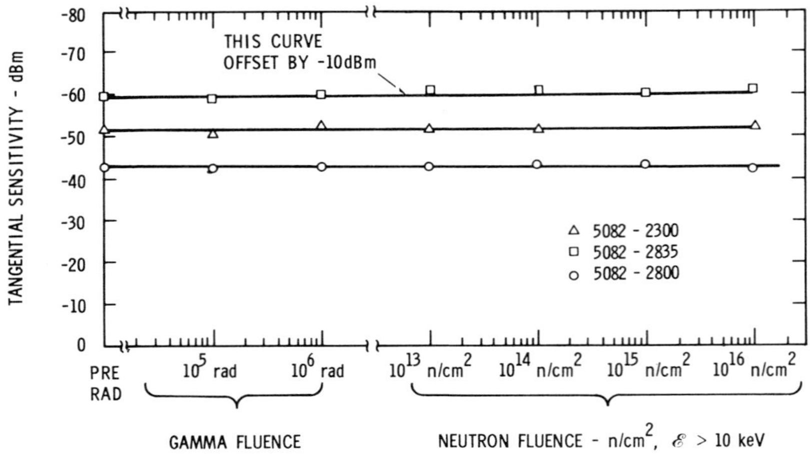

FIGURE 6.16
Effect of gamma and neutron radiation on three types of Schottky-barrier detectors.

for doing this are twofold: (1) It is easier to achieve large amplifier gains at the lower frequencies. (2) The 20 to 100 MHz frequency range is high enough to avoid most of the $1/f$ noise generated in the diodes (see Figure 6.14). The sensitivity of receivers made in this way can be good as 10^{-17} W (-140 dBm, [37]). Microwave mixers have been discussed and analyzed by several authors (see [3,8,11,33]). The commonly used figure of merit for a microwave receiver is its noise figure. The noise figure is defined [38] as

$$F = \frac{S_i/N_i}{S_o/N_o} \tag{6.23}$$

where S_i/N_i is the signal to noise ratio at the input of the receiver and S_o/N_o is the output signal to noise ratio of the receiver. Since lossless frequency conversion and noiseless amplification are not possible, F will be greater than unity. The closer to unity F is, the more sensitive the receiver will be. It is also possible to talk about the contribution of the individual parts of the receiver to the noise figure. Figure 6.17 shows a typical microwave receiver. The mixer converts the incoming signal, frequency f_s to the intermediate frequency, $f_{if} = f_s - f_{lo}$, with the aid of a local oscillator signal, f_{lo}. It can be shown [3,33] that the noise figure of the receiver may be expressed as

$$F\,(\text{in dB}) = 10 \log\,[L_m(F_{if} + t_m - 1)] \tag{6.24}$$

where L_m = mixer conversion loss expressed as a power ratio ($L_m \geq 1$),

F_{if} = noise figure of the IF amplifier, and t_m = effective noise temperature ratio of the mixer. The effective noise temperature ratio of the mixer is related to the diode noise temperature ratio [see Figure 6.14 and Eq. (6.21)] as follows [33]:

$$t_m = t_d\left(1 - \frac{2}{L_m}\right) + \frac{2}{L_m}. \tag{6.25}$$

For mixers with a fairly large conversion loss $t_m \to t_d$. The mixer shown in Figure 6.17 could consist of one, two, or four diodes, corresponding respectively to a single-ended, balanced, or double-balanced mixer. Only the double-balanced mixer will be discussed here, and a description of the other types may be found in [3,8,11,33]. The equivalent circuit of a double-balanced mixer [39] is shown in Figure 6.18A. The idealized operation of this

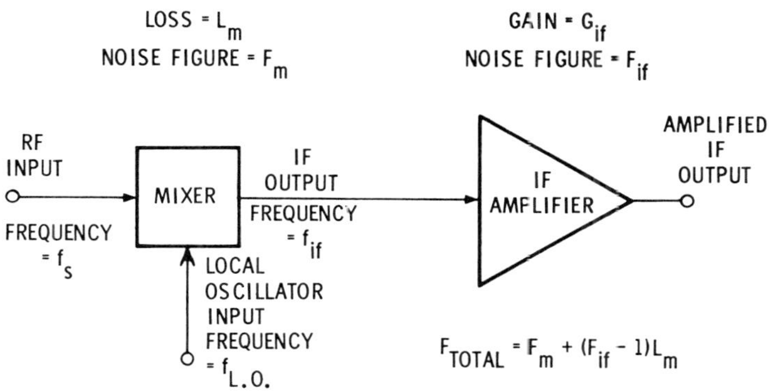

FIGURE 6.17
Typical microwave receiver.

type of mixer will be briefly explained using Figures 6.18B, C and 6.19. Figure 6.19 shows the effect of sine-wave voltage drive on the diode's conductance (dI/dV). We will assume that the diode's conductance varies between $G = \infty$ (a short circuit) at forward bias and $G = 0$ (an open circuit) at reverse bias. In actual practice, we only require $G \gg 1/R_0$ or $G \ll 1/R_0$, where R_0 = circuit impedance. Let us now return to Figure 6.18B and assume that the local oscillator amplitude is much larger than the signal amplitude (this is usually the case); then the local oscillator signal will determine which diodes (Figure 6.18A) are in the low impedance state and which are in the high impedance state. In Figure 6.18B the assumed local oscillator polarity is shown corresponding to the first half of the local oscillator rf cycle. The polarity is such that diodes 1 and 2 are effectively short circuits and diodes 3

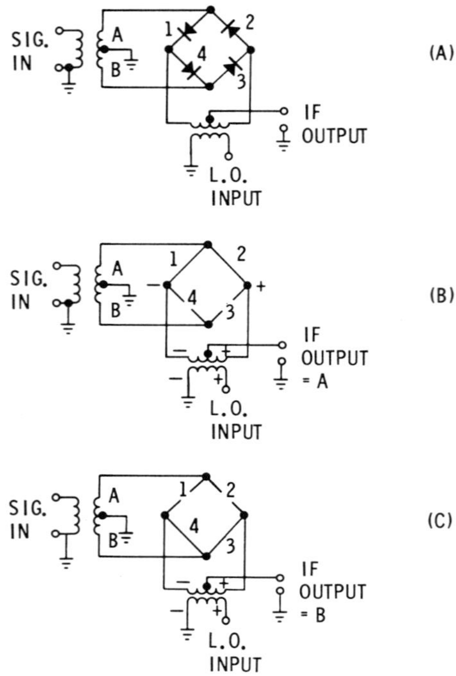

FIGURE 6.18
The double-balanced mixer.

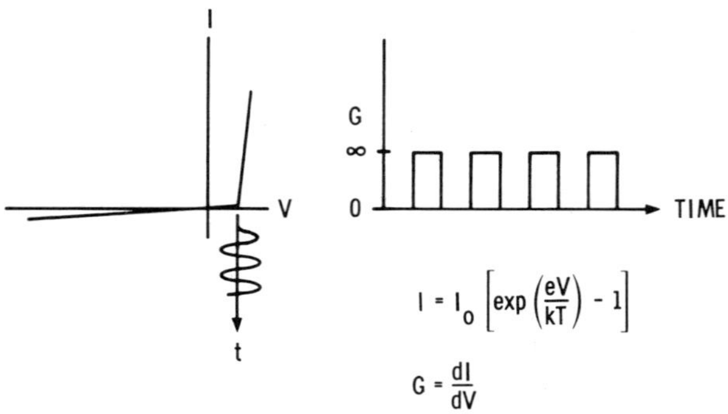

FIGURE 6.19
Conductance states of a Schottky-barrier diode with a sine-wave voltage drive.

and 4 are effectively open circuits, and the signal A appears at the IF terminal. During the second half of the local oscillator rf cycle (see Figure 6.18C) the polarity reverses; diodes 1 and 2 are now effectively open circuits, and 3 and 4 are short circuits. The signal B ($B = -A$) appears at the IF terminal. The net result is that input signal is transferred to the IF terminal with its polarity periodically reversed at the local oscillator frequency f_{lo}. The effect of this may be calculated by finding the Fourier transform of the quantity $s(t) \cdot f(t)$ where $f(t)$ = input signal, $s(t) = +1$ for $0 < t < 1/(2f_{lo})$ and $s(t) = -1$ for $1/(2f_{lo}) < t < 1/f_{lo}$ and $s(t)$ repeats with a period $T = 1/f_{lo}$. This can be expressed as

$$X(f) = \int_{-\infty}^{\infty} s(t)f(t)e^{-j\omega t}\,dt\,. \tag{6.26}$$

Hancock [40] has solved this case, and for the $f_s - f_{lo}$ frequency term the result is

$$X(f) = \frac{2}{\pi} F(f_s - f_{lo}) \tag{6.27}$$

where $F(f_s - f_{lo})$ is the Fourier transform of the input signal shifted to the frequency $f_s - f_{lo}$. The important point to be gained from Eq. (6.27) is that, under the assumptions made, a mixer of the type shown in Figure 6.18 should have an idealized† conversion loss of $L_m' = 20 \log (2/\pi) = -3.9$ dB. The value of L_m' could now be used directly in Eq. (6.24) to predict the noise figure of the microwave receiver. However, more accurate results are achieved when the effects of diode series resistance and junction capacitance are included. It was pointed out in Section 6.6 that only a fraction of the microwave power applied to a diode appears across the junction. This fraction is [33]

$$L_d = \frac{P_{in}}{P_{\text{Junction}}} = \left(1 + \frac{R_s}{R_J}\right)\left[1 + \left(\frac{f}{f_c}\right)^2\right] \tag{6.28}$$

where f_c is defined by Eq. (6.20) and R_s and R_J are the series resistance and junction resistance discussed previously. The actual value of conversion loss to be used in Eq. (6.24) and (6.25) is

$$L_m = L_m' \cdot L_d\,. \tag{6.29}$$

For this reason, it is now possible to discuss the noise figure of a microwave receiver in terms of basic mixer parameters as well as the effects of radiation on microwave mixers.

Thatcher et al. [42] have reported on studies made on point-contact mixers. They said that mixers made with the 1N23B, 1N21WE, 1N23WE, and

† Ignoring losses in the transformers and higher order frequency interaction terms.

1N25 point contact diodes showed poor performance above 10^{13} n/cm^2. Primarily they report a significant increase in noise figure and conversion loss with neutron fluences above 10^{13} n/cm^2.

Schnurr and Southward [24,25] were unable to observe any change in noise figure for silicon and gallium arsenide Schottky-barrier diode mixers at X-band after they were subjected to short intense x-ray pulses from a febetron (30-nsec pulse, 10^8 rad/sec dose rate).

Bristow, et al. [41] report no change in noise figure for a mixer made with a Microwave Associates MA-457R point contact diode for fast neutron fluences up to 4×10^{15} n/cm^2, $\mathscr{E} > 0.3$ keV, (Spectrum H). After 1.5×10^{16} n/cm^2 the noise figure of the mixer increased from 12 dB (pre-rad) to 20 dB. These measurements are for a mixer-IF amplifier together (such as shown in Figure 6.17) although all of the changes were attributed to mixer damage alone. Figure 6.20A shows the overall noise figure and crystal currents (rectified local oscillator current) versus irradiation time and Figure 6.20B shows the data obtained by Bristow, et al., which is a plot of the integrated neutron fluence (at the mixer) versus irradiation time. The authors attributed the damage to an increase in the series resistance, R_s, of the diode. The diode was made from silicon and carrier removal effects are blamed for increasing the R_s.

Chaffin [31] has measured noise figure and conversion loss for the three types of Schottky-barrier diodes shown in Figure 6.5. The diodes are the same as those discussed in Sections 6.5 and 6.6 and Table 6.1, i.e., the Hewlett-Packard 5082-2300, 5082-2835, and 5082-2800. The diodes were tested in a double balanced mixer similar to the one shown in Figure 6.18. The noise figure was measured using a Hewlett-Packard automatic noise figure meter, HP-340B. The local oscillator frequency was 250 MHz with a power level of 5 mW. The intermediate frequency was 60 MHz. The conversion loss was measured for both the upper sideband (310 MHz) and the lower sideband (190 MHz) and the figures that will be presented are an average of both. The noise figures given include the contribution from the 5.6 dB noise figure IF amplifier (see Figure 6.17). The values for both conversion loss and noise figure are DSB (double-sideband) which is the value obtained when one puts in energy at both the signal and image frequencies. The values usually used in a radar system is the SSB (single-sideband) value in which case the only useful signal input occurs at the signal frequency. The SSB values are 3 dB larger than the DSB values. The diodes were irradiated in both a Co^{60} gamma facility and in the Sandia ACPR reactor.† All neutron fluences listed are for $\mathscr{E} > 10$ keV. The results of the experiment are shown in Figures 6.21, 6.22, and 6.23 corresponding to the diode types shown in Figures 6.5A,

† It should be pointed out that the samples placed in the ACPR reactor receive a gamma dose of 10^{-9} rad/n/cm^2.

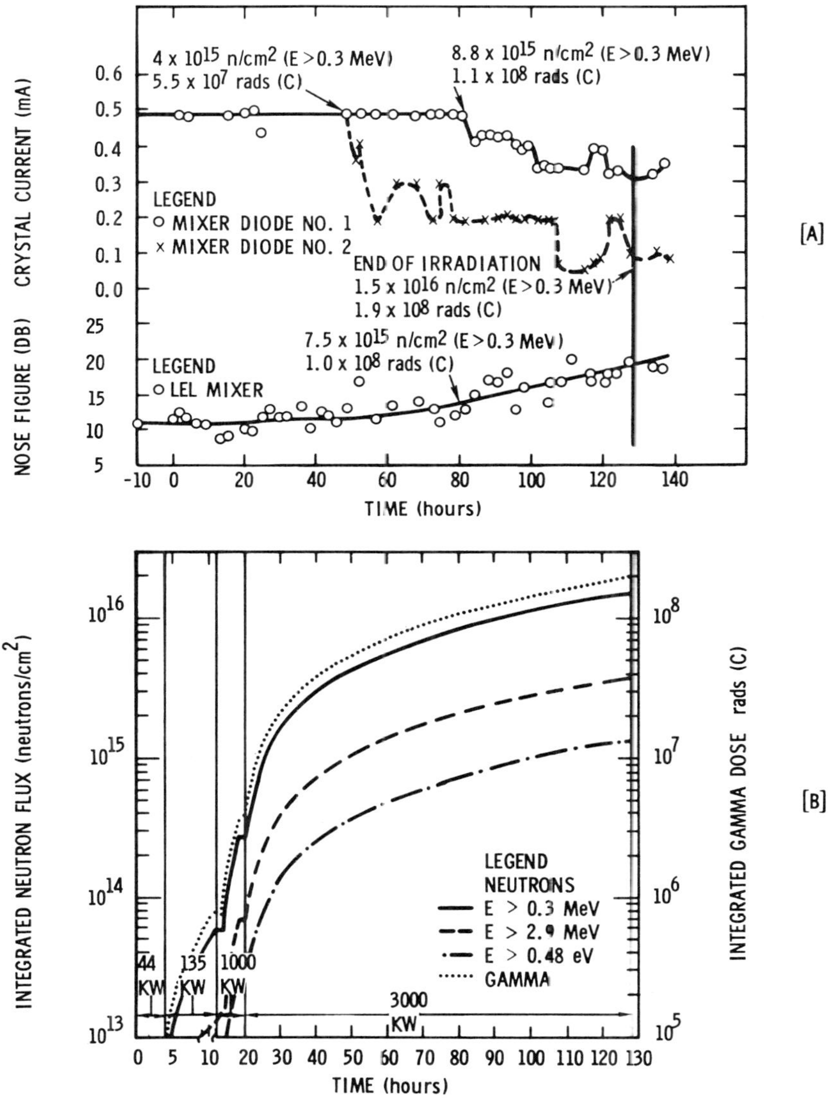

FIGURE 6.20
(A) Increase in noise figure of point-contact mixer with time. (B) Integrated neutron fluence versus time. (After Bristow et al. [41]).

6.5B, and 6.5C, respectively. From Figure 6.21 it can be seen that the 5082-2300 devices (see Figure 6.5A) show a slight increase in noise figure and conversion loss after 10^{16} n/cm^2. Figure 6.22 shows that the 5082-2835 devices (Figure 6.5B) show no change in conversion loss or noise figure at levels up to 10^{16} n/cm^2. From Figure 6.23 it can be seen that the 5082-2800 devices show considerable degradation in both noise figure and conversion loss after 10^{16} n/cm^2. This behavior is to be expected in light of the data presented in Table 6.1 and from Eq. (6.24), (6.25), (6.28), and (6.29). Equation (6.28) shows that loss due to diode series resistance, and consequently the mixer conversion

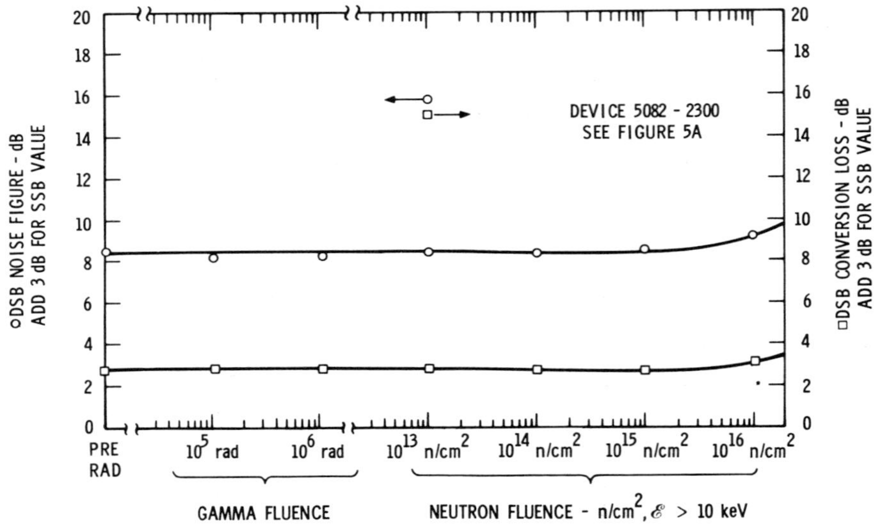

FIGURE 6.21
Noise figure and conversion loss for a Schottky-barrier mixer. Diode type shown in Figure 6.5A.

loss, Eq. (6.29), increases with an increase in R_s. The value of R_J is only a few ohms in the mixer because of the presence of the local oscillator drive signal. This makes the ratio R_s/R_J much more sensitive than it was in the case of the microwave detector where R_J was several $k\Omega$ (see Section 6.6). Therefore, while the increase in R_s at 10^{16} n/cm^2 for the 5082-2800 did not affect the TSS (Figure 6.16) of a microwave detector made with this diode, it is fatal to a mixer made with the same diode (Figure 6.23). Equation (6.24) shows that the noise figure will also increase with an increase in mixer conversion loss. Table 6.1 shows that the forward drop (proportional to R_s) increases slightly for the 5082-2300 device, not at all for the 5082-2835 device, and grossly for the 5082-

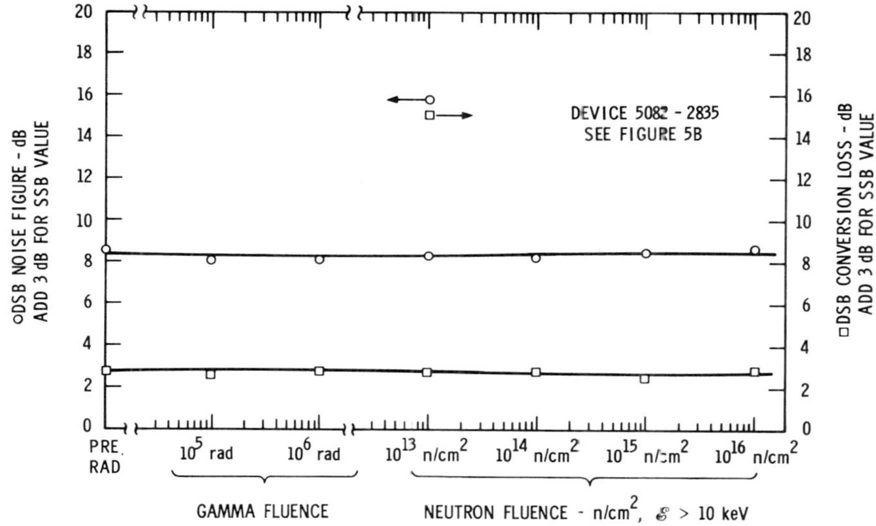

FIGURE 6.22
Noise figure and conversion loss for a Schottky-barrier diode mixer. Diode type shown in Figure 6.5B.

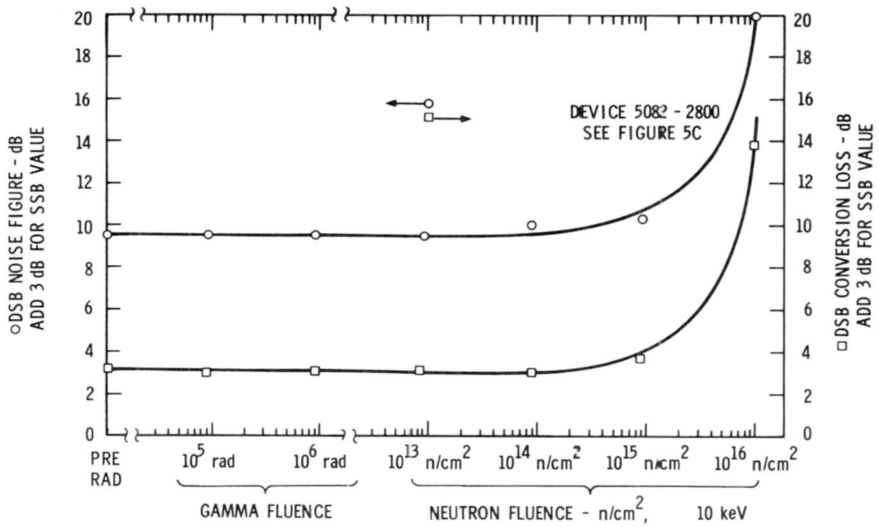

FIGURE 6.23
Noise figure and conversion loss for a Schottky-barrier diode mixer. Diode type shown in Figure 6.5C.

2800 device. An inspection of Figures 6.21, 6.22, and 6.23 shows that increases in noise figure and conversion loss are proportional to the increases in R_s seen in Table 6.1. This is to be expected from Eq. (6.24), (6.28), and (6.29). Another mechanism that could occur (but apparently does not) is that the effective noise diode temperature ratio, t_d [Figure 6.14 and Eq. (6.21)], could increase. This could increase the noise figure without changing the conversion loss [see Eq. (6.24) and (6.25)]. All of the changes seen in noise figure in Figures 6.21, 6.22, and 6.23 are directly relatable to changes in conversion loss through Eq. (6.24). The noise temperature ratio apparently remains constant with radiation up to 10^{16} n/cm^2 for all three device types tested. This can also be inferred from the TSS data given in Figure 16 and Eq. (6.22), i.e., if t_d changes the TSS should change as indicated by Eq. (6.22).

The conclusion that can be drawn from the works cited in this section is that silicon mixer degradation in a radiation environment is primarily† due to an increase in R_s, the series resistance of the device with radiation [see Eq. (6.16)]. This is caused by a decrease in conductivity of the silicon due to carrier removal effects. The degradation shows up as an increase in noise figure and conversion loss of the mixer. To improve radiation tolerance it is necessary to resist the effects of carrier removal in the device by using thin epitaxial layers and heavily doped substrates.

6.8 Transient Ionizing Effects

The effect of transient ionizing radiation on a Schottky-barrier diode is analogous to that in a simple *p-n* junction. The effects of transient ionizing radiation on a *p-n* junction were discussed in Chapter 4. In the present case, however, the pair generation in the metal can be neglected (due to the vanishingly short lifetime in the metal, 10^{-13} sec, see Section 6.3). This reduces Eq. (5.30) for the peak photocurrent to

$$i_{pp}(t) = eAG\{(W + L_p \operatorname{erf} \sqrt{t/\tau_p})U(t) - [W + L_p \operatorname{erf} \sqrt{(t - t_0)/\tau_p}]U(t - t_0)\} \quad (6.30)$$

where all of the symbols have been defined under Eq. (5.30), Chapter 5. One additional source of photocurrent that is not included in Eq. (6.30) is the photoemission of electrons from the metal into the semiconductor. Figure 6.24 depicts this effect as well as the pair generation in the depletion region. Photoemission occurs when an electron lying very close to the metal-semiconductor interface receives sufficient energy, either directly from a photon or indirectly from an energetic electron, to surmount the barrier and be injected

† This neglects the edge breakdown effects that occur in some devices (see Section 6.5) which would certainly increase the noise temperature ratio, t_d.

into the semiconductor. In general, the current contributed by this mechanism can be neglected because of the small effective volume in which it is generated.

The effect of ionizing radiation pulses on Schottky-barrier devices has been examined by Schnurr and Southward [24,25]. They exposed devices to 30 nsec pulses of intense x-ray irradiation from a febetron flash x-ray generator. The integrated dose per pulse was 20 rads. The peak photocurrent they observed

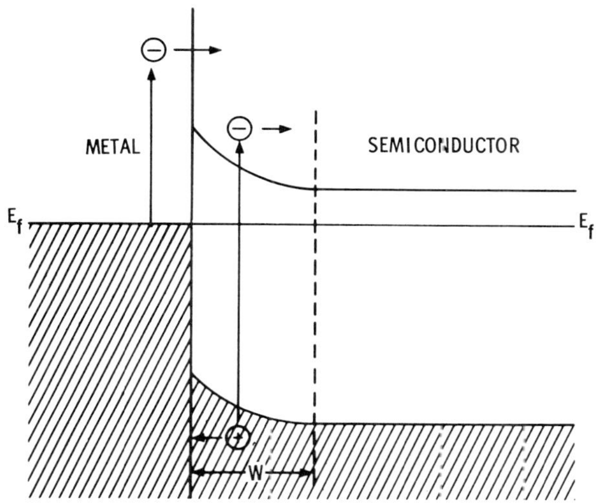

FIGURE 6.24
Zero-bias Schottky-barrier diode exposed to ionizing radiation.

was several orders of magnitude larger than that predicted by Eq. (6.30). The source of the excess current was found to be the diode package itself. When an empty diode package was substituted for the normal one, essentially the same peak photocurrent was measured. The authors were unable to detect any permanent changes in the diode parameters as a result of exposure to transient ionizing radiation pulses.

REFERENCES

[1] H. K. Henisch, *Rectifying Semiconductor Contacts*, Oxford at the Clarendon Press, Oxford, 1957, Section 1.2.

[2] W. Schottky, "Semiconductor Theory of the Barrier Film," *Naturwiss.*, **26**, 843 (1938).

[3] H. C. Torrey and C. A. Whitmer, *Crystal Rectifiers*, vol. 15, MIT Radiation Lab Series, McGraw-Hill, New York, 1948.

[4] Ibid., p. 71.

[5] A. M. Goodman, "Metal-Semiconductor Barrier Height Measurement by the Differential Capacitance Method—One Carrier System," *J. Appl. Phys.*, **34**, 329 (February 1963).

[6] C. A. Mead, "Metal-Semiconductor Surface Barriers," *Solid-State Electronics*, **9**, 1023 (1966).

[7] Torrey and Whitmer, op. cit., p. 77.

[8] H. A. Watson, *Microwave Semiconductor Devices and Their Circuit Applications*, McGraw-Hill, New York, 1969.

[9] M. Crane and H. Sorensen, "Using the Hot Carrier Diode as a Detector," *Hewlett-Packard J.*, **17**, 2 (December 1965).

[10] Torrey and Whitmer, op. cit., p. 81.

[11] H. V. Shurmer, *Microwave Semiconductor Devices*, John Wiley & Sons, New York, 1971.

[12] Watson, op. cit., p. 350.

[13] S. M. Sze, *Physics of Semiconductor Devices*, John Wiley & Sons, New York, 1969.

[14] Mead, op. cit., p. 1028.

[15] Torrey and Whitmer, op. cit., Chapter 10.

[16] Henisch, op. cit., p. 126.

[17] Watson, op. cit., p. 358.

[18] C. A. Burrus, Jr., "Millimeter-Wave Point-Contact and Junction Diodes," *Proc. IEEE*, **54**, 575 (April 1966).

[19] Shurmer, op. cit., p. 98.

[20] M. P. Lepselter and S. M. Sze, "Silicon Schottky Barrier Diode with Near-Ideal I–V Characteristics," *Bell System Tech. J.*, **47**, 195 (1968).

[21] G. N. Kaposhilin, "Hot-Carrier Diode Opens New Vistas," *Electronic Design*, p. 178 (March 15, 1966).

[22] W. E. Johnson and K. Lark-Horovitz, "Neutron Irradiated Semiconductor," *Phys. Rev.*, **76**, 442 (1949).

[23] D. K. Wilson et al., "Effects of Radiation on Semiconductor Materials and Devices," Final Report AFCRL-67-0068, prepared for Air Force Cambridge Research Laboratories, 1966.

[24] R. H. Schnurr and H. D. Southward, "Radiation Effects on Gallium Arsenide Devices and Schottky Diodes," Technical Report AFWL-TR-68-31, vol. 1, prepared for Air Force Weapons Lab, Albuquerque, N.M., 1968.

[25] R. H. Schnurr and H. D. Southward, "Radiation Effects upon Gallium Arsenide Devices," *IEEE Trans. Nucl. Sci.*, **NS-15**, 306 (December 1968).

[26] H. S. Lee and S. M. Sze, "Effects of Radiation on Silicon Schottky Barrier Diodes," presented at the IEEE Semiconductor Interface Specialists Conference, Las Vegas, Nevada, 1968.

[27] A. Y. C. Yu and E. H. Snow, "Radiation Effects on Silicon Schottky Barriers," *IEEE Trans. Nucl. Sci.*, **NS-16**, 220 (December 1969).

[28] E. H. Snow et al., "Radiation Effects on Novel Semiconductor Devices," Final Report Contract F19628-69-C-0118 USAF, *AD-691-044*.

[29] L. W. Aukerman, "Low Energy Proton Irradiation of Silicon Surface Barrier Detectors," *IEEE Trans. Nucl. Sci.*, **NS-17**, 245 (December 1970).

[30] K. L. Brower, "Structure of Multiple-Vacancy (Oxygen) Centers in Irradiated Silicon," (see Chapter 5, Ref. [12]).

[31] R. J. Chaffin (unpublished work).

[32] Torrey and Whitmer, op. cit., Chapter 11.

[33] Hewlett-Packard Application Note No. 907, "The Hot Carrier Diode, Theory and Design and Application," Hewlett-Packard Co., Palo Alto, Calif., 1967.

[34] Watson, op. cit., Chapter 12.

[35] Shurmer, op. cit., Chapter 1.

[36] A. M. Cowley and H. O. Sorenson, "Quantitative Comparison of Solid-State Microwave Detectors," *IEEE Trans. PGMTT*, **MTT-14**, 588 (December 1966).

[37] B. S. Siegal, "Schottky Diodes—Where We Stand Today," *Microwaves*, **10**, 44 (April 1971).

[38] W. W. Mumford and E. H. Scheibe, *Noise Performance Factors in Communication Systems*, Horizon House—Microwave, Inc., Dedham, Mass., 1968.

[39] R. B. Mouw and S. M. Fukuchi, "Broadband Double Balanced Mixer/Modulators," *Microwave J.*, **12** (March 1969).

[40] J. C. Hancock, *The Principles of Communication Theory*, McGraw-Hill, New York, 1961.

[41] W. D. Bristow et al., "An Experimental Investigation of Nuclear Radiation Effects on Radar Receiver Components," *IEEE Trans. Nucl. Sci.*, **NS-11**, 111 (November 1963).

[42] R. K. Thatcher et al., "The Effects of Nuclear Radiation on Electronic Components Including Semiconductors," Battelle Memorial Institute, REIC Report No. 36 (1964).

7

Radiation Damage in *p-i-n* Switches and Limiters

7.1 Introduction

The *p-i-n* diode was first introduced as a 60 Hz power rectifier. The conductivity modulated *i*-region permits a forward drop comparable with ordinary *p-n* junction devices while the wide intrinsic region results in a very high reverse breakdown voltage (as compared to a *p-n* junction device). This effect has been studied by many authors.†

Leenov [2] and others [3,4,5,7] have discussed the use of these devices as microwave switches and limiters. *p-i-n* diode microwave switches and limiters are finding increasing numbers of applications in military and space microwave systems. Typical applications are transmit-receive (T-R) switches for pulsed radars, digital phase shifters for phased-array radars and limiters to protect microwave mixers against accidental overload. Some of these military and space systems have radiation tolerance requirements.

This chapter is devoted to discussions of theoretical and experimental studies carried out on *p-i-n* microwave switches and limiters. Section 7.2 discusses the device physics of *p-i-n* diodes. The equivalent circuit of a packaged diode is also discussed in this section. Section 7.3 is devoted to an analysis of a typical *p-i-n* switch circuit, and Section 7.4 discusses the effects of radiation on both the diode and the overall circuit. Microwave limiters are discussed in Section 7.5. Section 7.6 presents experimental data on radiation-

† These papers have been reviewed by Rose [1].

induced changes in limiter performance. Section 7.7 briefly discusses the effect of transient ionizing radiation on *p-i-n* switches and limiters.

7.2 Device Physics of *p-i-n* Diodes

This section is devoted to deriving those diode parameters which will be of interest in a microwave *p-i-n* switch. The deviation is similar to that given by Leenov [1]. The parameters are derived from an abrupt junction model shown in Figure 7.1. The diode in Figure 7.1 is assumed to be a p^+-i-n^+ abrupt silicon structure. The width of the *i*-region, W, is 55 μm. The cross-sectional area is 2.6×10^{-4} cm^2. This resistance can vary over a wide range depending on the densities of injected carriers throughout the *i*-region at forward bias. Before attempting a solution for this resistance, let us first state the simplifying assumptions to be used: (1) mobility and lifetime are constant throughout the *i*-region and independent of carrier concentration; (2) $\mu_n = \mu_p = \mu$, and $\tau_p = \tau_n$ (equal mobility and lifetime of electrons and holes);

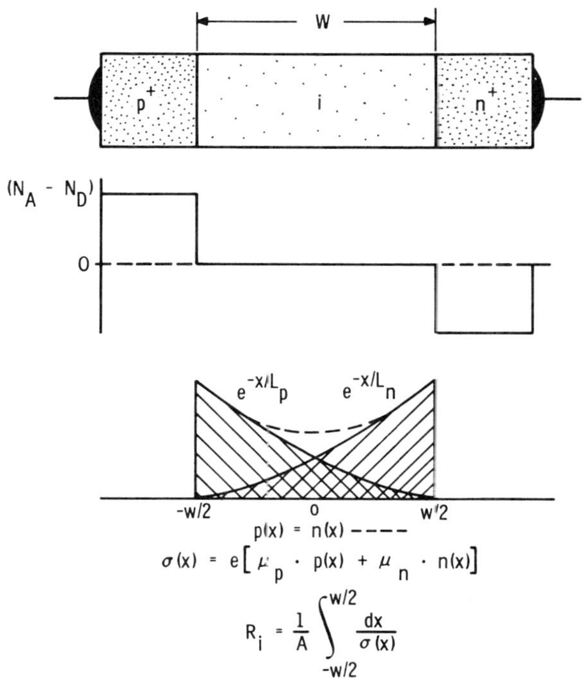

FIGURE 7.1
Abrupt junction diode model.

(3) quasi-neutrality, i.e., $p(x) = n(x)$, and hence $\partial n/\partial x = \partial p/\partial x$; (4) the current at the p^+-i interface consists entirely of hole current and a similar situation occurs for electrons at the n^+-i interface; and (5) injection of electrons into the p^+ region and holes into the n^+ region may be neglected (this is a requirement of 4).

From Section 2.8 of Chapter 2, the equations for the continuity of hole and electron current are

$$\frac{\partial n}{\partial t} = (g - R) + (\nabla \cdot J_n)/e \tag{7.1}$$

$$\frac{\partial p}{\partial t} = (g - R) - (\nabla \cdot J_p)/e \tag{7.2}$$

where $p = p(x)$ = hole density distribution
$n = n(x)$ = electron density distribution
$g - R$ = net generation-recombination rate (same for holes and electrons)
J_n, J_p = electron, hole current, respectively
e = electronic charge

Assume that the recombination is linear with excess carrier concentration, then from Chapter 2, Section 2.8

$$g - R = -\frac{(n - n_i)}{\tau} \quad \text{for electrons} \tag{7.3}$$

and

$$g - R = -\frac{(p - p_i)}{\tau} \quad \text{for holes} \tag{7.4}$$

where n_i, p_i = intrinsic carrier concentrations.

From Eq. (2.43A and B) the hole and electron currents in one dimension are:

$$J_n = ne\mu E + eD_n \frac{dn}{dx} \tag{7.5}$$

$$J_p = pe\mu E - eD_p \frac{dn}{dx} \tag{7.6}$$

where E = electric field and $D_p = D_n = D$ [a result of assumption (2) above]. Combining Eq. (7.1) to (7.6), we can write

$$\frac{\partial n}{\partial t} + \frac{\partial p}{\partial t} = -\frac{(p - p_i)}{\tau} - \frac{(n - n_i)}{\tau} + D\frac{\partial^2 p}{\partial x^2} + D\frac{\partial^2 n}{\partial x^2} + \mu\frac{\partial}{\partial x}[(n - p)E]. \tag{7.7}$$

Since E will be small in a heavily injected region and also since $n = p$, the last term on the R.H.S. of Eq. (7.7) may be dropped. Since $n = p$ we can also convert Eq. (7.7) to:

$$\frac{\partial n}{\partial t} = -\frac{n - n_i}{\tau} + D\frac{\partial^2 n}{\partial x^2} \tag{7.8A}$$

or

$$\frac{\partial p}{\partial t} = -\frac{p - p_i}{\tau} + D\frac{\partial^2 p}{\partial x^2} \tag{7.8B}$$

and, in the dc steady state, $\partial n/\partial t = \partial p/\partial t = 0$. Hence, in the steady state and if n and p are much larger than n_i and p_i (which is the definition of conductivity modulation), we may write:

$$\frac{\partial^2 n}{\partial x^2} + \frac{n}{L^2} = 0 \tag{7.9A}$$

$$\frac{\partial^2 p}{\partial x^2} + \frac{p}{L^2} = 0 \tag{7.9B}$$

where $L = \sqrt{D\tau}$ = diffusion length. Equations (7.9A) and (7.9B) are the so-called steady-state diffusion equations, and their solution (satisfying a given set of boundary conditions) defines the carrier distributions in the *i*-region. The coordinate system to be used is shown in Figure 7.1. Since $n(x) = p(x)$, it is sufficient to solve either Eq. (7.9A) or (7.9B). Equation (7.9A) is a simple second-order differential equation, and the solution is known to be of the form:

$$n(x) = A_1 \exp\left(+\frac{x}{L}\right) + A_2 \exp\left(-\frac{x}{L}\right) \cdot \tag{7.10}$$

Due to the double injection, finite diffusion length, and assumption (2), we expect $n(x)$ to be a minimum at $x = 0$, i.e.,

$$\left.\frac{dn(x)}{dx}\right|_{x=0} = 0\,. \tag{7.11}$$

This requires that $A_1 = A_2$. Therefore, Eq. (7.10) can be written

$$n(x) = 2A_1 \cosh\left(\frac{x}{L}\right) \cdot \tag{7.12}$$

At $x = W/2$ we require

$$\left.n(x)\right|_{x=W/2} = n\left(\frac{W}{2}\right) \qquad \text{a boundary condition.}$$

Applying this boundary condition to Eq. (7.12), we get:

$$n(x) = \frac{n(W/2)\cdot\cosh(x/L)}{\cosh(W/2L)}\,. \tag{7.13}$$

Therefore, once the value of n at $x = W/2$ (n^+-i interface) is known the carrier distributions in the i-region will be given by Eq. (7.13).

The solution for $n(W/2)$ proceeds as follows, at $x = +W/2$, $J_p = 0$ [assumption (4)]. Hence, from Eq. (7.6)

$$J_p = 0 = e\mu pE - eD\frac{dp}{dx} \tag{7.14}$$

or

$$\mu pE = D\frac{dp}{dx}\,. \tag{7.15}$$

From Eq. (7.15) and (7.5), the continuity of current, and recalling that $p(x) = n(x)$, it follows that

$$I_0 = J_n\cdot A\Big|_{x=W/2} = 2eDA\frac{dn}{dx} \tag{7.16}$$

where I_0 = bias current and A = diode area. Differentiating Eq. (7.13) to get dn/dx and solving Eq. (7.16) for $n(W/2)$, we get

$$n\left(\frac{W}{2}\right) = \frac{I_0 L\cosh(W/2L)}{2eDA\sinh(W/2L)}\,. \tag{7.17}$$

Substituting this result into Eq. (7.13), we have:

$$p(x) = n(x) = \frac{I_0 L\cosh(x/L)}{2eDA\sinh(W/2L)}\,. \tag{7.18}$$

This is the carrier distribution we have been seeking. The carrier densities are seen to depend on the bias current density, I_0/A, and the ratio $W/2L$ (i-region width divided by twice the diffusion length). A normalized plot of the carrier distributions for two values of $W/2L$ is obtained by changing the carrier lifetime, the values given on Figure 7.2 are obtained from:

$$\frac{W}{2L} = \frac{55\times 10^{-4}\text{ cm}}{2\sqrt{D\tau}} \tag{7.19}$$

where D = 18 cm^2/sec (diffusion coefficient) and τ = 1 μsec (upper curve), 0.03 μsec (lower curve). The resistance of the i-region may be calculated from

$$R_i = \frac{1}{A}\int_{-W/2}^{+W/2}\frac{dx}{\sigma(x)}\,. \tag{7.20A}$$

From Eq. (2.39) and assumption (2),

$$\sigma(x) = e\mu[n(x) + p(x)] = 2e\mu n(x)\,. \tag{7.20B}$$

R_i may be found by substituting Eq. (7.18) into Eq. (7.20) and carrying out the integration [6]. The result is:

$$R_i = \frac{2(kT/e)}{I_0}\cdot \sinh\left(\frac{W}{2L}\right)\cdot \tan^{-1}\left[\sinh\left(\frac{W}{2L}\right)\right]\cdot \tag{7.21}$$

As an example, take $I_0 = 50$ mA, $W/2L = 0.65$, $kT/e = 0.026$. The resulting value of R_i is $R_i = 0.44$ ohms.

The i-layer is shunted by a capacitance C_i which is essentially that of a parallel plate capacitor formed by the p^+ and n^+ regions. This capacitance may be estimated from:

$$C_i = \frac{\epsilon A}{W}\cdot \tag{7.22}$$

For a silicon diode with an area of 2.6×10^{-3} cm^2 and $W = 55$ μm, $C_i \cong 0.5$ pF.

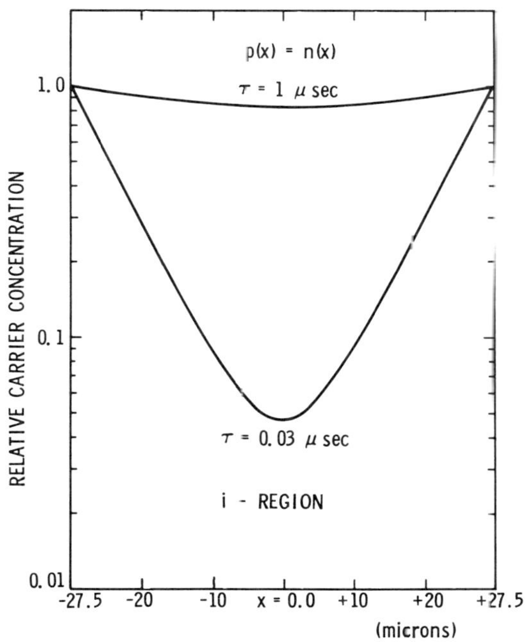

FIGURE 7.2
Normalized carrier distributions.

At forward bias there are also junction effects which must be included in the diode impedance model. The depletion capacitance must be considered at forward bias. From Section 3.4

$$C_J = \left[\frac{eN_A N_D \epsilon}{2V_B(N_A + N_D)}\right]^{1/2} \cdot A \tag{7.23}$$

where N_A, N_D = donor and acceptor concentrations, ϵ = dielectric constant and V_B = potential barrier height. In practice it is not a true p^+-i-n^+ diode, but usually a p^+-π-n^+ structure is utilized. Therefore, if we take $N_A = 10^{12}/\text{cm}^3$ and $N_D \gg N_A$ and $V_B \cong 0.25$ V, we obtain $C_J \cong 2$ pF.

At high frequencies ($\omega\tau \gg 1$) and forward bias, the presence of the complex diffusion capacitance must also be considered. Using Eq. (3.16), (3.17), (3.33), and (3.34), it can be shown that at any frequency the junction admittance due to carrier diffusion is given by:

$$Y = G + jB = (J_p \cdot \sqrt{1 + j\omega\tau} + J_n \cdot \sqrt{1 + j\omega\tau}) \cdot \frac{e}{kT} \cdot A \tag{7.24}$$

where J_p, J_n = hole and electron current density at the juntion
ω = frequency
τ = lifetime
A = device area.

If it is assumed that one type of current dominates (i.e., J_p or J_n), this current must then be the same as I_0 (bias current $I_0 = J_0 \cdot A$) from the continuity of current. Also, at very high frequencies $\omega\tau \gg 1$

$$\sqrt{1 + j\omega\tau} \cong \sqrt{j\omega\tau} = \frac{1 + j}{\sqrt{2}} \sqrt{\omega\tau}\,. \tag{7.25}$$

In the light of the above arguments Eq. (7.24) can be rewritten as

$$Y = G + jB = \frac{eI_0}{kT} \cdot \sqrt{\frac{\omega\tau}{2}} \cdot (1 + j)\,. \tag{7.26}$$

If $I_0 = 0.05$ amp, $T = 290$°K, $\tau = 1$ μsec, and $\omega = 1.25 \times 10^{10}$ (2 GHz), then

$$G = 1.6 \times 10^2 \text{ mhos}$$
$$B = 1.6 \times 10^2 \text{ mhos}\,.$$

The effective capacitance corresponding to the susceptance is $B/\omega = 12{,}500$ pF, and the resistance associated with G is $R = 1/G = 0.006$ ohms. So, the equivalent circuit of the junction alone at 50 mA forward bias consists of two capacitances ($C_J = 2$ pF, $C_D = 12{,}500$ pF) in parallel with a resistance, $R = 0.006$ ohms. Hence, at 2 GHz the junction effects can be neglected in comparison with the *i*-region resistance (0.44 Ω) given by Eq. (7.21).

At both forward and reverse bias we must also consider the series resistance of the p^+ and n^+ regions (see Figure 7.1) as well as any contact resistance. A value including all three contributions is estimated to be $R_S = 0.1\ \Omega$.

At reverse bias (punch-through) the *i*-region is completely depleted of mobile carriers. The effects of G and B [Eq. (7.24)] disappear; the *i*-region resistance [Eq. (7.21)] $R_i \to \infty$ (no carriers); C_J becomes C_i and is given by Eq. (7.22).

In practice the diode chip is mounted in a package so that it may be handled. At microwave frequencies this package contributes parasitic impedances which must be considered. There is a capacitance, C_c, which is in shunt with the chip (this capacitance is between the two contacts to the chip). There is a series inductance, L_s, due to the package leads, and also a package capacitance, C_p, shunting the outside of the package.

Taking all of the parameters mentioned here, due both to the device physics and to the package, it is possible to draw a reasonably accurate lumped constant equivalent circuit of a complete packaged device. (This can be done as long as all dimensions are much less than a wavelength.) This is done in Figure 7.3.

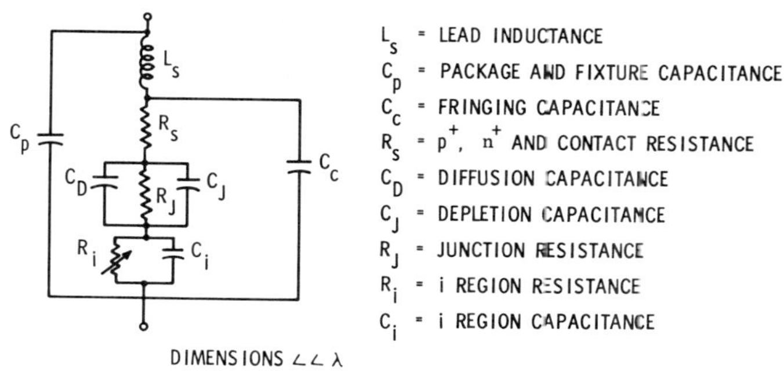

FIGURE 7.3
Exact diode equivalent circuit (forward bias).

It was pointed out earlier that at microwave frequencies and forward bias the junction effects may be neglected in comparison to R_i. Also, at reverse bias the equivalent circuit is simplified to the point that the chip just contributes R_s and C_J. These simplified models are shown in Figure 7.4 for both forward and reverse bias. The equivalent circuits shown in Figure 7.4 are the ones that will be used for the remainder of this chapter.

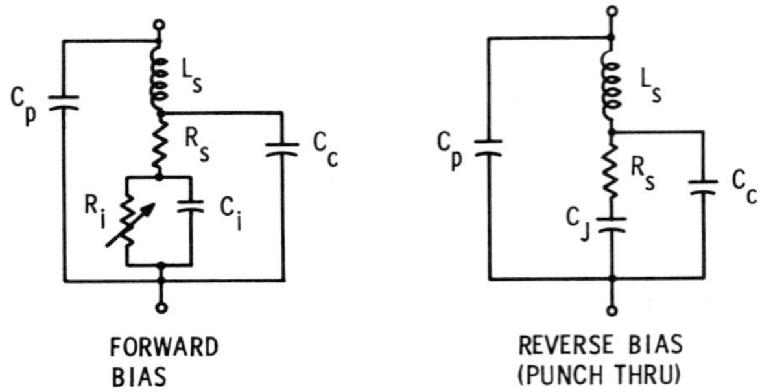

FIGURE 7.4
Approximate equivalent circuits of a packaged *p-i-n* diode at (A) forward bias and (B) reverse bias.

7.3 The *p-i-n* Microwave Switch

To make a microwave switch out of the *p-i-n* diode it is necessary to connect the diode across a transmission line, in shunt. (A series connection may also be used, but it is not as common and will not be discussed here.) The diode is then switched between dc forward bias (where it presents a low microwave impedance and reflects nearly all of the incident rf power) and reverse bias (where it presents a high microwave impedance and allows free transmission of rf power along the transmission line). The effect of an impedance or admittance in shunt with a transmission line will be considered

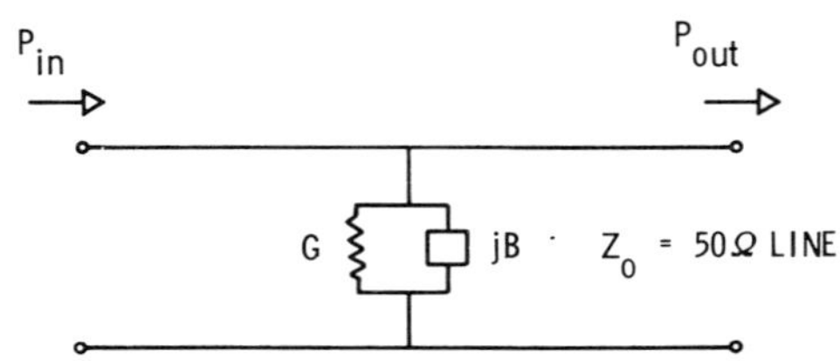

$$\alpha = 10 \log_{10}\left(\frac{P_{in}}{P_{out}}\right)$$

$$\alpha = 10 \log_{10}\left[\left(1 + \frac{GZ_0}{2}\right)^2 + \left(\frac{BZ_0}{2}\right)^2\right]$$

FIGURE 7.5
Transmission line with shunt admittance.

with the aid of Figure 7.5. The insertion loss on the transmission line (reflected and absorbed power) is defined as

$$\alpha = 10 \log_{10} \left(\frac{P_{\text{in}}}{P_{\text{out}}}\right). \quad (7.27)$$

Where α is given by [7]

$$\alpha = 10 \log_{10} \left[\left(1 + \frac{GZ_0}{2}\right)^2 + \left(\frac{BZ_0}{2}\right)^2\right]. \quad (7.28)$$

If the diode is connected directly across the line, then the isolation (at forward bias) and the insertion loss (at reverse bias) can be obtained by finding the G

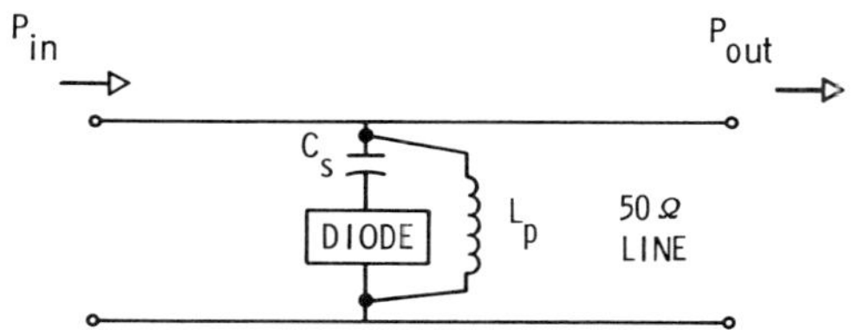

FIGURE 7.6
Simple shunt *p-i-n* switch configuration.

and B of the equivalent circuits shown in Figure 7.4 (A and B) and inserting those values into Eq. (7.28). In practice, the diode is not connected directly across the line, but rather the package inductance, L_S, is series-resonated at forward bias to improve the isolation, and the diode capacitance, $C_J + C_c + C_p$, is parallel-resonated at reverse bias to reduce the insertion loss. This is shown in Figure 7.6. In practice L_p and C_s cannot be chosen independently, but they interact. If the proper values are chosen, a typical result that would be obtained is shown in Figure 7.7. From this figure it can be seen that the isolation peaks and the insertion loss dips at some design frequency, f_0 (where the circuits are resonant). The Q of the resonance and general shape of the curves may be calculated from the frequency dependences of G and B in Eq. (7.28). These frequency dependences are in turn obtained from L_p and C_s combined with the equivalent circuits shown in Figure 7.4. (The parameters given in Figure 7.4 are assumed to be frequency independent.)

For this reason, the task of predicting the effect of radiation on the microwave *p-i-n* switch reduces to finding the effect of radiation on the parameters of the equivalent circuits given in Figure 7.4.

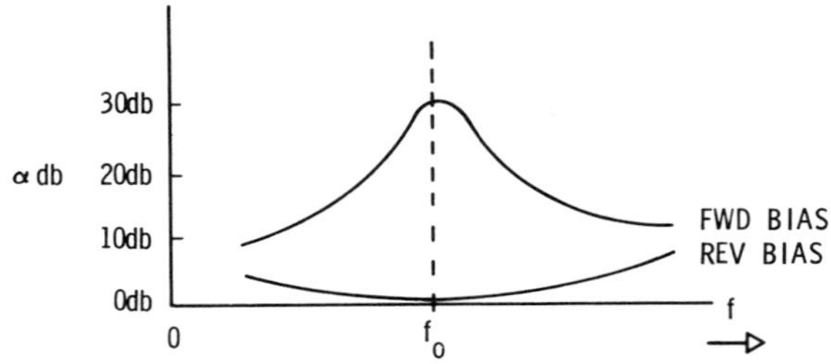

FIGURE 7.7
Typical performance of a shunt *p-i-n* switch versus frequency.

7.4 Permanent Radiation Effects in *p-i-n* Switches

This section discusses and analyzes the effect of neutron irradiation on the device and on the combined device and circuit. The package parasitics and transmission line structure are assumed not to change under the radiation levels discussed here. The parameters that will change are those associated with the diode chip. Neutron damage to silicon has been discussed in Chapter 5. The most lightly doped regions will be affected first (the *i*-region) while damage in the p^+ and n^+ regions will require much higher fluences. In the *i*-region we can expect changes in mobility, lifetime, and thermal equilibrium carrier concentrations. In silicon, at room temperature and for low doping levels, the primary carrier scattering mechanism is interactions with lattice vibrations. Defect scattering does not become important until defect concentrations of the order of $3 \times 10^{15}/\text{cm}^3$ are reached. Assuming three defects per incident neutron/cm³ (see Figures 5.10 and 5.11), this would require a fluence of 10^{15} n/cm² to alter the mobility. Reduction of thermal equilibrium carrier densities by carrier removal in the *i*-region is not significant since the major contribution to the conductivity of the *i*-region is due to injected excess carriers. Lifetime of these injected carriers is the most sensitive parameter to neutron irradiation in these devices, in fact, all of the observed changes in the diodes are due to a change in the carrier lifetime. From Eq. (5.23) we know that the carrier lifetime as a function of neutron fluence can be expressed as:

$$\tau = \frac{\tau_0}{1 + K_\tau \Phi \tau_0} \tag{7.29}$$

where τ_0 = initial lifetime, Φ = neutron fluence (n/cm²), and K_τ = lifetime damage constant. If we assume $\tau_0 = 1\ \mu\text{sec}$, $K_\tau = 1.75 \times 10^{-7}$, we see that the lifetime will be reduced by a factor of 2 at $\Phi = 5.7 \times 10^{12}$ n/cm². The diffusion length, $L = \sqrt{D\tau}$ will be reduced by $1/\sqrt{2}$ at this fluence. The carrier distributions [Eq. (7.18) and Figure 7.2] are critical functions of the ratio $W/2L$. This implies also that the *i*-region resistance [Eq. (7.21)] will also depend strongly on lifetime. As will be shown later in this chapter, this is the failure mechanism of this device (an increase in R_i with fluence). Swartz and Thurston [8] examined the effect of neutron irradiation on low frequency *p-i-n* power rectifiers and concluded that there will be an increase in forward voltage drop due to the decrease in lifetime.

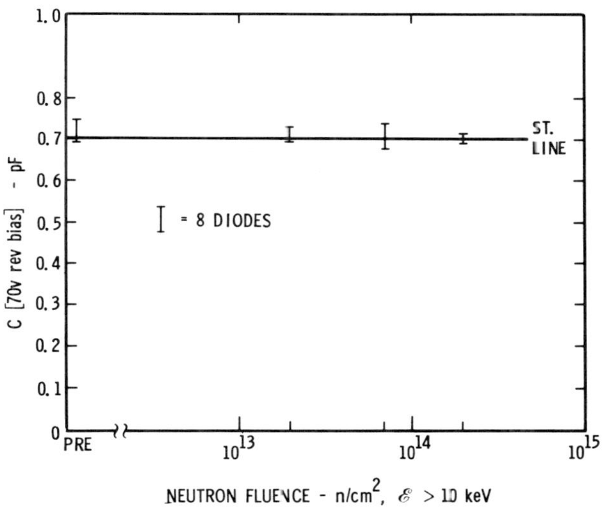

FIGURE 7.8
Punch-through capacitance versus neutron fluence for *p-i-n* diodes.

An experimental study was carried out using the diode described in Section 7.2 of this chapter. The irradiations were performed at the Sandia ACPR (Annular Core Pulse Reactor) facility.† The diodes were characterized by the following measurements: (1) reverse bias punch-through capacitance; (2) carrier lifetime measurement (forward bias); (3) forward V-I characteristic; and (4) microwave impedance measurements from 200 MHz to 4 GHz (forward and reverse bias). These measurements were repeated on all diodes (irradiated and controls) after each irradiation.

The capacitance measurements were made on a 1 MHz Boonton digital bridge, and the results are summarized in Figure 7.8. From Figure 7.8 it can

† All neutron fluences given are $\mathscr{E} > 10$ keV, Spectrum B.

be seen that the reverse bias punch-through capacitance, C_J in Figure 7.4B, does not vary appreciably with neutron fluence. This is because the depletion width is essentially fixed by the n^+-i and i-p^+ boundaries and neutron radiation does not have a pronounced affect on the p^+ and n^+ regions. The invariance of C_J with neutron fluence implies that there will be very little change in the low insertion loss (reverse-bias) state of the switch described in

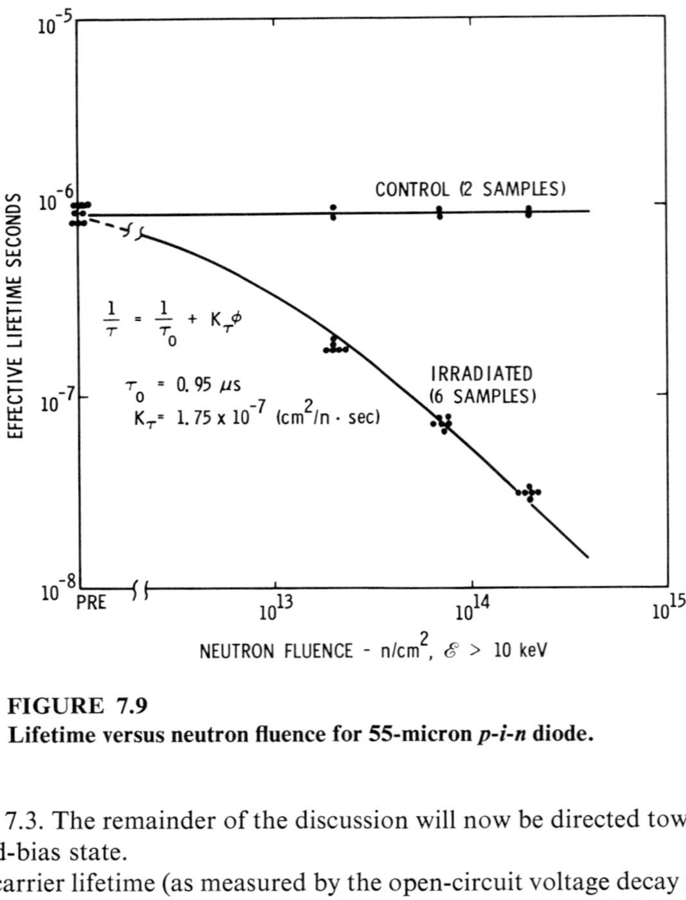

FIGURE 7.9
Lifetime versus neutron fluence for 55-micron *p-i-n* diode.

Section 7.3. The remainder of the discussion will now be directed towards the forward-bias state.

The carrier lifetime (as measured by the open-circuit voltage decay method, see Figure 4.14) was observed to decrease with fluence. This is plotted in Figure 7.9 which shows that the experimentally measured values of lifetime are in good agreement with Eq. (7.29) for $K_\tau = 1.75 \times 10^{-7}$ (cm^2/n-sec). This value of K_τ is also in good agreement with Figure 5.25 for high injection ratios. (The injection ratio in forward biased *p-i-n* diode approaches unity.)

The forward V-I characteristic of the diodes was measured, and the voltage across the diode at a given current was found to increase with fluence. This result is shown in Figure 7.10 where it can be seen that the forward voltage drop at a given current increases dramatically with neutron fluence. This is due primarily to an increase in R_i brought on by a reduction in diffusion

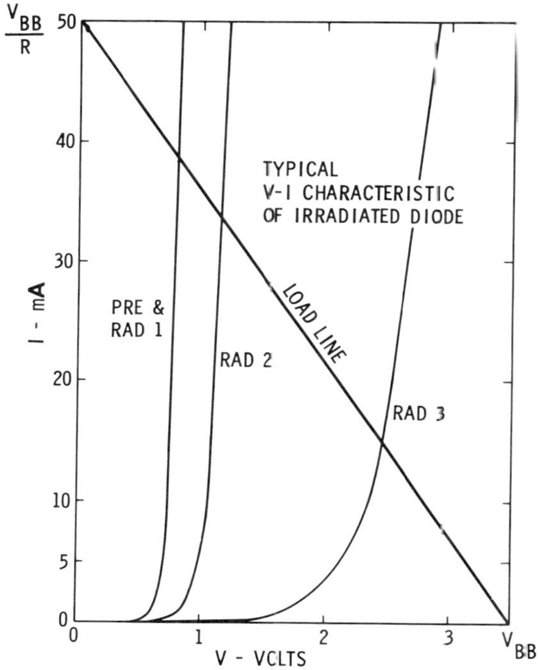

FIGURE 7.10
Typical V-I characteristic of *p-i-n* diode: rad 1 = 2 × 10^{13} n/cm²; rad 2 = 7 × 10^{13} n/cm²; and rad 3 = 2 × 10^{14} n/cm² (accumulated dose).

length. The behavior seen in Figure 7.10 can be qualitatively explained as follows: The voltage across a *p-i-n* diode may be written as

$$V = V_B + V_J \tag{7.30}$$

where V_B = voltage drop across base (i) region and V_J = junction voltage. For diodes with wide i regions the V_B dominates and V_B increases with decreasing τ (poorer conductivity modulation). Therefore, it would be expected that diodes with a wide *i*-region would have an increasing forward drop with fluence [τ decreases, see Eq. (7.29)]. For diodes with narrow *i*-regions, the V_J term dominates. From Eq. (2.77), (3.12), and (3.20), and for a one-sided junction, we can write

$$J = \frac{en_i^2\sqrt{D_n}}{\sqrt{\tau_n}N_A}\left[\exp\left(\frac{eV_J}{kT}\right) + 1\right]. \tag{7.31}$$

Equation (7.31) may be rewritten as (assuming $V_J \gg e/kT$)

$$V_J = \frac{kT}{e}\ln\left[\frac{J\cdot N_D\sqrt{\tau_n}}{en_i^2\sqrt{D_p}}\right]. \tag{7.32}$$

Taking the derivative of V_J with respect to τ_n, we obtain

$$\frac{dV_J}{d\tau_n} = \frac{kT}{2e\tau_p}. \tag{7.33}$$

This is a positive slope which implies that for a constant current, V_J increases with increasing τ. Or, inversely, as τ decreases (as it does with neutron fluence), V_J decreases. The behavior seen in Figure 7.10 clearly shows that the V_B term rather than the V_J term dominates in the case of the 55 μm *p-i-n* diode. Choo [9] has studied this problem in detail and shows that for a *p-i-n* diode there is an optimum lifetime which gives a minimum forward drop.

The microwave impedance measurements on the diode were made on a digital microwave impedance analyzer [10] which incorporated the Hewlett-Packard 8410 network analyzer. The measurement accuracy is typically $\pm 0.4\ \Omega$ (for values below 50 Ω) at any frequency from 200 MHz to 10 GHz. The microwave impedance of the diode was measured at several frequencies between 200 MHz and 4 GHz at forward and reverse bias. In all, eight diodes were measured and there was very little impedance variation from diode to diode (typically $\pm 5\%$) in the measured results. The parameter of interest was R_i, and in order to separate it from the parasitic effects of the other elements (shown in Figure 7.4A), it was necessary to transform the measured complex impedances back through the network shown in Figure 7.4A. To do this a simple computer program was used. Before the transformation could be accomplished, however, it was first necessary to evaluate the remainder of the parameters in the equivalent circuit (Figure 7.4A). This was done through high and low frequency measurements, use of a shorted diode, and through calculation where possible. The results obtained for R_i as a function of fluence are plotted in Figure 7.11. Figure 7.11 also shows the result predicted by Eq. (7.21) for R_i, where it is assumed that $\tau(\Phi)$ is given by Eq. (7.29). The experimental and theoretical values of R_i are seen to be in good agreement. The measured values of R_i may also be plotted versus the measured lifetimes;

this is done in Figure 7.12, which also shows the value of R_i calculated from Eq. (7.21). The experiment and theory are again seen to be in good agreement. It should be added here that all measurements of R_i on Figures 7.11 and 7.12 are for a constant current, $I_0 = 50$ mA.

From Eq. (7.21) it can be seen that the value of R_i should depend linearly on $(I_0)^{-1}$. This implies that it is possible to reduce the value of R_i by increasing the bias current. This effect was also measured using the digital microwave impedance system. The results for $f = 2$ GHz, before and after 2×10^{14}

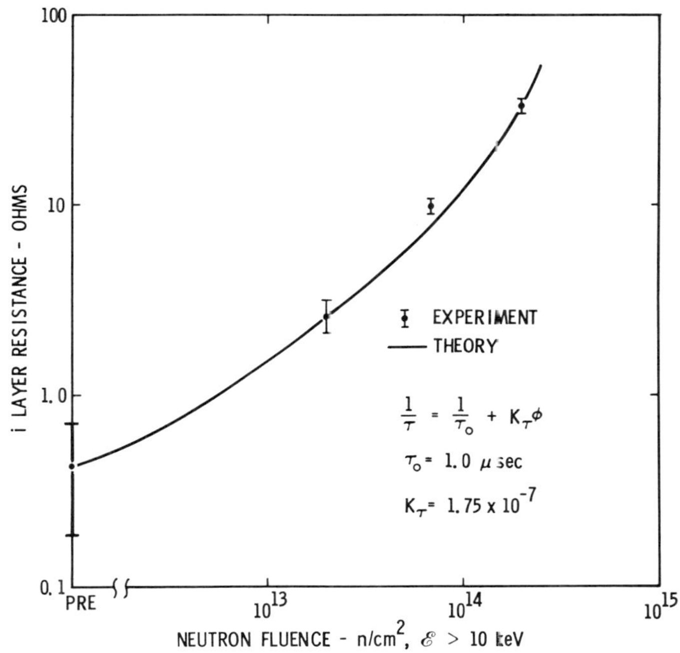

FIGURE 7.11
R_i versus neutron fluence for a *p-i-n* diode biased at $I_0 = 50$ mA.

n/cm^2 are shown in Figure 7.13. From Figure 7.13 it can be seen that R_i is indeed decreased by increasing I_0; however, the dependence is not linear but rather is proportional to $(I_0)^{-0.57}$ in the prerad case and to $(I_0)^{-0.70}$ in the postrad case. The discrepancy between this result and Eq. (7.21) is thought to be due, in part, to assumption of an abrupt-junction model and in part to injection into the p^+ and n^+ regions. This injection is reduced in the postradiation case (due to a shortened diffusion length) and in the postradiation case the dependence is closer to the $(I_0)^{-1}$ law given by Eq. (7.21).

At this point we have covered the primary radiation effects to the device, namely: (1) an increase in R_i (due to a reduction of τ); and (2) a change in the forward V-I characteristic. Now consider the effect these changes would have on an operational circuit such as the one shown in Figure 7.6. Figure 7.14 shows how a switch of the type shown in Figure 7.6, using the diode which has been discussed, would react to neutron fluence. The curves on Figure 7.14 were calculated from Eq. (7.28) for a design frequency of 2.5 GHz. The

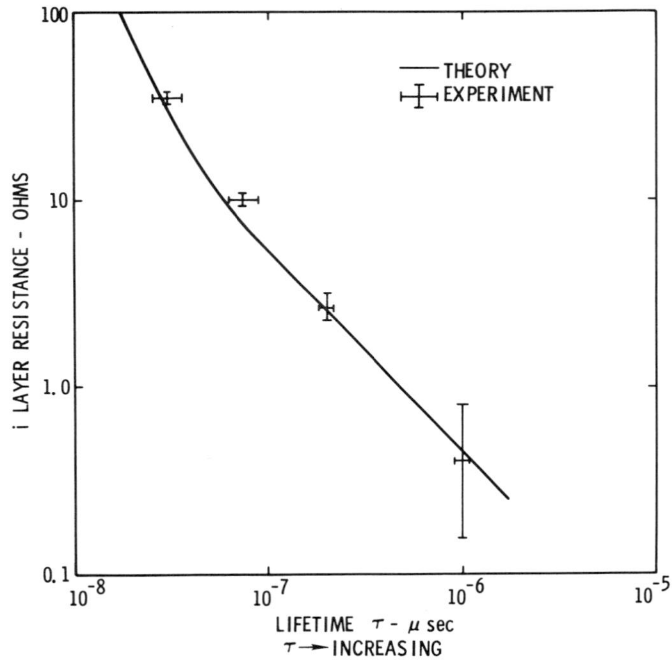

FIGURE 7.12
R_i versus lifetime for a *p-i-n* diode biased at I_0 = 50 mA.

equivalent circuits in Figure 7.4 were changed after each radiation to match the experimental impedances measured. (R_i is the only parameter that changes.) Figure 7.14 shows that the switch initially provided nearly 40 dB of isolation [see Eq. (7.27)], and after 2×10^{14} n/cm^2 the isolation is only 8 dB. (All forward-bias curves are for $I_0 = 50$ mA.) The Q of the series resonance is also seen to decrease; this is also due to the increase in R_i. Initially, it was thought there could possibly be a detuning effect (or shift in the resonance), but as can be seen from Figure 7.14 this effect is swamped out by the decrease

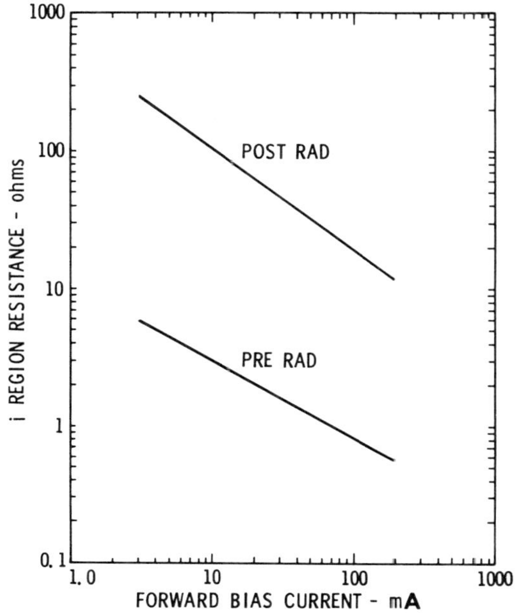

FIGURE 7.13
***i*-Region resistance versus bias current prerad and after 2 × 10^14 n/cm^2.**

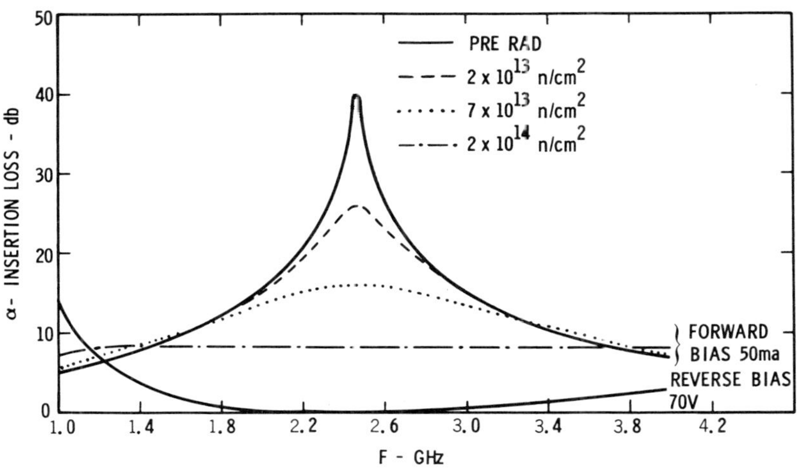

FIGURE 7.14
Effect of neutron fluence on *p-i-n* switch performance.

in Q. The curve for reverse bias (on Figure 7.14) does not change with irradiation (as was discussed earlier).

The actual failure point in terms of lost isolation is relative to the system in which the switch is to be used. If it is arbitrarily assumed that the switch has failed when $R_i = 1\ \Omega$, then it is possible to generalize the results of this study to other *p-i-n* diodes. In particular, let us assume that Eq. (7.29) with $K_\tau = 1.75 \times 10^{-7}$ holds for all microwave *p-i-n*'s and neglecting $1/\tau_0$ in comparison

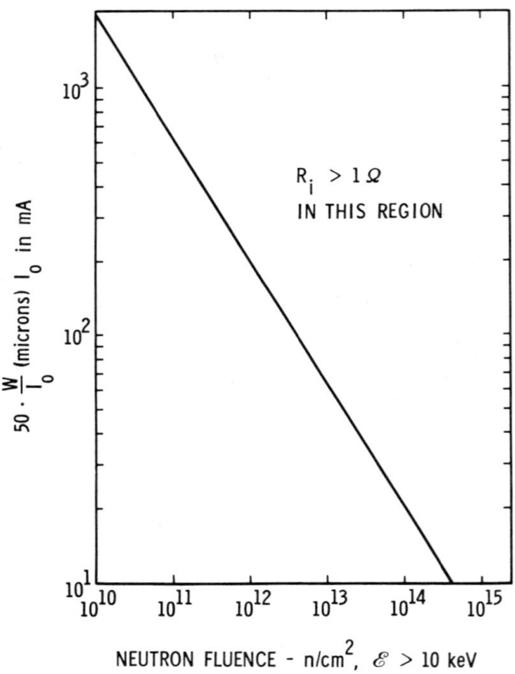

FIGURE 7.15
Failure fluence versus *i*-region width for a *p-i-n* microwave switch.

to $1/\tau$, we then have an expression for τ (neutron fluence). Let us further assume $I_0 = 50$ mA (a typical value for commercial *p-i-n*'s). Then it is possible, for a given *i*-region width, W, to calculate R_i as a function of neutron fluence through Eq. (7.21) and, inversely, to calculate the fluence where $R_i = 1\ \Omega$. In this way it is possible to plot the fatal fluence versus *i*-region width. This is done in Figure 7.15 where it can be seen that the radiation hardness increases with decreasing *i*-region width. This, however, is countered by the fact that the avalanche breakdown voltage decreases with decreasing *i*-region width [11]. Thus, the minimum *i*-region width needed to hold off the peak rf voltage without going into avalanche breakdown must be determined.

There are two possible ways around this breakdown problem: The first is to design the switch (if in a radar T-R) so that high rf powers are only present during the forward-bias state; then since received powers (reverse bias state) are usually very low, W can be made small without concern about avalanche breakdown. If this cannot be done, then it is still possible to put several devices with narrow *i*-regions (for hardness) in series so that the reverse voltage is divided up between the diodes. This would involve some loss in available isolation due to increased series resistance.

Finally, while this work has been done for neutron irradiation, the results can be extended to other forms of radiation if Eq. (7.29) is used with an appropriate damage constant (see Chapter 5). In particular, for γ-rays with energy of 1.25 MeV the damage constant is $K_\tau = 2.8 \times 10^{-10}$ (cm^2/photon-sec) for room-temperature irradiated silicon. Hence, the damage done by 2×10^{13} n/cm^2 (Spectrum B) should be equivalent to about 10^{16} photons/cm^2 or approximately 5.4 Mrad [see Eq. (5.4)]. This should be of interest to space satellite system designers.

In addition to the radiation-induced loss in isolation, there is a secondary problem. Since R_i is increased (with irradiation) there comes a point when $R_i \approx 50\ \Omega$. If the diode is at the end of a transmission line stub (used in some switches), then it can easily be seen that all of the incident power will be absorbed instead of reflected (when $R_i = 50\ \Omega$). If the stub is used as a T-R switch in a radar, then it is not unreasonable to have 10 kW of pulsed rf power (10 watts average) incident on the device. This device cannot dissipate 10 watts of average power and for this reason, it will fail catastrophically (i.e., burn out—shorted). This failure mode would then disable the receive function of the radar and thus fail the whole system. Therefore, this secondary failure mode can be extremely important. The ratio of $P_{\text{dissipated}}/P_{\text{incident}}$ for the particular diode studied in this chapter (at the end of a 50 Ω stub) is plotted versus fluence on Figure 7.16 for $I_0 = 50$ mA (the curve will fall off as R_i goes above 50 Ω).

Another secondary failure mode that should be mentioned can be best seen by reexamining Figure 7.10. Forward-bias current is usually achieved for a *p-i-n* switch with a voltage source, V_{BB}, and a series resistor, R. If this scheme were used, the operating point would be found from the intersection of the load-line with the diode V-I curve. (The load-line is a straight line drawn between the voltage V_{BB} on the abscissa and the current V_{BB}/R on the ordinate (see Figure 7.10)). It can then be seen from Figure 7.10 that the operating point will shift to lower currents with increasing neutron fluence. Since R_i is inversely related to I_0, this reduction in current will further increase the value of R_i. The solution to this problem is to bias the diode with a constant current generator rather than a voltage source and a resistor. If this is done, then the "load-line" will be a horizontal line (in Figure 7.10), and the bias current will not change with neutron fluence.

One final comment about diode bias is that Figure 7.13 shows that some of the increase in R_i can be eliminated by increasing the bias current after

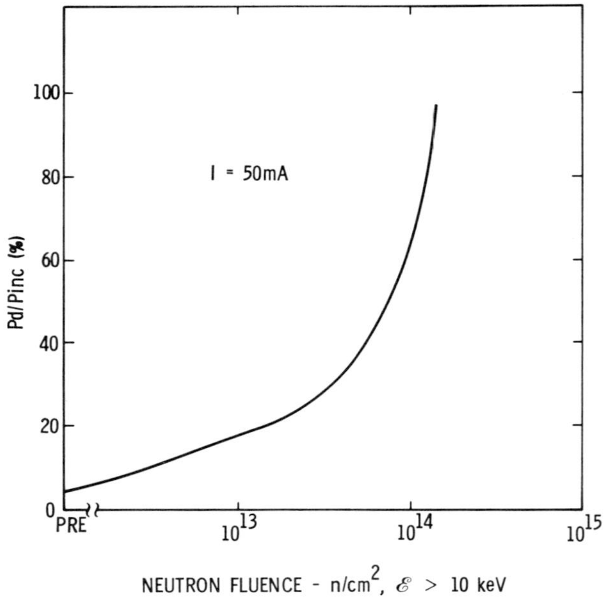

FIGURE 7.16
Absorbed rf power versus neutron fluence for a *p-i-n* switch.

irradiation. However, the forward voltage drop also increases after radiation, and care must be taken so that the heating caused by dc power dissipation in the diode ($V_{dc} \cdot I_{dc}$) does not cause catastrophic failure.

7.5 *p-i-n* Diode Microwave Limiters

Microwave limiters are usually used as a passive protective device. They are placed between the antenna and the mixer in a microwave receiver to protect the mixer from burnout due to too large an rf input to the antenna.

The essential characteristic of a limiter is that it have little or no insertion loss at low-signal input levels (so that it will not adversely affect the receiver noise figure) and a large insertion loss at high-signal input levels (to attenuate the signal and protect the mixer). This characteristic can be achieved in practice by shunting a transmission line with a diode such as is shown in Figure 7.17. Figure 7.17A shows two diodes shunted across a balanced stripline transmission line. The line is also shunted by an inductive stub (shorted line segment, less than $\lambda/4$ long). The inductive stub provides a dc return for the diode. For small-signal levels (where the peak rf voltage is less than the diode

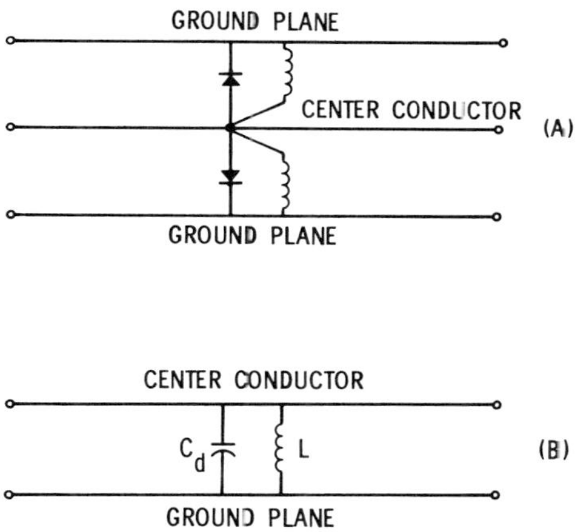

FIGURE 7.17
Microwave diode limiter: (A) stripline cross section; (B) equivalent circuit.

knee $\sim$0.5 V) the equivalent circuit of the configuration is given in Figure 7.17B where C_d represents the zero-bias depletion capacitance of the diodes and L represents the inductance of the stub. Since L may be changed by varying the length of the stub, it is possible to parallel resonate the diode capacitance with the inductive stub at any desired frequency, f_0. The resonant circuit presents a very high impedance in shunt with the line and to first-order may be neglected. Therefore, ideally, the insertion loss at small-signal levels is zero. When the signal level increases (such that the peak rf voltage gets above the diode knee), appreciable injection of minority charge occurs. This increases the diffusion capacitance of the junction so that it is several orders of magnitude larger than the depletion capacitance [see Eq. (3.23) and (3.37)]. This in turn shifts the resonant frequency down to the point where at the original frequency the line is essentially shorted by the large capacitance, which results in a large insertion loss. This can be calculated using Eq. (7.28). The diodes used in limiters usually have very narrow *i*-regions (typically a few μm), or they may even be abrupt *p-n* junction diodes. The reason for this is that narrower devices produce better limiting action at lower power levels than do wide devices.

7.6 Permanent Radiation Effects in Microwave Limiters

To measure the effects of energetic radiation on a limiter, a test was performed using silicon p^+-n-n^+ diodes with a 3 μm *n*-region width. The *n*-region doping density was 10^{14} per cm^3. These diodes were packaged in pairs and mounted across a balanced 50 Ω stripline circuit like the one shown in Figure 7.17.

The following parameters were measured on the diodes, and on the diodes and rf test fixture combined, as a function of neutron fluence: (1) forward V-I characteristic; (2) carrier lifetime; (3) zero-bias capacitance; (4) reverse breakdown voltage; and (5) limiting characteristic between −5 dbm and +40 dbm. The irradiations were performed in the Sandia ACPR reactor, and all fluences given are for $E > 10$ keV (Spectrum B). The forward V-I characteristic of each diode was measured, and the voltage across the diode at a

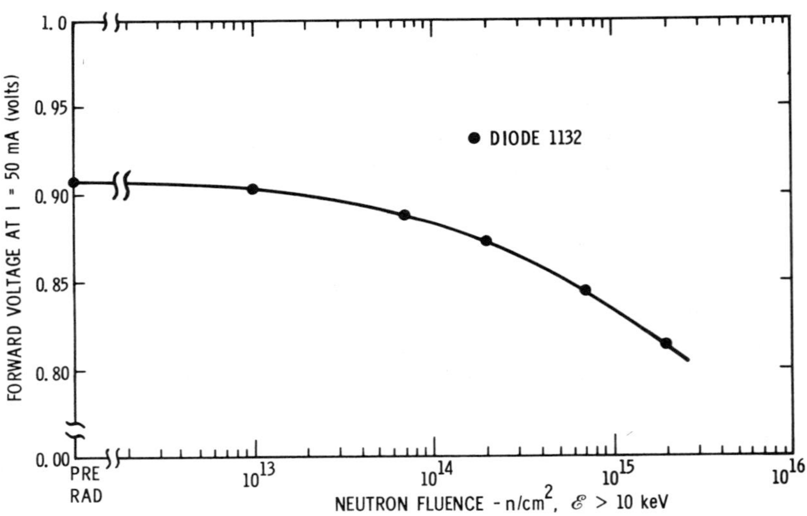

FIGURE 7.18
Forward voltage versus neutron fluence for a 3 μm *p-i-n* diode.

given current was found to decrease after irradiation. Typical values are shown in Figure 7.18. This drop in forward voltage with neutron fluence is due to the change in junction parameters discussed earlier [see Eq. (7.30) to (7.33)] and implies that the V_J term dominates over the V_B term for this narrow *p-i-n* diode.

The minority carrier lifetime was measured using the charge removal method (see Figure 4.15). A plot of the data obtained versus neutron fluence is shown in Figure 7.19, where it can be seen that the minority carrier lifetime can again be expressed in the standard form given by Eq. (7.29). The best fit value for K_τ is $K_\tau = 1.5 \times 10^{-7}$ cm^2 sec^{-1} which is very close to the 1.75 $\times 10^{-7}$ value found earlier for the 55 μm *p-i-n* diode.

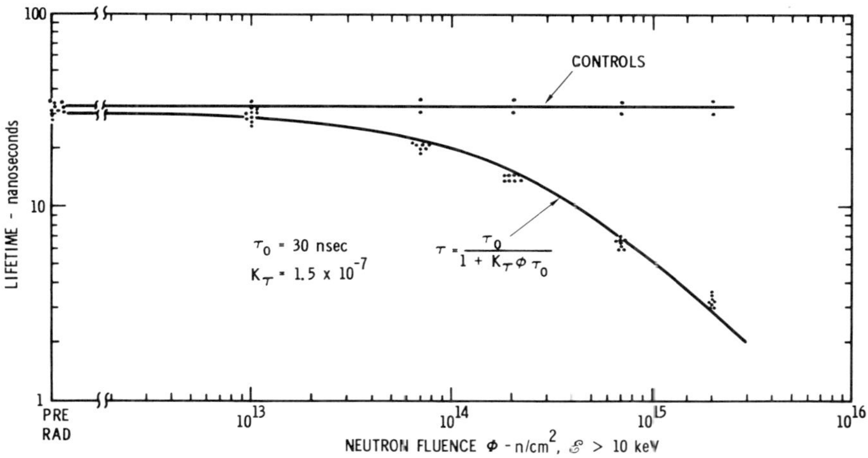

FIGURE 7.19
Lifetime versus neutron fluence for a 3 μm *p-i-n* diode.

The zero-bias capacitance was measured after each irradiation. The capacitance was found to decrease slightly after irradiation. An example of this is shown in Figure 7.20. This decrease in capacitance could cause problems since this capacitance determines the resonant frequency of the limiter (see Section 7.5); any change in its value will change the insertion loss of the limiter in the small-signal case. This was in fact seen and will be discussed later. The change in capacitance with radiation is attributed to an increase in zero-bias depletion width with radiation. This is due to carrier removal effects (see Chapter 5).

The limiting characteristic is simply a plot of power input versus power output for the diodes and fixture combined. This characteristic was measured

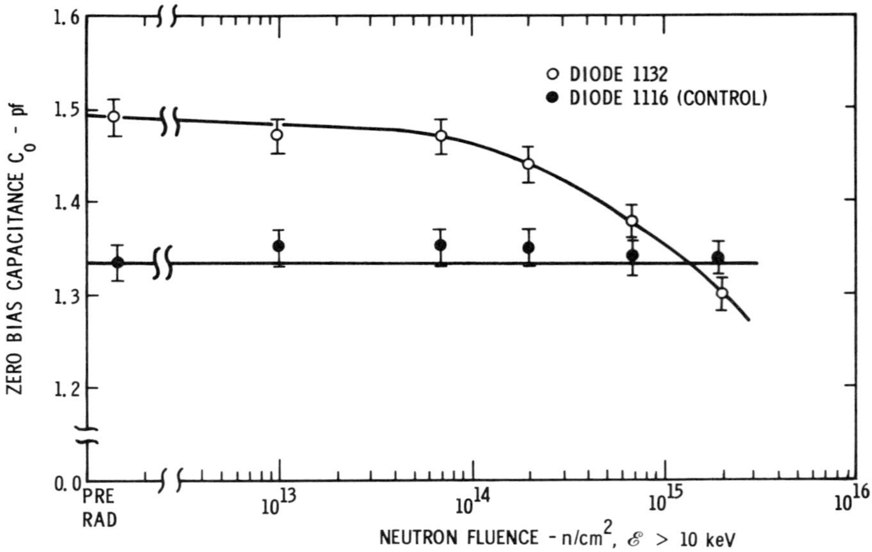

FIGURE 7.20
Zero-bias diode capacitance versus neutron fluence for a 3 μm *p-i-n* diode.

over an input power range of −5 dbm to +40 dbm using the setup shown in Figure 7.21. The measurements above +20 dbm were made on a pulse basis to avoid thermal destruction of the diodes. Figure 7.22 shows typical data taken from a limiter before and after irradiation. The unit shows no change at high input levels and an improvement in the postirradiation case at small

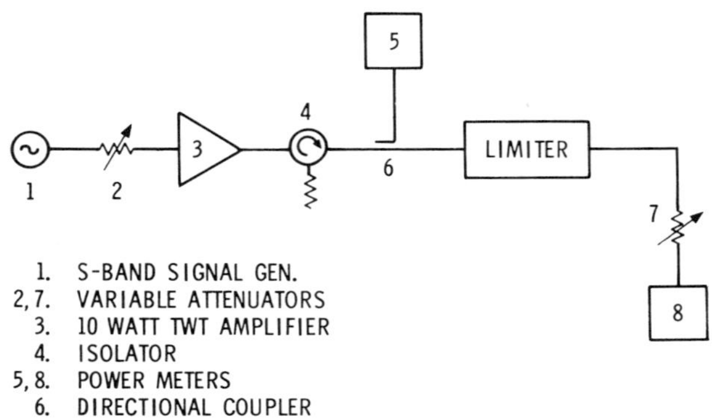

FIGURE 7.21
Limiter test setup.

signal levels. The reduction in insertion loss at low-signal levels postrad can be attributed to a shift in the resonant frequency upwards caused by a decrease in diode capacitance with irradiation (see Figure 7.20). It was observed initially (using swept frequency methods) that the diodes and fixture actually resonated below the test frequency. However, the resonance had a sufficiently low Q that reasonable performance could still be obtained at the test fre-

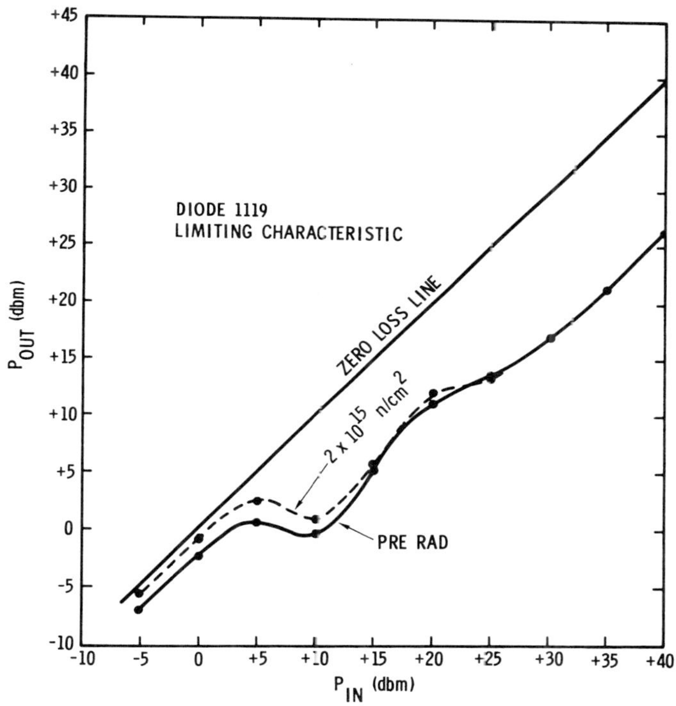

FIGURE 7.22
Limiting characteristic of irradiated *p-i-n* diode limiter.

quency and an upward shift in resonance would be beneficial. Also plotted in Figure 7.22 is the "zero loss line"—that input-output characteristic which would be obtained if no limiter were present. The insertion loss of the limiter is simply the vertical distance between the zero loss line and the limiter curve. So, from Figure 7.22 we see an insertion loss of about 2 dB at 0 dbm and an insertion loss of 14 dB at +40 dbm which is the type of variation expected of a limiter.

Because of a design error, the limiter resonance was low to begin with, and the upward shift was beneficial. An actual improvement in performance was obtained with irradiation. Had the limiter resonance been too high to begin with, the low-level insertion loss would have increased rather than decreased with radiation, and this could have been a problem.

7.7 Transient Ionizing Effects in Switches and Limiters

Chapter 5 pointed out that ionizing radiation produces additional electron hole-pairs in semiconductor crystals through pair generation. This effect will be most important when the density of carriers in the device is at its smallest value. For the devices discussed in this chapter the smallest carrier concentration occurs in the reverse-biased, wide-base *p-i-n* switching diode. The carrier concentration is essentially zero in this case, and there may be as much as 75 V of reverse bias across the diode. If no current limiting is provided, and an ionizing radiation dose rate of $\dot{\mathrm{R}}$ (rads per second) is present then the current that will flow is [from Eq. (5.35) and (5.36)]

$$I \cong \frac{100 \cdot A \cdot \rho \cdot \dot{\mathrm{R}} \cdot W \cdot e}{\mathscr{E}_p} \tag{7.34}$$

where L_n and L_p are $\ll W$, and $W =$ *i*-region width $\approx$ depletion width. If the $V \cdot I$ product is sufficiently large, diode burnout could occur. Therefore, the use of a current limiting resistor in the reverse bias supply of the *p-i-n* diode switch is recommended. In the other cases: (1) forward-biased *p-i-n* switch, and (2) *p-i-n* limiters, the ionizing radiation will add to the isolation because of the increased carrier concentration, but since only small (1 volt or less) voltages are present, heat dissipation is not a severe problem.

REFERENCES

[1] A. Rose, "Comparative Anatomy of Models for Double Injection of Electrons and Holes into Solids," *J. Appl. Phys.*, **35**, 2664 (1964).

[2] D. Leenov, "The Silicon PIN Diode as a Microwave Radar Protector at Megawatt Levels," *IEEE Trans. PGED*, **ED-11**, 53 (1964).

[3] A. Uhlir, "The Potential of Semiconductor Diodes in High Frequency Communications," *Proc. IRE*, **46**, 1099 (1958).

[4] R. V. Garver, "Theory of TEM Diode Switching," *IRE Trans. PGMTT*, **MTT-9**, 224 (1961).

[5] J. K. Hunton and A. G. Ryals, "Microwave Variable Attenuators and Modulators using *p-i-n* Diodes," *IRE Trans. MTT*, **MTT-10**, 262 (1962).

[6] H. B. Dwight, *Tables of Integrals and Other Mathematical Data*, Macmillan & Co., New York, 1957, p. 158.

[7] H. Watson, *Microwave Semiconductor Devices and Their Circuit Applications*, McGraw-Hill, New York, 1969, p. 305.

[8] J. Swartz and M. Thurston, "Analysis of the Effect of Fast Neutron Bombardment on the Current-Voltage Characteristic of a Conductivity Modulated *p-i-n* Diode," *J. Appl. Phys.*, **37**, 745 (1966)

[9] S. Choo, "Effect of Carrier Lifetime on the Forward Characteristics of High Power Devices," *IEEE Trans. PGED*, **ED-17**, 647 (1970).

[10] R. J. Chaffin and T. E. Gerst, *Semiautomated, Digital Microwave Impedance Analyzer*, Sandia Laboratories, SC-RR-70-608, September 1970.

[11] S. M. Sze, *Physics of Semiconductor Devices*, John Wiley & Sons, New York, 1969, p. 126.

8

Radiation Damage in IMPATT and TRAPATT Devices

8.1 Introduction

IMPATT and TRAPATT devices are fairly new in the microwave field. Devices of the IMPATT (IMpact, Avalanche, Transit-Time) type were originally proposed by Read [1] in 1958. However, it was not until 1965 that successful device operation was demonstrated by Johnston et al. [2]. The first TRAPATT (TRApped, Plasma, Avalanche, Triggered Transit) oscillator was demonstrated in 1967 by Prager et al. [3]. However, it was not until 1969 that the theory behind the operation was developed [4,5].

IMPATT devices are usually operated CW, and repeatable powers of a few hundred milliwatts are obtained in the 3 GHz to 50 GHz range. These devices can be used for low-power doppler radars, pumps for parametric amplifiers, or as local oscillators for radars.

TRAPATT devices are usually operated in the pulsed mode with peak power outputs of several hundred watts in the 0.5 GHz to 5 GHz range. Typical duty cycles range from 0.001 to 0.01. These devices have been used as rf sources for radar transmitters and IFF transponders.

Some of the military and space microwave systems that use these devices have radiation tolerance as a requirement. This chapter is devoted to an investigation of the behavior of these devices in a radiation environment. Section 8.2 qualitatively discusses the operation of IMPATT oscillators, pointing out those key semiconductor parameters which determine device performance. Section 8.3 gives a similar treatment for TRAPATT oscillators. Section 8.4 examines permanent radiation damage effects in IMPATT oscillators, and Section 8.5 does the same for TRAPATT oscillators. Section 8.6 discusses transient ionizing effects in IMPATT devices.

8.2 IMPATT Oscillator Theory

The basic concept of the IMPATT class of devices was first described in a farsighted paper by Read [1] in 1958. Since the first successful demonstration of an actual operating device [2] in 1965, a great deal of literature has been written on the theoretical and experimental aspects of these devices (see for example Ref. [6–12]). Therefore, a brief, quasi-theoretical description of device operation will be given here. The description to be given will be adequate, however, to point out those critical device parameters which may change in a radiation environment.

The IMPATT oscillator device is a two-terminal, *p-n* junction device that operates at reverse bias well into avalanche breakdown (see Figure 3.6). The device uses avalanche injection and transit-time effects to produce an ac component of terminal current approximately 180° out of phase with the ac terminal voltage, and the result is an ac negative resistance. The device structures that have been successfully used are the *p-n*, *p-i-n*, p^+-*n*-n^+ (called *p-n* IMPATTs); the p^+-*n*-*i*-n^+ and n^+-*p*-*i*-p^+ (called the Read structure); the p^+-*p*-*n*-n^+ (or double-drift structure [13]); and Schottky-barrier IMPATT devices. Devices have been made successfully from Ge, Si, and GaAs materials.

For illustration, the p^+-*n*-n^+ silicon IMPATT (one of the most commonly used types) will be discussed. Two basic concepts which should be recalled at this point are avalanche multiplication (see Section 3.5) and velocity saturation (see Section 2.6). Figure 8.1A shows the device being considered. It is assumed that the device has a dc bias voltage, V_A, applied to it so that it is biased just below the critical field for breakdown, E_A (see Figure 8.1B and Figure 3.8). The doping density of the *n*-region is chosen so that the *n*-region is fully depleted or "punched-through" at breakdown (see Figure 3.8). The doping concentration is also picked so that the smallest field, E_S, (see Figure 8.1B) that occurs in the *n*-region at breakdown is sufficiently large to cause velocity saturation to occur (see Figure 2.10). The transit time of electrons through the *n*-region is then defined as

$$T = \frac{L}{v_s} \tag{8.1}$$

where v_s = electron saturated drift velocity and L = *n*-region width (see Figure 8.1A). Assume that a very small ac voltage of frequency, ω_0 (where $\omega_0 T = \pi$), is added to the dc bias voltage, V_A (see Figure 8.1H). The resulting internal carrier distributions at several different instants of time during the rf

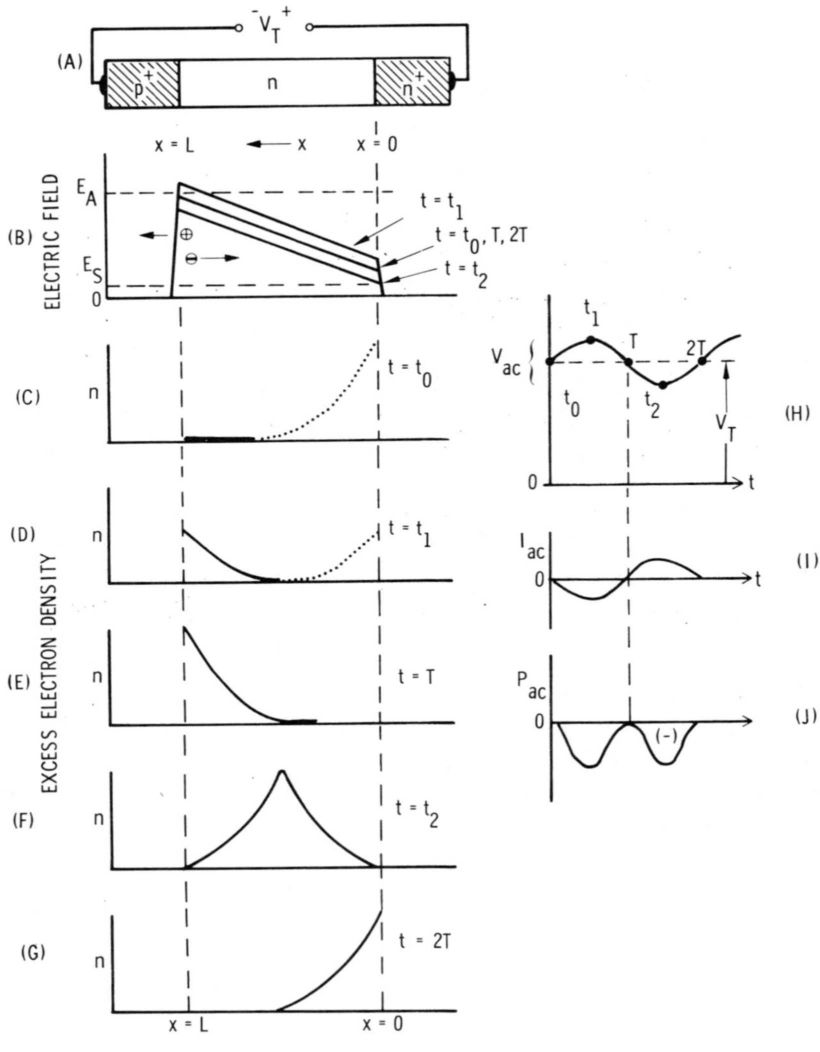

FIGURE 8.1
Explanation of IMPATT operation: (A) device geometry, (B) electric field profile, (C)–(G) carrier distribution at different times, (H) terminal voltage, (I) terminal current, (J) small-signal power.

cycle are shown in Figures 8.1C–G. The broken lines in Figures 8.1C–G represent carriers from the previous rf cycle. At $t = t_0^+$, the terminal voltage swings slightly above the avalanche value, and avalanche multiplication of the reverse saturation current begins in the high-field region near $x = L$ (see Figure 8.1A). For each impact ionization that occurs, one hole and one electron are created near $x = L$. The applied field (Figure 8.1B) causes the hole to drift into the p^+ region where it recombines almost immediately. The electron drifts to the right (Figure 8.1B) through the drift region at a velocity, v_s. All the time that the terminal voltage is above V_A, the carrier density at $x = 0$ continues to grow exponentially and the avalanche-generated electrons continually drift to the right. This is shown in Figure 8.1C, D, E. When $t = T$ (see Figure 8.1H), the voltage drops below the avalanche value and the carrier density at $x = L$ begins to decay towards zero while the generated electrons continue to drift to the right. This is shown in Figure 8.1F. Finally, when $t = 2T$ (see Figure 8.1H), half of the generated electrons have drifted out of the n-region (into the n^+ region where they recombine), and a new cycle starts over at Figure 8.1C. If recombination in the n-region can be neglected, it can be shown [1] that the terminal current due to the drifting electrons is given by:

$$I(t) = A \cdot \int_0^L \frac{\rho(x, t)\, dx}{T} \tag{8.2}$$

where A = device area, $\rho(x, t) = en(x, t)$, the excess charge density in n-region. The terminal current waveform (Figure 8.1I) can be qualitatively determined from Figures 8.1C–G and Eq. (8.2). The minimum value of $I(t)$ occurs at $t = t_1$, that is when the area under the curves (both solid and broken) in Figures 8.1C–G is a minimum. Correspondingly it can easily be seen that the integral of $\rho(x, t)$ is a maximum at $t = t_2$, and therefore $I(t)$ is also maximum at $t = t_2$. The small signal power flow can be determined by graphically multiplying Figure 8.1H by Figure 8.1I. The result is shown in Figure 8.1J. From Figure 8.1J it can be seen that the average ac power per cycle is negative! This implies that once the oscillations at a frequency, $\omega_0(\omega_0 T = \pi)$, get started, they can continue and the device can generate ac power.† The usual assumption made is that at a finite temperature the high frequency components of the thermal noise voltage start oscillations, and, once started, they are self-sustaining.

Analytical theories for the behavior of IMPATT oscillators are, at the very best, only a crude approximation, and at their worst they are grossly in-

† The ac power is actually converted from the dc bias power input with a typical efficiency of 5%.

adequate. Most of these theories begin with three basic equations: Poisson's equation [see Eq. (3.1)]

$$\frac{\partial E}{\partial x} = \frac{e}{\epsilon}(N_D - N_A + p - n)\,, \tag{8.3}$$

and the continuity equation [Eq. (2.68)] written separately for holes and electrons (neglecting recombination)

$$e\frac{\partial n}{\partial t} - \frac{\partial J_n}{\partial x} = \alpha_n J_n + \alpha_p J_p\,, \tag{8.4}$$

$$e\frac{\partial p}{\partial t} + \frac{\partial J_p}{\partial x} = \alpha_n J_n + a_p J_p\,, \tag{8.5}$$

where α_n, α_p are discussed† in Section 3.5, and $J_n = -env_{sn}$ and $J_p = +epv_{sp}$. Assume that $v_{sn} = v_{sp}$ (equal electron and hole saturated drift velocities) and also that $\alpha_n = \alpha_p$ (equal electron and hole ionization coefficients). If we then do a small-signal ac expansion [similar to the one discussed below Eq. (3.18)] using the above equations and neglecting diffusion effects, we can show [6] that for an n^+-p-i-p^+ structure (Read diode) the small signal terminal admittance is given by

$$Y_{ac} = j\omega C\left[\frac{\omega T(1 - \omega^2/\omega_a^2)}{\omega T(1 - \omega^2/\omega_a^2) + j(1 - e^{-j\omega T})}\right] \tag{8.6}$$

where $C = \epsilon A/L$ and ω_a is called the avalanche resonance frequency.

$$\omega_a^2 = \frac{2(m + 1)J_{dc}}{\epsilon E_A T_1} \tag{8.7}$$

where J_{dc} = dc bias current density, and T_1 = transit time through the avalanche region; T_1 is normally $0.1T$. Figure 8.2 shows a plot of Eq. (8.6) for the normalized values $C = 1$, $\omega_a = 1$, $T = 1$. From Figure 8.2 it can be seen that in this case the negative conductance peaks near $\omega = 2$ and falls off, and secondary peak is seen at very high frequencies. This secondary peak in negative conductance corresponds to spending more than one rf cycle in the drift region and is not usable in the practical case due to diffusion effects.

To build an oscillator, all that is needed is to build a circuit that provides a conjugate match for the admittance predicted by Eq. (8.6). This, however, usually does not yield the optimum performance. Equation (8.6) is based on a small-signal derivation and is not valid for the large-signal, real-world case. In the large-signal case, the differential equations become nonlinear, space-charge effects need to be included, and nonsaturated drift velocities can

† The definition of α used here is different from Eq. (3.48). Here it is assumed $\alpha = \alpha_0 E^m$ (where $m \cong 6$ for Si).

occur. The only adequate method for modeling the device in this case is a device physics computer model [13]. These computer models are quite complex and require fairly large amounts of computer time, but they are very accurate [13].

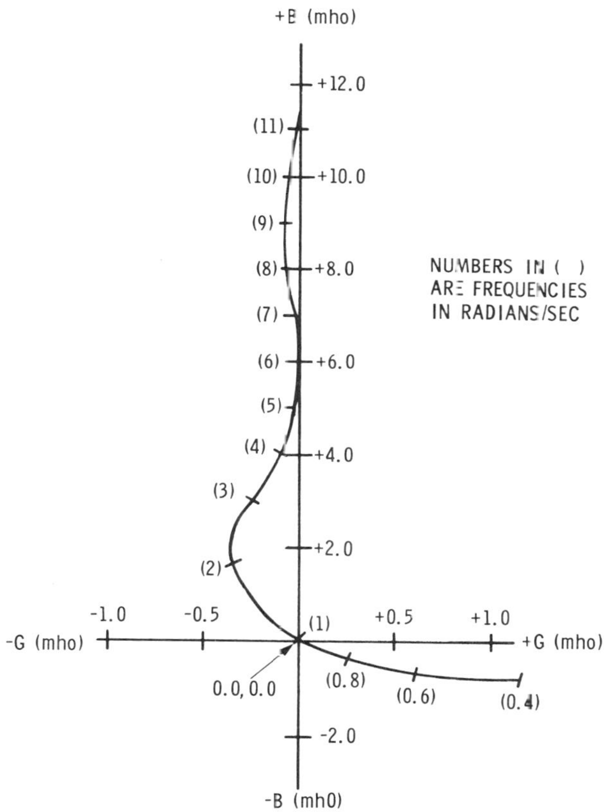

FIGURE 8.2
Plot of Read diode admittance, $Y = G + jB$ for $C = 1$, $\omega_a = 1$, $T = 1$ in Eq. (8.6).

The key parameters which determine device operation are seen to be the electric field distribution, the ionization coefficients, the reverse leakage current, the saturated drift velocity, and the carrier lifetime (it must be greater than T). One further important point is the positive temperature coefficient of avalanche breakdown [27] (i.e., the breakdown voltage increases with increasing temperature). This prevents hot spotting or filament formation from occurring in avalanche diode oscillators and tends to distribute the heat evenly across the cross-sectional area of the device.

8.3 TRAPATT Oscillator Theory

After the discovery [3] of high-power, high-efficiency TRAPATT oscillations, a full two years elapsed before an adequate theory for the mode was developed [4,5]. This is because there is no small-signal theory for the operation of the TRAPATT oscillator. TRAPATT operation only occurs in the large-signal case, and hence a small-signal expansion of the device's differential equations is simply not adequate. The TRAPATT oscillator device is *similar* to an IMPATT device in that it is usually a p^+-n-n^+ device and operates in reverse breakdown. The most distinctive *differences* between the IMPATT and TRAPATT are that the latter has a much higher efficiency (up to 75% has been reported [14]) and the latter also operates at a much lower frequency for a given device (n-region) width. The optimum frequency for TRAPATT operation is defined [5] as $\omega_0 T = 0.14\pi$ [T defined by Eq. (8.1)] as compared to $\omega_0 T = \pi$ for an IMPATT oscillator. The high efficiency and low frequency lead to the device being named the "Anomalous Mode" by its discoverers [3]. However, after the device theory was worked out, the acronym TRAPATT was applied to this device. The analytic theory [4,5] for the operation of a TRAPATT was preceded by a device physics computer simulation [15]. This simulation showed that during a portion of the rf cycle, large densities of holes and electrons were present, and the electric field was suppressed to fairly small values. This meant that the electron-hole plasma was "trapped," i.e., it takes much longer to remove the carriers when the field is small and the carrier velocity is well below the saturated value. The mechanism for the formation of this "trapped" plasma will be discussed with the aid of Figure 8.3.

Figure 8.3A shows a typical n^+-n-p^+ abrupt junction TRAPATT device. Figure 8.3B ($t = 0$ curve) shows the electric field in the device assuming that it is biased to just punch-through but that the peak field is well below the critical field for avalanche breakdown, E_A (see also Figure 3.8). The shape of the electric field may be found from Poisson's equation [Eq. (3.1)]

$$\frac{dE}{dx} = \frac{\rho(x)}{\epsilon} = \frac{eN_D}{e}, \tag{8.8}$$

i.e., the slope of the electric field is due to the ionized donor density N_D in the punched-through device (recall that, at least for the moment, it is reverse

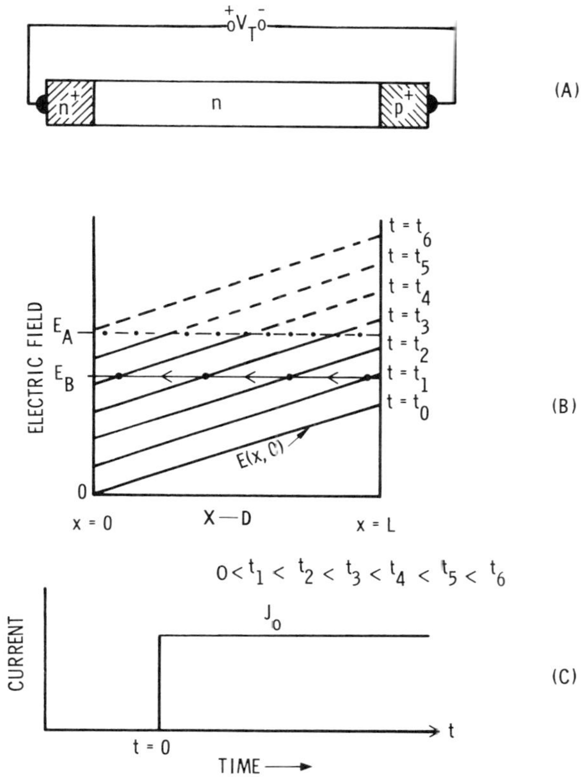

FIGURE 8.3
Explanation of shock front propagation in a TRAPATT oscillator device.

biased and below breakdown and so essentially no mobile carriers are present). If we assume $E(x = 0) = 0$, then integrating Eq. (8.8) we obtain:

$$E(x) = \frac{eN_D}{\epsilon} x . \tag{8.9}$$

Let us assume that at $t = 0$ a current generator of value J_0 is suddenly applied to the device terminals (see Figure 8.3C). As long as $E < E_A$, there are no mobile carriers present, and the only current that can flow is a displacement current, given by

$$J_0 = \frac{\partial D}{\partial t} = \epsilon \frac{\partial E}{\partial t} . \tag{8.10}$$

Integrating Eq. (8.10), we find that

$$E(x, t) = E(x, 0) + \frac{J_0}{\epsilon}\cdot t\,, \tag{8.11}$$

i.e., the electric field everywhere increases with time. Inserting Eq. (8.9) for $E(x, 0)$ into Eq. (8.11), we obtain

$$E(x, t) = \frac{eN_D}{\epsilon}\cdot x + \frac{J_0}{\epsilon}\cdot t\,. \tag{8.12}$$

Let us examine the behavior of an arbitrary point of constant field $E = E_B$. From Figure 8.3B and Eq. (8.12) we see that the point of constant field, E_B, moves to the left at a velocity

$$v_x = -\frac{J_0}{eN_D}\,. \tag{8.13}$$

If J_0 is large enough and N_D is small enough, the velocity v_x can exceed the saturated drift velocity! (This is like the classic problem of the intersection of a wave with the beach.) The significance of the field velocity can, however, be demonstrated with the aid of Figure 8.4. For values of $E < E_A$ (avalanche) Figure 8.4 is really a blowup of Figure 8.3B. However, when E is greater than E_A, the field will not continue to increase with time because of the sudden increase in local carrier density due to avalanche multiplication (note that if E gets appreciably above E_A, the avalanche growth is very rapid). On Figure 8.4 it has been assumed that v_x [see Eq. (8.13)] is greater than the saturated drift velocity of the holes and electrons. This implies that the avalanche generated carriers do not have time to redistribute themselves, and this in turn momentarily allows the field to get considerably above the avalanche value. The tremendous avalanche-induced carrier densities suppress the field to a very low value behind the crest of the electric field wave as this wave "shoots through" the device from right to left. This wave front which propagates through the device is called the *avalanche shock front* [4]. Typical propagation times (L/v_x) are on the order of 100 picoseconds for a TRAPATT device. The terminal voltage across the device is the integral of the electric field shown in Figure 8.4. Thus, at $t = t_3$ the terminal voltage is fairly high and is slightly above the avalanche breakdown voltage (which is typically a few hundred volts). Shortly after $t = t_6$, the terminal voltage is quite small (several volts). Therefore, the terminal voltage drops from a few hundred volts to just a few volts in roughly 100 picoseconds. This is shown on Figure 8.5A as the line *A-B*. Disregard the applied current step (Figure 8.3C) for the moment and assume that the TRAPATT diode is connected in a circuit of the type [16] shown in Figure 8.6. The transmission line is charged to a voltage $V \cong V_B$ at point A in Figure 8.5A. Suddenly (100 ps later), the diode appears as short circuit across the line (point B in Figure 8.5A), and the resulting current flow is

$$I_p \cong \frac{V_B}{Z_0} \tag{8.14}$$

where Z_0 = transmission line impedance. Before reaching point A in Figure 8.5A the diode current is very small (equal to the reverse saturation current of

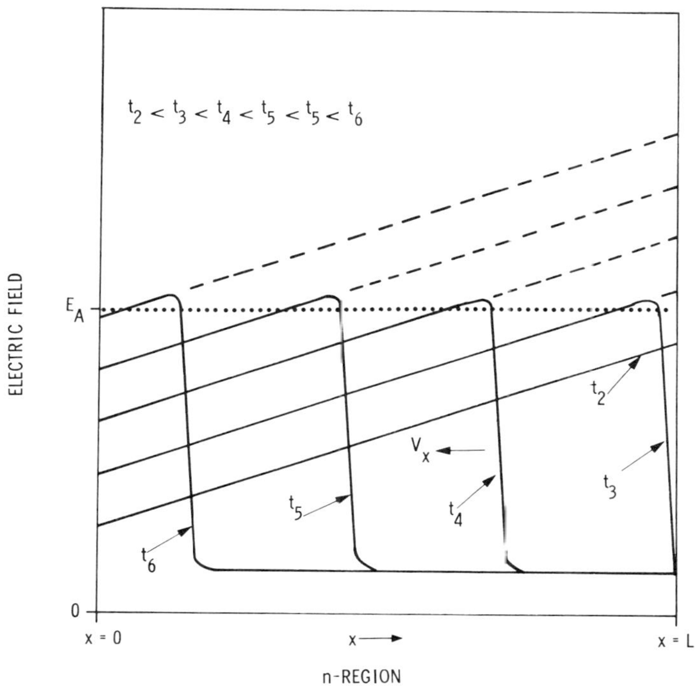

FIGURE 8.4
Propagation of an avalanche shock front through a TRAPATT device.

the diode). Therefore, the terminal current changes very rapidly from a small value to a large value. Figure 8.5B shows this current waveform. Eventually, all of the "trapped" charges are extracted by the current I_p, and the current falls back to a small value. Then the diode and line voltage charge back up to $V \cong V_B$ using charge supplied through bias choke (see Fig. 8.6). The sudden voltage drop (line A-B, Figure 8.5A) can be thought of as a large negative voltage pulse being launched at $x = 0$ on the transmission line in Figure 8.6.

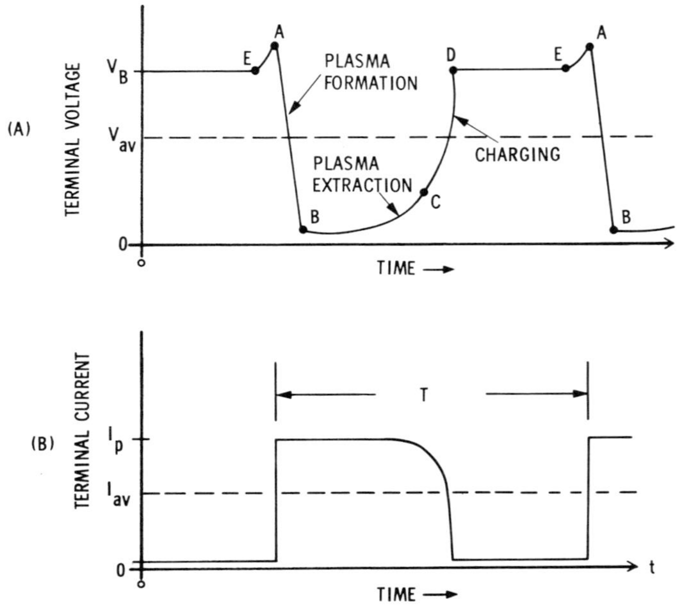

FIGURE 8.5
TRAPATT waveforms for a diode connected into the circuit shown in Figure 8.6.

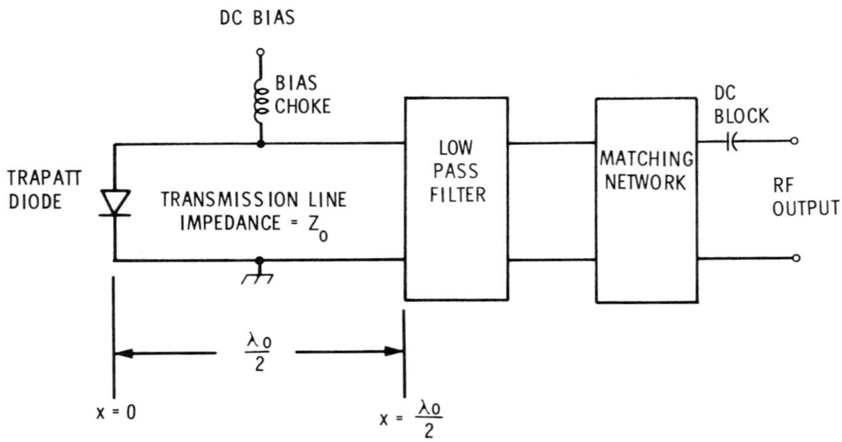

FIGURE 8.6
TRAPATT oscillator circuit

Assume that at a distance $\lambda_0/2$† from the diode there is a low pass filter which will pass only the fundamental TRAPATT frequency, f_0, and reflect all higher harmonics with a reflection coefficient of -1 (a high frequency short). Then the leading edge of the negative pulse just described will be inverted and reflected back towards the diode by the low pass filter. In the TRAPATT oscillator this positive pulse arrives back at the diode after the plasma has been removed and the diode and line have charged back up to $V \cong V_B$ (point E in Figure 8.5A). At this time there are no mobile carriers present in the device and the returning pulse induces a rising voltage (line E-A in Figure 8.5A) which produces a displacement current $J_0 = C \cdot (dV/dt)$ in the device. This current can then initiate another shock front and cause the cycle to repeat again. For this reason, once the oscillations get started, they can continue indefinitely, converting the dc bias power into rf power. Since these oscillations do not exist in the small signal case, thermal noise will not start them. They can be started directly [17] if a fast-rise bias voltage pulse is applied (so that the $J_0 = C \cdot (dV/dt)$ current initiates the shock front). This, however, is impractical since the pulse generator and coupling network are difficult to fabricate for the typical 100 V/nanosecond risetime rates required [17]. Instead, nearly all TRAPATT oscillators in existence today are initiated by IMPATT oscillations in a high Q IMPATT circuit. In other words, when the bias is first turned on (or at the beginning of each bias pulse in the pulsed case), IMPATT oscillations grow exponentially (assuming that a high Q circuit at the IMPATT frequency has been provided) until the peak-to-peak swing becomes large enough so that the $C \cdot (dV/dt)$ current becomes large enough to trigger an avalanche shock front, and then TRAPATT oscillations begin. This has been observed experimentally; see Figure 8.7 for an example.

From Figure 8.7 and the preceding discussion it can be seen that for most applications the IMPATT mode is a critical part of the TRAPATT starting mechanism. Since the TRAPATT frequency is typically one-third to one-tenth that of the IMPATT frequency, several people have referred to the TRAPATT mode as a subharmonic IMPATT mode; however from Figure 8.7 it can be seen that once TRAPATT mode gets started, the IMPATT oscillations lose their identity, and this terminology is not correct.

For the waveforms shown in Figure 8.5, Clorfeine et al. [5] have shown that the negative resistance at the fundamental TRAPATT frequency is given by

$$R \cong -0.4 \frac{V_B}{I_{av}} \cdot \tag{8.15}$$

The matching network in Figure 8.6 must conjugate match the fundamental TRAPATT impedance. The real part should match Eq. (8.15), while the

† Where $\lambda_0 = v_p/f_0$, where v_p = propagation velocity on the transmission line and f_0 = TRAPATT frequency.

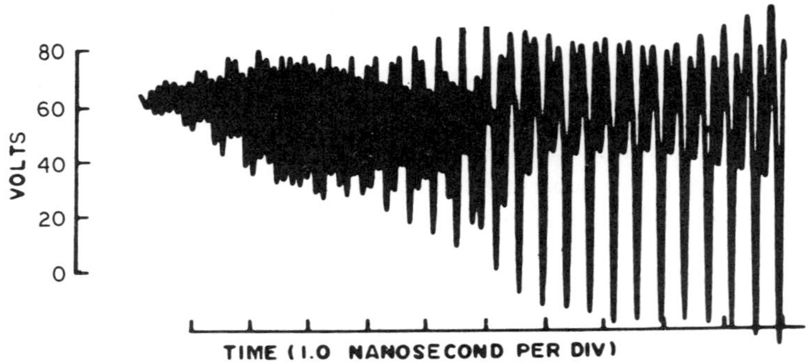

FIGURE 8.7
Diode voltage versus time for pulsed 2.6 GHz TRAPATT oscillator (after Johnston et al. [15]).

imaginary part will determine the operating frequency. Since in the TRAPATT mode the device alternates between a high current–low voltage state and a low current–high voltage state (see Figure 8.5), the efficiency is very high, as noted earlier. It will be found later that the radiation-sensitive parameters in TRAPATT operation are those associated with the IMPATT mode used to initiate the TRAPATT mode.

8.4 Permanent Radiation Effects in IMPATT Oscillators

Several authors [18–26] have studied the effects of radiation on IMPATT oscillators. The critical parameters for IMPATT operation are the electric field distribution, the ionization coefficients, the reverse saturation current, the saturated drift velocity, and the carrier lifetime. The behavior of many of these properties with radiation has been discussed in Chapter 5. However, no data is available for radiation effects on the ionization coefficients or the saturated drift velocities. They will be assumed not to change for fast-neutron fluences to 10^{16} n/cm^2 ($\mathscr{E} > 10$ keV). Wilson [18] points out that both the saturated drift velocity and the avalanche multiplication phenomenon are collision dominated, and the mean free path between collisions at room temperature is still short compared to the mean distance between defects even after 10^{16} n/cm^2. The discussion of permanent radiation damage to IMPATT oscillators will be broken into two parts: first, a discussion of the dc and low-frequency changes in the diode itself and, second, an examination of the changes in performance of microwave IMPATT oscillators subjected to a radiation environment. The discussion here will be limited to neutron-irradiation effects in silicon devices. The extension to other types of radiation, or other semiconductor materials can be made from the data presented in Chapter 5.

Dc and Low Frequency Radiation Effects. Wilson et al. [18] have made a thorough study of X-band, silicon, p^+-n-n^+ IMPATT devices which had an n-region doping density of 10^{16} per cm^3 and an n-region width of 4 μm. The devices were exposed to integrated neutron fluences (of both Spectrums A and B)† to 10^{16} n/cm² ($\mathscr{E} > 10$ keV). In this study the carrier lifetime was found to follow the form of Eq. (5.23) (see also Figure 7.9) with a value of $K_\tau = 7 \times 10^{-7}$ cm² n^{-1} sec^{-1} (for Spectrum A† irradiations). This means that after a fluence of 10^{16} n/cm² ($\mathscr{E} > 10$ keV, Spectrum A) the lifetime would be approximately 10^{-10} sec. The electron transit time for a 4 μm n-region at saturated drift velocity (see Figure 2.10) may be found from Eq. (8.1) as 0.4×10^{-10} sec. This means an appreciable fraction of the transiting electrons may recombine at 10^{16} n/cm². The effects of this on IMPATT oscillator performance will be discussed later.

EerNisse and Chaffin [23] have reported the results of a theoretical and experimental study on silicon p^+-n-n^+ devices. The doping profile for the devices used is shown in Figure 8.8. In this study, a device physics computer model (like the one described in Ref. [13]) is used to accurately model the device behavior both before and after neutron irradiation. The radiation effects were brought in through the use of the two-level divacancy defect (see Section 5.3). The punch-through capacitance (measured at 1 MHz) is found to decrease slightly with neutron irradiation (for all fluences in this study, $\mathscr{E} > 10$ keV, Spectrum B). This is shown in Figure 8.9. The reduction in depletion capacitance with neutron fluence is due to a slight widening of the depletion region due to carrier removal at the edges of the p^+-n and n-n^+ regions. The calculated electric field distribution at 1000 A/cm² bias and for several neutron irradiation produced trap densities is shown in Figure 8.10 where it can be seen that a large trap density increases the depletion width (discussed earlier) and it also distorts the electric field. The reason for this is that large numbers of both electrons and holes are created by avalanche generation at the peak in electric field (Figure 8.10). The direction of the electric field is such that the holes drift to the left and the electrons drift to the right (in Figure 8.10). The divacancy trap has the property that it traps the most populous carrier species, and thus to the left of the field maximum, holes will be trapped (+ charge) and, to the right of the field maximum, electrons will be trapped (− charge). These trapped charges will modify the net space charge (due to ionized donor and acceptor atoms as well as drifting charges) present before irradiation. This is shown in Figure 8.11. The changes

† See Appendix C.

in net space charge shown in Figure 8.11 lead directly to the distortions in the electric field shown in Figure 8.10. The effects of this electric field distortion on oscillator behavior will be discussed later.

Chaffin [24] has experimentally measured the effects of neutron irradiation (Spectrum A) on a p^+-n-n^+ silicon IMPATT device which had an n-region doping density of 5×10^{15} per cm^3 and an n-region width of 4.5 μm. Figure

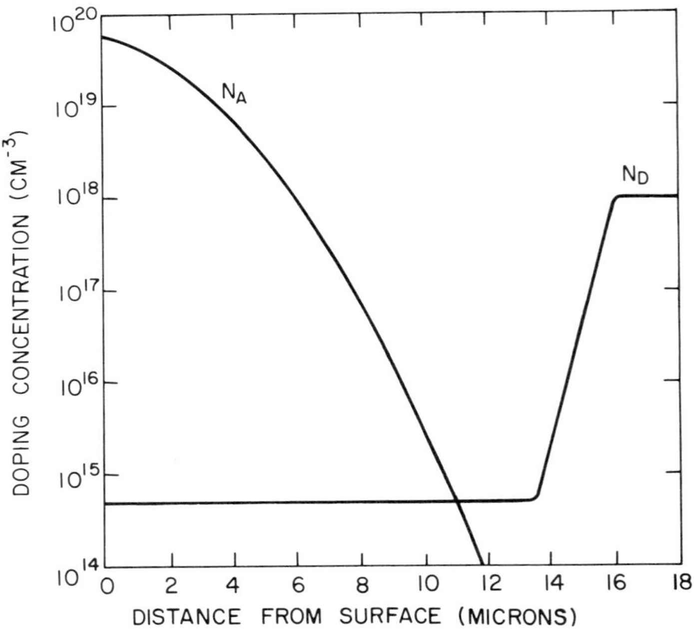

FIGURE 8.8
Doping profiles for IMPATT and TRAPATT diode study (after EerNisse and Chaffin [23]).

8.12 shows that the forward voltage for a given current decreases with neutron fluence. The reasons for this have been discussed in Section 7.4. Equation (3.12) may be rewritten as

$$I = I_0 \cdot \left[\exp\left(\frac{eV_A}{kT}\right) - 1 \right] \tag{8.16}$$

where I_0 = reverse saturation current and V_A = voltage applied to the device terminals. A comparison of Figure 8.12 and Eq. (8.16) shows that I_0, the reverse saturation current, must increase with radiation. The increase in I_0

from prerad to 5×10^{15} n/cm^2 corresponds to an increase in I_0 by a factor of 100 (assuming $e/kT = 38$ V^{-1}). Therefore, the neutron irradiation can be seen to increase the reverse saturation current. (Note that the indirect approach used above to calculate I_0 is necessary since reverse-bias measurements are dominated by surface leakage effects.) The effect of this increase in reverse saturation current on oscillator performance will be discussed later. This study [24] also showed that the reverse breakdown voltage of the devices increased with neutron fluence. This is shown in Figure 8.13. The main reason

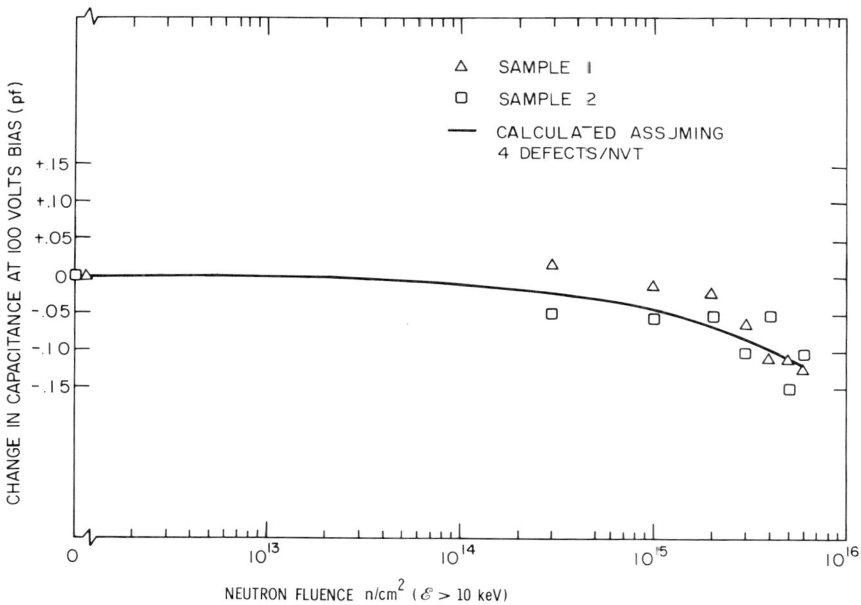

FIGURE 8.9
Measured and calculated punch-through depletion capacitance versus neutron fluence for the diode in Figure 8.8 (after EerNisse and Chaffin [23]).

for the increase in the breakdown voltage is that carrier removal at the edges of the space-charge region causes a widening of the depletion region for the same applied voltage. The result is a lower peak field for the same terminal voltage which implies the breakdown voltage will increase. Another factor affecting the breakdown voltage is the distortion in the electric field caused by trapped charges in the depletion region. Figure 8.10 shows both effects: a widening of the space-charge region and a distortion of the electric field profile.

Both Wilson [18] and EerNisse [28] state that for neutron fluences above 4×10^{15} n/cm^2 the temperature coefficient of the avalanche breakdown voltage ($V_B^{-1} \cdot \partial V_B/\partial T$) becomes *negative* and this leads to unstable operation and possible dc burnout. Recall that earlier in this chapter we discussed the importance of the normally *positive* temperature coefficient of avalanche breakdown in preventing hot spotting and thermal runaway. The negative temperature coefficient after 4×10^{15} n/cm^2 is attributed to an increased

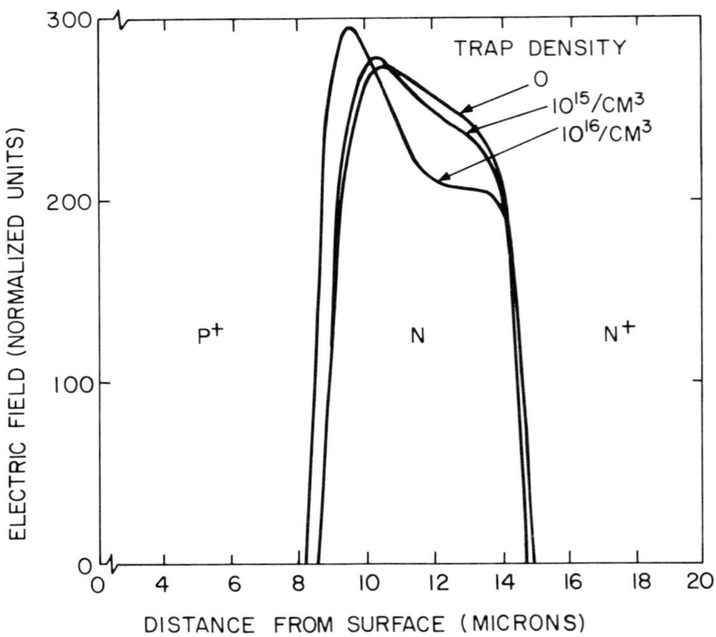

FIGURE 8.10
Calculated electric field versus distance for the doping profile shown in Figure 8.8 biased into avalanche breakdown (1000 A/cm²) for several different neutron-induced trap densities (after EerNisse and Chaffin [23]).

trap release rate for trapped charges with increasing temperature [28]; i.e., Figure 8.13 shows that the avalanche voltage increases with irradiation due to trapping (carrier removal) effects. At elevated temperatures the effect is reduced, and thus the avalanche voltage decreases. If this happens on a localized basis, a hot spot and device burnout can occur. This effect places a survivability upper limit on avalanche devices exposed to neutron fluence of about 10^{16} n/cm^2, independent of other mechanisms that may be present.

Effects of Radiation on Microwave Performance. Early research on radiation effects in IMPATT oscillators [18,19] showed that the microwave output power for a constant current input continued to increase with increasing neutron fluence up to the point of dc failure† near 10^{16} n/cm^2. The power

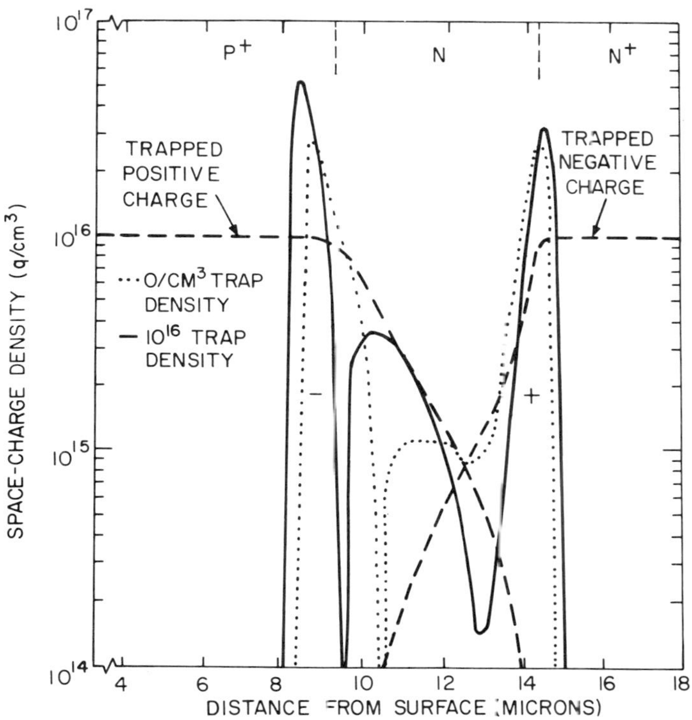

FIGURE 8.11
Calculated net space-charge density (in electronic charges/cm^3) versus distance for the junction of Figure 8.8 biased into avalanche breakdown to 1000 A/cm^2 for a trap density of 10^{16}/cm^3. The trapped charge in the two-level trap is also shown to demonstrate how it contributes to the net space charge.

output increases for two reasons. The first is because as the breakdown voltage increases with irradiation (see Figure 8.13), the power input at a constant current increases, and if the efficiency remains constant, the power output must also increase. Secondly, it has been observed [18,19] that the

† This failure is due to negative temperature coefficient of the avalanche breakdown.

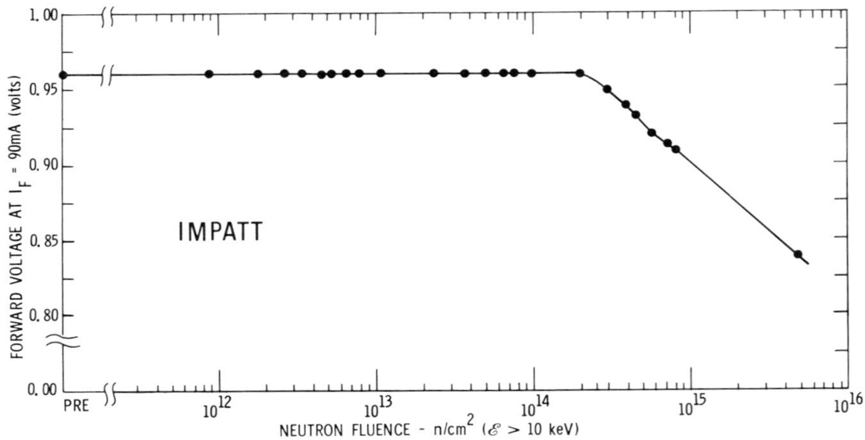

FIGURE 8.12
Forward voltage drop versus neutron fluence for a p^+-n-n^+ silicon IMPATT.

efficiency actually increases with radiation so the output power would increase even if the breakdown voltage remained constant. Figure 8.14 shows the measured oscillator efficiency as a function of neutron fluence. This data shows that after a slight initial decrease, the oscillator efficiency increases with neutron fluence. The explanation of this effect is as follows: Early

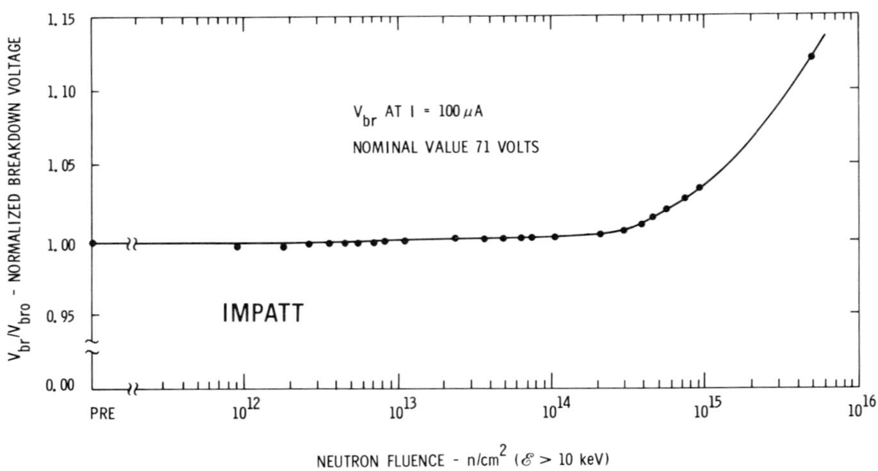

FIGURE 8.13
Normalized reverse breakdown voltage versus neutron fluence for a p^+-n-n^+ silicon IMPATT device.

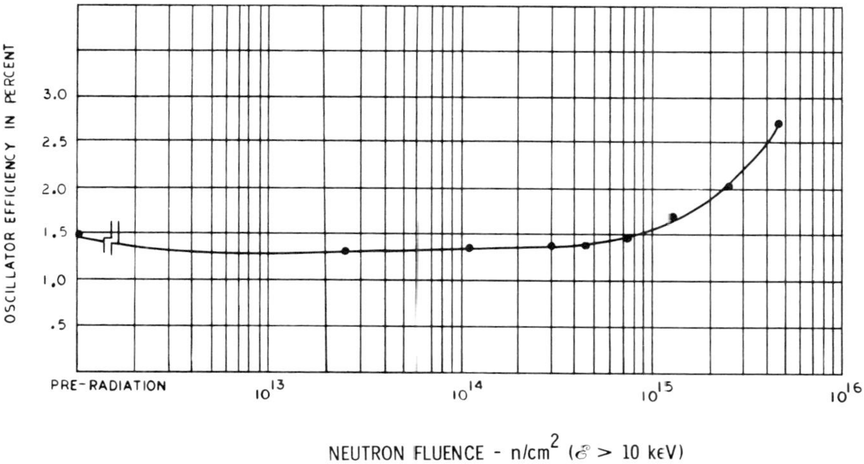

FIGURE 8.14
IMPATT oscillator efficiency versus neutron fluence for an early IMPATT device (Spectrum E).

IMPATT devices were designed so that they were not quite "punched-through" at breakdown, and consequently a thin layer of high resistivity silicon (typically 1 Ω-cm) appeared in series with the device as a parasitic loss element. This is shown in Figure 8.15. It has been shown that this layer degrades the oscillator efficiency [29]. As the device is irradiated, two opposing effects occur. First, the resistivity of the lightly doped, undepleted region increases with neutron fluence due to carrier removal effects. This increases

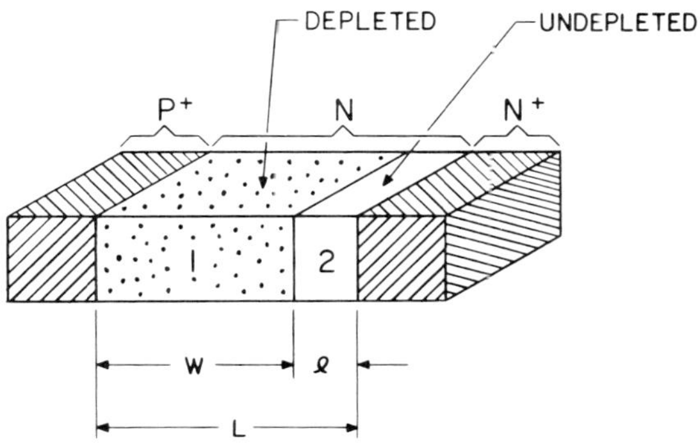

FIGURE 8.15
IMPATT device which is not punched-through at breakdown.

the parasitic loss term. Secondly, the depletion region widens due to the same carrier removal effects, and this reduces l (see Figure 8.15). Therefore, this is a parasitic resistor whose resistivity increases with irradiation but whose length decreases with irradiation. Consequently, the parasitic resistance $R = \rho l/A$ can either increase or decrease depending on whether changes in the ρ term or the l term dominate in a radiation environment. This in turn is determined by the specific values of ρ and l for the device being used [19]. For the device represented by Figure 8.15, the l-term dominates at high fluence levels and thus the efficiency increases with irradiation. The signifi-

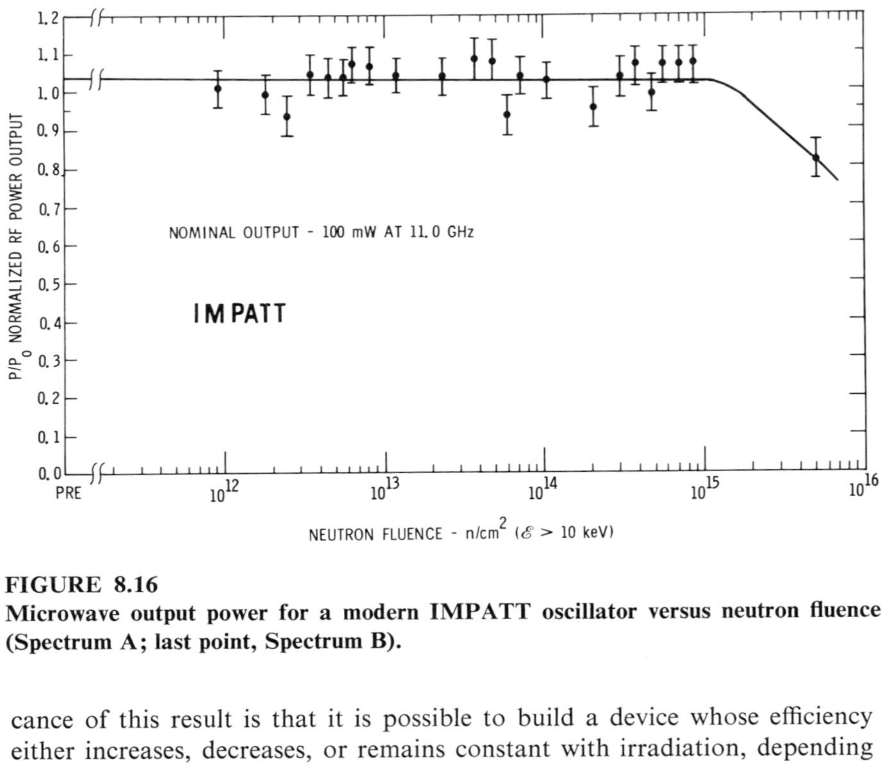

FIGURE 8.16
Microwave output power for a modern IMPATT oscillator versus neutron fluence (Spectrum A; last point, Spectrum B).

cance of this result is that it is possible to build a device whose efficiency either increases, decreases, or remains constant with irradiation, depending on how the device is initially fabricated. (Note that since the series parasitic resistance reduces the efficiency, this effect can only be achieved at the expense of building a less than optimum device initially.)

Modern IMPATT devices are designed so that they have no undepleted epitaxial material in series with the active layer. This results in higher initial efficiencies (3% to 7%). These devices show a degradation in output power at high neutron fluences. Figure 8.16 shows the microwave output power of a device described earlier (see Chaffin [24] discussed earlier in this section) versus neutron fluence. From Figure 8.16 it can be seen that the microwave output power for an IMPATT device of this type does indeed degrade with neutron fluence. The primary reasons for this degradation are twofold and were mentioned at the beginning of this section. First, recombination effects must be considered. The carrier lifetime degradation data [18] was used to calculate device small-signal negative conductance [23]. The results of this calculation are shown in Figure 8.17. where it can be seen that the calculated

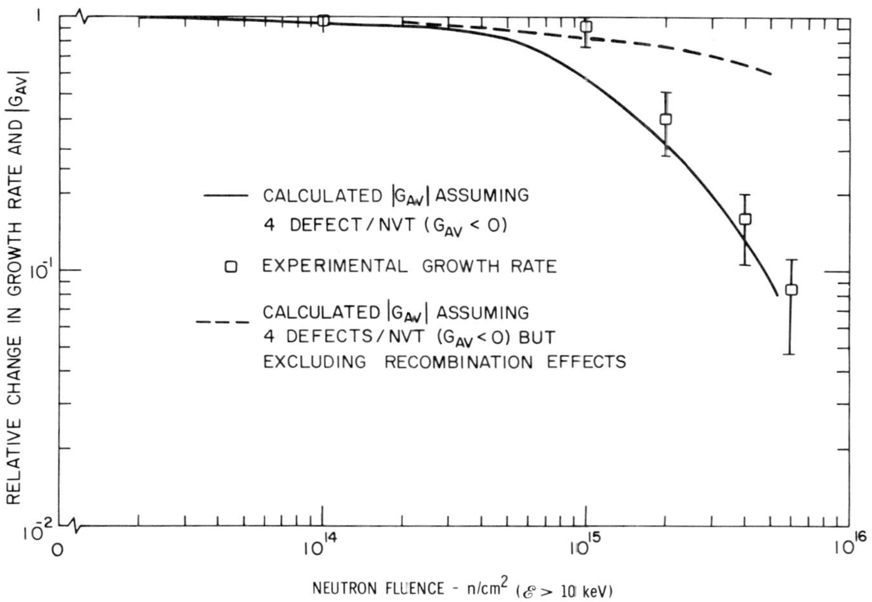

FIGURE 8.17
Normalized small-signal negative conductance (proportional to growth rate) measured and calculated for the device shown in Figure 8.8 (Spectrum B). (After EerNisse and Chaffin [23]).

small-signal negative conductance including (solid line) and neglecting (dashed line) recombination effects leads to quite different results, and the results which include recombination effects fit the experimental data [23] much better. The reduction of the negative conductance due to recombination effects is one mechanism which leads to a falloff in power output with neutron fluence, such as is shown in Figure 8.16. The second mechanism which degrades the output power is the increase in reverse saturation current discussed earlier in conjunction with Figure 8.12 and Eq. (8.16). Misawa [30] has shown that an increase in reverse saturation current reduces the phase delay associated with the avalanche process (see Figure 8.1) which severely reduces the microwave power output and efficiency.

Another important microwave characteristic of IMPATT oscillators is the oscillation frequency. The oscillation frequency is determined by resonance of the cavity-supplied† inductive susceptance with the diode-supplied capacitive susceptance [see Eq. (8.6)]. From Figure 8.9 it can be seen that the depletion capacitance at punch-through decreases with neutron fluence (due to a widening of the depletion region), and this effect tends to decrease the diodes' capacitive susceptance [see Eq. (8.6)]. The distortion in the electric field profile (Figure 8.10) with neutron irradiation leads to a narrowing of the

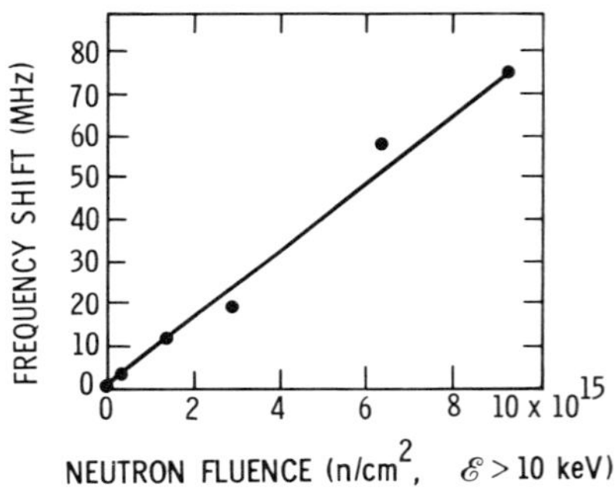

FIGURE 8.18
Frequency shift versus neutron fluence for a typical 9.5 GHz IMPATT oscillator ($\mathscr{E}$ > 10 keV, Spectrum B). (After Anderson [25].)

avalanche region width (and avalanche region transit time, T_1). This in turn leads to an increase in the avalanche resonance frequency ω_a [see Eq. (8.7)] which in turn decreases the diodes' capacitive susceptance [see Eq. (8.6)]. The reduction in capacitive susceptance due to the reduction in depletion capacitance and increase in ω_a, leads to an increase in operating frequency. The operating frequency is the point where $\omega C' = 1/\omega L$ ($\omega C'$ = diode capacitive susceptance and $1/\omega L$ = inductive susceptance fixed by the cavity). The resulting frequency shift is usually less than 1% even after 10^{16} n/cm^2 ($\mathscr{E}$ > 10 keV). Typical experimental data [25] showing this effect are shown in Figure 8.18 where the frequency shift versus neutron fluence for a 9565 MHz, silicon IMPATT oscillator is plotted.

† The diode is assumed to be mounted in a microwave cavity for all oscillator measurements.

Wilson [18] found that the AM and FM noise levels produced by an IMPATT oscillator changed by less than 3 dB for neutron fluences up to 10^{16} n/cm^2 ($\mathscr{E} > 10$ keV, Spectrum B).

8.5 Permanent Radiation Effects in TRAPATT Oscillators

Only a small number of radiation studies of TRAPATT oscillators have been conducted [23,31,32]. These devices appear to be very tolerant to radiation fluence, and the most radiation sensitive parameters are those associated with the IMPATT mode used to initiate the TRAPATT mode (see Section 8.3). Once the TRAPATT mode gets started, very little degradation is seen even at 6×10^{15} n/cm^2 ($\mathscr{E} > 10$ keV).

The dc and low-frequency diode parameters have been measured versus radiation [23]. The diode represented by Figures 8.8 and 8.9 is a TRAPATT diode. The previous section illustrated that this diode (Figure 8.8) shows a slight decrease in punch-through capacitance, a slight increase in breakdown voltage, and a negative temperature coefficient for the avalanche breakdown after a sufficiently high neutron fluence.

The effects of neutron irradiation on the TRAPATT microwave oscillator are summarized in Figures 8.19 to 8.22. Figure 8.19 shows that the TRAPATT oscillator frequency increases slightly with neutron fluence. This is attributed to the decrease in punch-through depletion capacitance that occurs with irradiation (see Figure 8.9). As was mentioned previously, the major change in TRAPATT performance with neutron irradiation is due to a degradation of the IMPATT mode which is used to start the TRAPATT oscillations (see Figure 8.7). From the previous section (Section 8.4) we recall that the small-signal negative conductance, and hence the negative Q ($Q = \omega C'/G$) and the growth rate ($\omega/Q = G/C'$), of the oscillations are changed by irradiation. Figure 8.17 shows that the growth rate of the IMPATT oscillations decreases with fluence, and this implies that the starting delay† of TRAPATT oscillations should increase with fluence. This behavior has been observed experimentally [31] and is shown in Figure 8.20 which shows a pulsed TRAPATT oscillator before and after 6×10^{15} n/cm^2. From Figure 8.20 it can be seen that the starting delay before irradiation is approximately 50 ns, and after 6×10^{15} n/cm^2 it is approximately 200 ns. The peak power output, once the oscillations get started, can be seen not to change appreciably even after 6×10^{15} n/cm^2. The reason for this is the alternating on-off nature of the TRAPATT mode (see Figure 8.5). During the high voltage–low current portion of the rf cycle, the device is completely depleted of free carriers, and the effects of radiation-induced traps are small. During the high current–low

† The starting delay is the time required for the initiation of TRAPATT oscillations after bias is applied (see Figure 8.7).

voltage part of the rf cycle, the carrier density is very large (typically 10 times the *n*-region doping density [5]). These large densities allow significant recombination to occur, and yet the carrier density remains large enough to keep the field low during the extraction part of the rf cycle (see Figure 8.5). Although the peak power output per pulse does not change with irradiation, the average power output per pulse does change. This is because the device is generating microwave power for a smaller fraction of the pulse width, due to

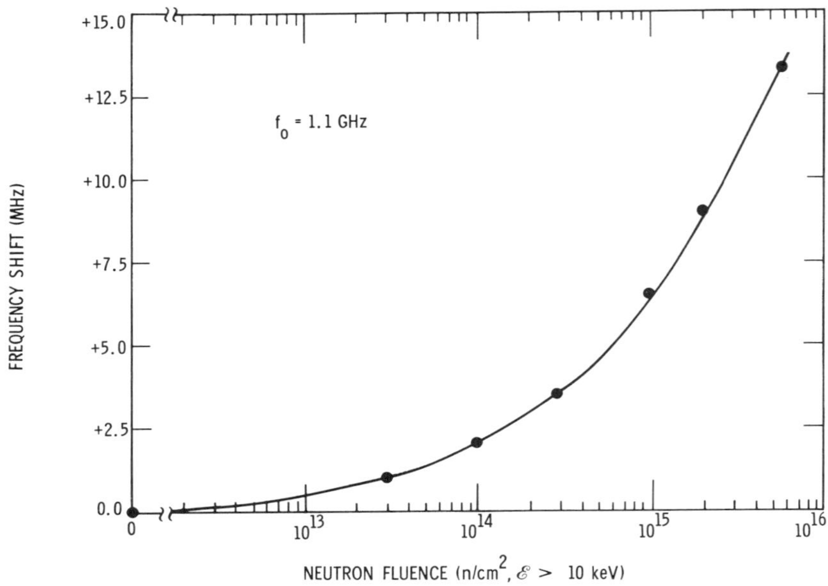

FIGURE 8.19
Frequency shift versus neutron fluence for a typical TRAPATT oscillator (Spectrum B).

the increased starting delay (see Figure 8.20). This effect is shown in Figure 8.21. The starting delay shown in Figure 8.20 *cannot* be compensated for by simply increasing the pulse width. If the pulse width, is increased, the average power input to the diode per pulse would increase proportionally and could lead to diode burnout. (From Figure 8.20 it can be seen that the diode draws a large current even before oscillations start). This increase in starting delay, due to degradation of the IMPATT mode with irradiation, constitutes the primary failure mechanism for TRAPATT devices. (Also recall the dc burnout mechanism due to the negative temperature coefficient discussed earlier.) In Section 8.4 it was pointed out that the degradation of the IMPATT mode

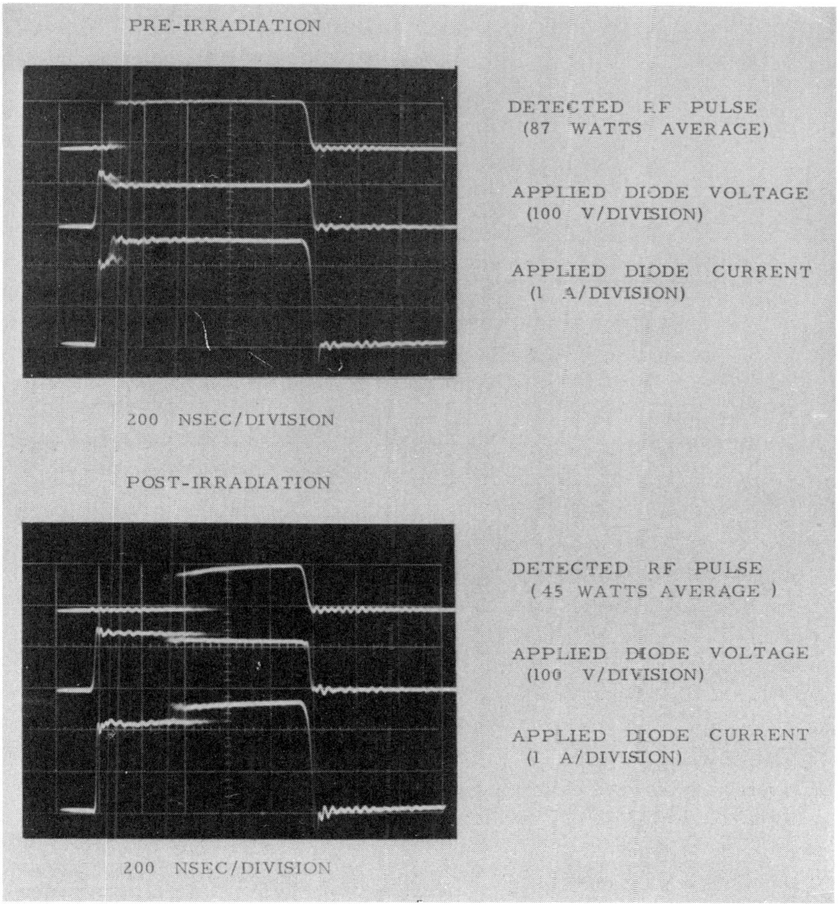

FIGURE 8.20
Starting delay effect after neutron irradiation (6×10^{15} n/cm^2, $\mathscr{E} > 10$ keV, Spectrum B) of a 1 GHz TRAPATT oscillator.

growth rate is due to increased generation and recombination in the depletion region. In practice it is found that these effects become important when $\tau \leq 7T$ (lifetime $\leq$ 7 times the transit time). Let us assume $K_\tau = 7 \times 10^{-7}$ $cm^2\ n^{-1}\ sec^{-1}$ and $1/\tau_0 \ll 1/\tau$. Then from the preceding discussions, and Eq. (5.23) and (8.1), and using the fact that for TRAPATT operation $\omega_0 T = 0.14\pi$ (Ref. [5]), we can derive the failure fluence due to increased starting delay for TRAPATT devices in general. The result is

$$\Phi_f = 3 \times 10^6 f_0 \tag{8.17}$$

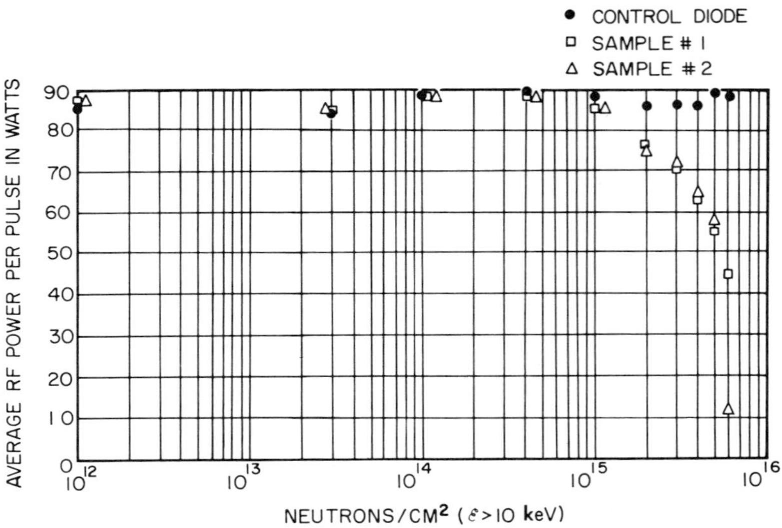

FIGURE 8.21
Average microwave power (per pulse) versus neutron fluence (Spectrum B) for a 1 GHz TRAPATT oscillator.

where f_0 = TRAPATT design frequency. Equation (8.17) is plotted in Figure 8.22 which shows the regions where rf failure (starting delay) and dc failure (negative temperature coefficient) are likely to occur. Figure 8.22 also shows the results of measurements on two TRAPATT oscillators with different design frequencies.

8.6 Transient Ionizing Radiation Effects in IMPATT Oscillators

The effects of transient ionizing radiation on an operating IMPATT oscillator have been investigated [22,24,33]. The primary effect seen is an interruption of the oscillations at sufficiently high ionizing radiation dose rates [rads per second, see Eq. (5.4)]. The oscillations return to normal shortly (a few transit times) after the ionizing radiation ceases. If the ionizing radiation pulse lasts considerably longer than 1 rf cycle (10^{-10} sec at 10 GHz), then it becomes a steady-state problem [24]. For steady-state conditions, we know from Eq. (5.36) that photocurrent densities are given by

$$J = g \cdot e \cdot (L_p + L_n + W_t) \tag{8.18}$$

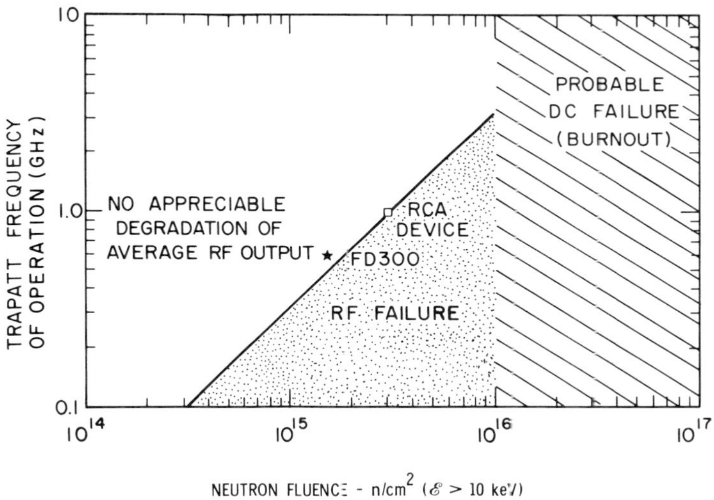

FIGURE 8.22
Graph of TRAPATT frequency of operation versus neutron fluence, which outlines the areas for no appreciable average rf output degradation, for rf failure (buildup time > 250 nsec), and for probable dc burnout. The unity slope line represents expected behavior for rf failure with neutron fluence. (After EerNisse and Chaffin [32]).

assuming that $L_p = L_n \cong 0$ and $W_t = L$ (depletion width) one finds

$$J = g \cdot e \cdot L\,. \tag{8.19}$$

For example, if we consider a silicon device with $L = 4.5\ \mu\text{m}$ which dissipates 7×10^7 rads per sec of ionizing radiation, Eq. (8.19) and (5.35) predict a photocurrent of $J = 0.2\ \text{A/cm}^2$. Since this is a steady-state problem, this current may be simply added to the reverse saturation current. Typical reverse saturation currents for silicon IMPATT devices are on the order of 10^{-3} to $10^{-4}\ \text{A/cm}^2$. For this reason, the photocurrent effectively increases

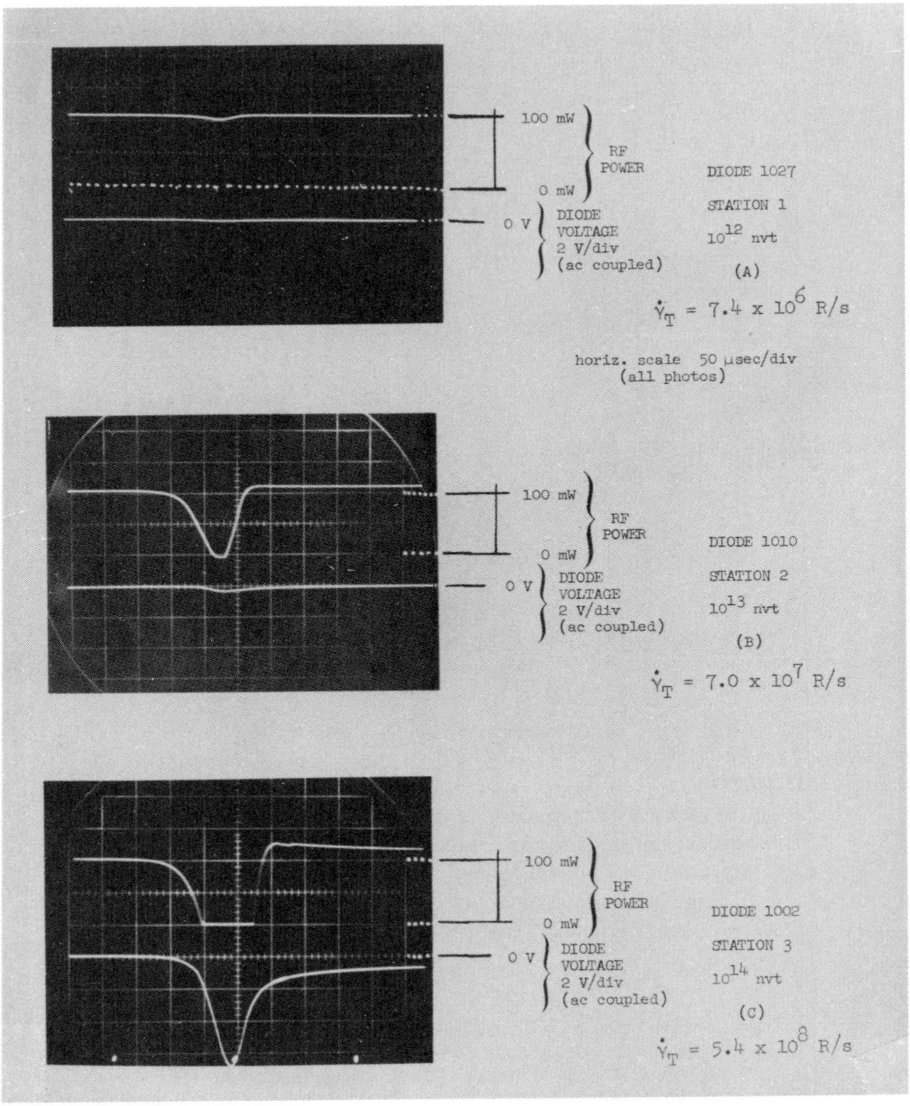

FIGURE 8.23
Transient ionizing radiation effects in a silicon IMPATT diode.

the reverse saturation current by several orders of magnitude. As discussed earlier, Misawa [30] has shown that too large a reverse saturation current leads to a premature buildup of the avalanche current and reduces the phase delay due to the avalanche process. If the phase delay is reduced sufficiently, the negative conductance will be insufficient to overcome circuit losses, and oscillations will cease.

Experimental data [24] showing this effect is presented in Figure 8.23. The data shown in Figure 8.23 was taken using a cavity-mounted, silicon p^+-n-n^+ IMPATT device operating at 11 GHz. The n-region had a doping density of 5×10^{15} per cm^3 and a width of 4.5 μm. The nominal operating point was 80 V and 50 mA. The device was exposed† to combined neutron-gamma pulses of approximately 40 μsec duration (measured at half-height). The integrated neutron fluence per pulse is low enough so that permanent damage can be neglected, and only transient ionizing effects need be considered. Figure 8.23 shows the effects of ionizing radiation of differing intensities on similar IMPATT oscillators. From Figure 8.23 it can be seen that ionizing radiation dose rates below 7×10^6 rads per sec do not appreciably affect the diode operation. At and above ionizing radiation dose rates of 7×10^7 rads per sec, the IMPATT oscillations are temporarily interrupted. The reason for this was discussed earlier in this section. The decrease in diode voltage seen at high dose rates is due to an increased voltage drop across the 300-Ω current limiting resistor when the photocurrent flows. (A current limiting resistor is used even in the absence of ionizing radiation environment because of the steepness of the reverse breakdown curve.)

At ionizing radiation levels much above 10^8 rads per sec, the radiation effects on the air in microwave cavity must also be considered. This is discussed in Appendix B.

Anderson [22] has observed an anomalously long (15-minute) interruption of IMPATT oscillations when an IMPATT oscillator was exposed to a 15 ms pulse of fast neutrons and gamma rays in a Triga-type reactor. The ionizing radiation dose rate was 3×10^6 rads per sec, and the integrated neutron fluence was 2×10^{15} n/cm^2 ($\mathscr{E} > 10$ keV) per pulse. This long delay effect is in disagreement with Ref. [24] and [33]. The only explanation that can be offered is that the device remains in a fairly intense gamma field in the reactor core in Ref. [22] due to the decay of fission fragments produced during the radiation pulse.

REFERENCES

[1] W. T. Read, "A Proposed High-Frequency, Negative Resistance Diode," *BSTJ*, **37**, 401 (1958).

† At the Sandia SPR-II reactor (Spectrum A).

[2] R. L. Johnston, B. C. DeLoach, and B. G. Cohen, "A Silicon Microwave Diode Oscillator," *BSTJ*, **44**, 369 (1965).
[3] H. J. Prager, K. K. N. Chang, and S. Weisbrod, "High Power, High Efficiency Silicon Avalanche Diodes at Ultra High Frequencies," *Proc. IEEE*, **55**, 586 (1967).
[4] D. J. Bartelink and D. L. Scharfetter, "Avalanche Shock Fronts in *p-n* Junctions," *Appl. Phys. Letters*, **14**, 320 (1969).
[5] A. S. Clorfeine et al., "A Theory for the High-Efficiency Mode of Oscillation in Avalanche Diodes," *RCA Review*, **30**, 397 (1969).
[6] H. A. Watson, ed., *Microwave Semiconductor Devices and Their Circuit Applications*, McGraw-Hill, New York, 1969.
[7] J. E. Carroll, *Hot Electron Microwave Generators*, Arnold Publishers, London, 1970.
[8] H. V. Shurmer, *Microwave Semiconductor Devices*, Wiley-Interscience, New York, 1971.
[9] Special Issue on Semiconductor Bulk Effect and Transit-Time Devices *IEEE Trans. PGED*, **ED-13** (January 1966).
[10] Special Issue on Microwave Circuit Aspects of Avalanche Diode and Transferred Electron Devices, *IEEE Trans MTT*, **MTT-18** (November 1970).
[11] Special Issue on Microwave Semiconductors, *IEEE Proc.*, **59** (August 1971).
[12] B. C. DeLoach, "Modes of Avalanche Diodes and Their Associated Circuits," *IEEE J. of Sol. St. Cts.*, **SC-4**, 376 (1969).
[13] D. L. Scharfetter and H. K. Gummel, "Large Signal Analysis of a Read Diode Oscillator," *IEEE Trans. ED*, **ED-16**, 64 (1969).
[14] D. F. Kostishack, "UHF Avalanche Diode Oscillator Providing 400-Watts Peak Power and 75 Per Cent Efficiency," *IEEE Proc.*, **58**, 1282 (1970).
[15] R. L. Johnston et al., "High Efficiency Oscillations in Germanium Avalanche Diodes Below the Transit-Time Frequency," *Proc. IEEE*, **56**, 1611 (1968).
[16] W. J. Evans, "Circuits for High Efficiency Avalanche Diode Oscillators," *IEEE Trans. MTT*, **MTT-17**, 1060 (1969).
[17] H. Yanai et al., "Large Amplitude High Efficiency Oscillation Using Si-Avalanche Diode and Its Experimental Analysis," *MOGA-70*, Kluwer-Deventer (Pub.), The Netherlands, 1970, p. 1971.
[18] D. K. Wilson *et al.*, "Radiation Effects on Silicon Avalanche (IMPATT) Diodes," *IEEE Trans. Nuc. Sci.*, **NS-15**, 6, 114 (1968).
[19] R. J. Chaffin, "The Effect of Neutron Radiation on an IMPATT Diode," *IEEE Trans. MTT*, **MTT-17**, 119 (1969).
[20] W. A. Anderson, "The Performance of IMPATT Diodes under the Influence of Gamma Radiation," *Proc. IEEE*, **57**, 852 (1969).
[21] W. A. Anderson, "The Performance of IMPATT Diodes under the Influence of High Intensity Gamma and Neutron Irradiation," *Proc. IEEE*, **57**, 1441 (1969).
[22] W. A. Anderson, "A Radiation Pulse Interrupts IMPATT Diode Oscillation," *Proc. IEEE*, **58**, 807 (1970).

[23] E. P. EerNisse and R. J. Chaffin, "Carrier Trapping and Recombination in Avalanche Diodes," *IEEE Trans. PGED*, **ED-17**, 520 (1970).

[24] R. J. Chaffin, *Transient and Permanent Radiation Effects in Gunn and IMPATT Oscillators*, Sandia Laboratories, Albuquerque, N.N., Report No. SC-RR-710378, June 1971.

[25] W. A. Anderson, "Comparing Radiation Effects in Gunn and IMPATT Diodes," *IEEE Trans. Nuc. Sci.*, **NS-18**, 5, 404 (1971).

[26] W. A. Anderson, "Computer Studies of Radiation Effects in Avalanche Diodes," *IEEE Trans. Nuc. Sci.*, **NS-19**, 894 (February 1972).

[27] S. M. Sze, *Physics of Semiconductor Devices*, Wiley-Interscience, New York, 1969.

[28] E. P. EerNisse, "Temperature Coefficients of Avalanching p^+-n-n^+ Junctions with Carrier Trapping," *Appl. Phys. Letters*, **16**, 506 (1970).

[29] S. R. Kovel and G. Gibbons, "The Effect of Unswept Epitaxial Material on the Microwave Efficiency of IMPATT Diodes," *Proc. IEEE*, **55**, 2066 (1967).

[30] T. Misawa, "Saturation Current and Large Signal Operation of a Read Diode," *Sol, St. Elec.*, **13**, 1363 (1970).

[31] R. J. Chaffin et al., "The Effect of Neutron Irradiation on the TRAPATT Diode," *IEEE Proc.*, **56**, 2063 (1969).

[32] E. P. EerNisse and R. J. Chaffin, "Design of Neutron Radiation Tolerant High Efficiency Microwave Avalanche Diode Sources," *IEEE Trans. Nuc. Sci.*, **NS-17**, 227 (December 1970).

[33] J. M. Borrego et al., "Transient Ionizing Radiation Effects on IMPATT Diode Oscillators," presented at 1972 IEEE Conference on Nuclear and Space Radiation Effects, Seattle, July 1972.

9

Radiation Damage in Gunn and LSA Devices

9.1 Introduction

The possibility of obtaining a bulk semiconductor microwave negative resistance was first predicted theoretically in 1961 by Ridley and Watkins [1] and Hilsum [2]. The first experimental evidence of such behavior was reported in 1963 by Gunn [3]. The mode observed by Gunn is one in which negative differential resistance (NDR) causes the formation of a dipole domain which drifts through the device at nearly saturated drift velocity. In 1966, Copeland [4] predicted and observed the existence of the LSA (*L*imited *S*pace-charge *A*ccumulation) mode of operation for these devices. In the LSA mode, careful control of the device voltage waveform is used to assure that dipole domains do not form. This means that the whole device can contribute to the NDR and that the LSA device is not transit-time limited.

The Gunn class of devices has found applications as radar local oscillators and as transmitters for low-power radars. These devices work best in the frequency range from 6 GHz to 20 GHz and produce continuous powers of up to a few hundred milliwatts. If the domain formation is suppressed, these devices may also be used as small-signal, broad-band, reflection-type microwave amplifiers [5]. The Gunn device produces less noise (typically 10 dB to 20 dB less) than an IMPATT device and, for this reason, is more suitable as a local oscillator source. The LSA devices are used to produce pulsed rf powers of up to several hundred watts peak in the 3 GHz to 15 GHz range. LSA-pulsed oscillators find applications as beacon transponders and as moderate-power pulsed radar transmitters.

This chapter is devoted to a study of radiation effects in Gunn and LSA devices. Sections 9.2, 9.3, and 9.4 will discuss the basic device physics of Gunn and LSA devices. Sections 9.5 and 9.6 are devoted to an examination of permanent radiation damage to Gunn and LSA devices, respectively, and Section 9.7 discusses the effects of transient ionizing radiation on Gunn oscillators.

9.2 Bulk Negative Differential Resistance (NDR)

The next three sections of this chapter will discuss the NDR, Gunn, and LSA mechanisms. This subject has been treated thoroughly [6–9] in the literature, and therefore discussion here will be brief, emphasizing only those parameters needed to understand radiation effects in these devices.

This section will describe the mechanism responsible for the NDR used in Gunn and LSA devices. The discussion will be restricted to n-type GaAs. The mechanism to be described has been observed in other materials, InP, InAs, etc.; however, at present GaAs is the only material used commercially to make devices. The NDR effect will be explained with the aid of Figure 9.1 which shows an expanded view of Figure 2.4B, which is the energy-momentum relation for GaAs. The important things to note about Figure 9.1 are that GaAs has an upper valley in the conduction band which has an effective mass that is much larger than that of the lower valley. This can be understood by recalling Equation (2.9)

$$m^* = \frac{\hbar^2}{d^2\mathscr{E}/dk^2} \cdot \tag{9.1}$$

From Figure 9.1 it can be seen that $d^2\mathscr{E}/dk^2$ is much larger for the lower valley than for the upper valley, and therefore

$$m_1^* \ll m_2^* \, . \tag{9.2}$$

This implies that the mobility of the electrons in the upper valley must be much less than that of the lower valley electrons. This can be seen by recalling the approximate formula, Eq. (2.34)

$$\mu_n \cong \frac{e\tau}{2m_{ce}^*} \tag{9.3}$$

where τ = collision time, m_{ce}^* = conductivity effective mass. From Eq. (9.2) and (9.3) it can be seen that $\mu_1 \gg \mu_2$. (Typical values are $\mu_1 = 8000$, $\mu_2 = 150$ cm^{2}V^{-1}sec^{-1}.) Let us next consider the relative populations of the two valleys. First, it is important to note that the two valleys are separated by 0.36 eV, an energy which is less than the bandgap (1.43 eV) but is much larger than kT/e (0.26 eV) at room temperature. For small applied fields at room temperature the electrons will all have low energies, and most of them will be found in the lower valley, i.e., $n_1 = n_0$, $n_2 \cong 0$ ($n_0 = n_1 + n_2$, total electron density). For very large applied fields the electrons will be accelerated and gain energy, thus moving upwards on Figure 9.1. As this happens, it becomes possible for some of the electrons to be scattered into the upper valley by phonon collisions. The calculation of the exact population of each valley requires a knowledge of the density of states of both valleys [8]. Let us

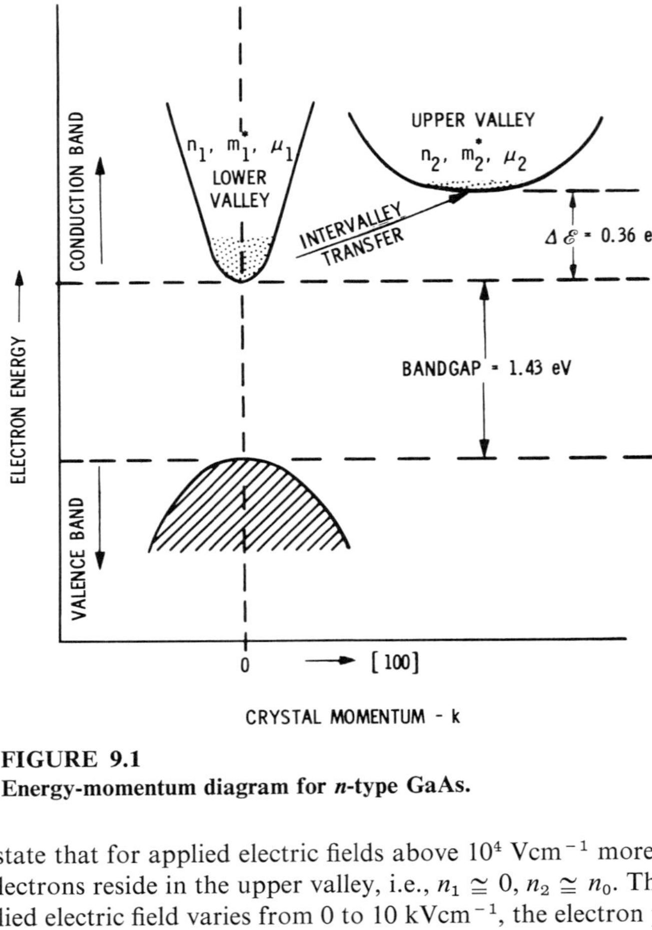

FIGURE 9.1
Energy-momentum diagram for *n*-type GaAs.

simply state that for applied electric fields above 10^4 Vcm^{-1} more than 90% of the electrons reside in the upper valley, i.e., $n_1 \cong 0$, $n_2 \cong n_0$. Therefore, as the applied electric field varies from 0 to 10 kVcm^{-1}, the electron population essentially migrates from the lower valley to the upper valley. If an average mobility is defined as

$$\mu_{av} = \frac{n_1\mu_1 + n_2\mu_2}{n_0}, \tag{9.4}$$

then, the average electron velocity is given by [see Eq. (2.33)]

$$\bar{v}_{av} = -\mu_{av}\bar{E}\,. \qquad (9.5)$$

We have already stated that as E increases, n_1 decreases and n_2 increases. Since $\mu_2 \ll \mu_1$, Eq. (9.4) implies that μ_{av} decreases as E increases. In fact it can be shown [8] that the net effect is that $|\bar{v}_{av}|$ actually decreases with increasing E in a certain range. This is shown in Figure 9.2. Figure 9.2 illustrates that for $E_T < E < E_V$ the average electron velocity actually decreases

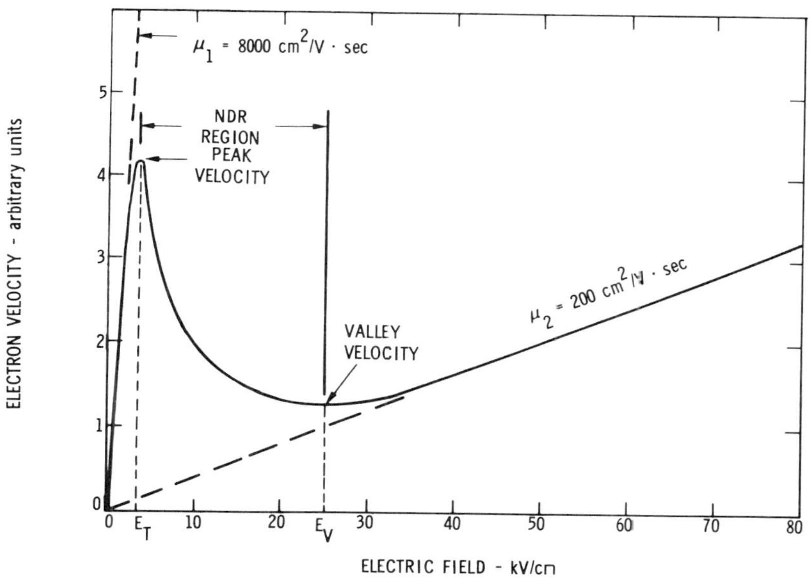

FIGURE 9.2
Idealized velocity-field relationship for *n*-type GaAs.

with increasing E. After all of the electrons have transferred to the upper valley, the velocity might again be expected to increase with field corresponding to the upper valley mobility, as is shown in Figure 9.2. Due to velocity saturation effects however (see Section 2.6), the velocity does not increase again, but rather it levels out as is shown in Figure 2.10. The region of interest is the NDR region in Figure 9.2. An average conductivity may also be defined as

$$\sigma = n_0 e v_{av}\,. \qquad (9.6)$$

Using the point form of Ohm's law ($\bar{J} = \sigma\bar{E}$), we may define a one-dimensional differential conductivity as

$$\sigma = \frac{dJ_x}{dE_x} = n_0 e \frac{dv_{\text{av}}}{dE_x} \cdot \tag{9.7}$$

From Eq. (9.7) and Figure 9.2 it can be seen that for $E_T < E < E_V$ the differential conductivity (and thus the differential resistance) *will be negative*! This is the mechanism utilized by Gunn and LSA devices to produce microwave oscillations. One further note is that in order for this effect to be useful at microwave frequencies, the intervalley transfer time (see Figure 9.1) must be very short. The estimated value is 2×10^{-12} sec [8].

9.3 The Gunn Mode

The NDR mechanism can be used to produce microwave oscillations. This was first observed by Gunn [3]. Before discussing the Gunn effect, we will consider the effect of negative differential conductivity on the dielectric relaxation phenomena. It can be shown [10] that an excess (nonneutral) charge distribution in a material will disperse due to mutual repulsion forces following the relation

$$\rho = \rho_0 \exp\left(-\frac{\sigma}{\epsilon}t\right) \tag{9.8}$$

where σ, ϵ = material's conductivity, dielectric constant and ρ_0 = initial charge density. This decay of nonequilibrium charge density with time is called the dielectric relaxation effect. The characteristic relaxation (or decay) time is given by

$$T_d = \frac{\epsilon}{\sigma} \cdot \tag{9.9}$$

For the case when $E_T < E < E_V$, the differential conductivity is negative. This implies that the dielectric relaxation phenomena becomes a dielectric growth phenomena, i.e., any small charge perturbation grows instead of decaying. This growth will continue as long as $E_T < E < E_V$. Consider a hypothetical example of this case: Figure 9.3 shows a one-dimensional section of nearly uniform n-type GaAs which instantaneously has the field distribution shown in Figure 9.3A. (This cannot represent a dc solution since it is not stable.) This figure shows that the entire region is biased between E_T and E_V. The electric field is in the $-x$ direction; thus the electrons drift in the $+x$ direction. For the electric field distribution shown in Figure 9.3A, the velocity-field curve shown in Figure 9.2 predicts that those electrons in the lowest field will move the fastest; therefore, electrons will accumulate on the left-hand edge of the field bump in Figure 9.3. Those electrons at the field peak move the slowest, and those to the right of the peak move somewhat faster; consequently, a depletion layer is created just to the right of the field maximum. The resulting dipole layer adds to the electric field bump resulting in a feedback effect, i.e., the larger the field bump grows, the greater the velocity differences become. This discontinuity will continue to grow with

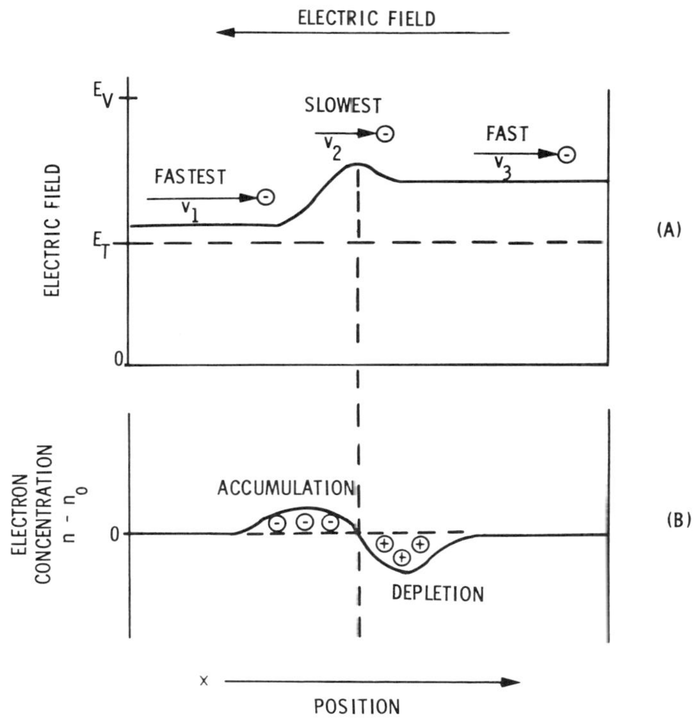

FIGURE 9.3
Hypothetical demonstration of dielectric growth phenomena (transient solution only).

time until the dipole layer drifts out of the device or until the field outside the dipole layer decreases below E_T. This happens because $\int E \cdot dx$ is limited to the applied terminal voltage. The actual shape of the discontinuity is not critical to the preceding arguments as long as the electric field discontinuity has both positive and negative slopes.

The Gunn device consists of a block of n-type GaAs with an ohmic contact on both ends of the device. One of these contacts is referred to as the cathode and the other as the anode. (The electrons drift from the cathode to the anode.) In this sense we may speak of a Gunn "diode," although the device does not have a *p-n* junction. The operation of the Gunn oscillator will be explained with the aid of Figure 9.4 which shows an n^+-n-n^+ GaAs device (the n^+

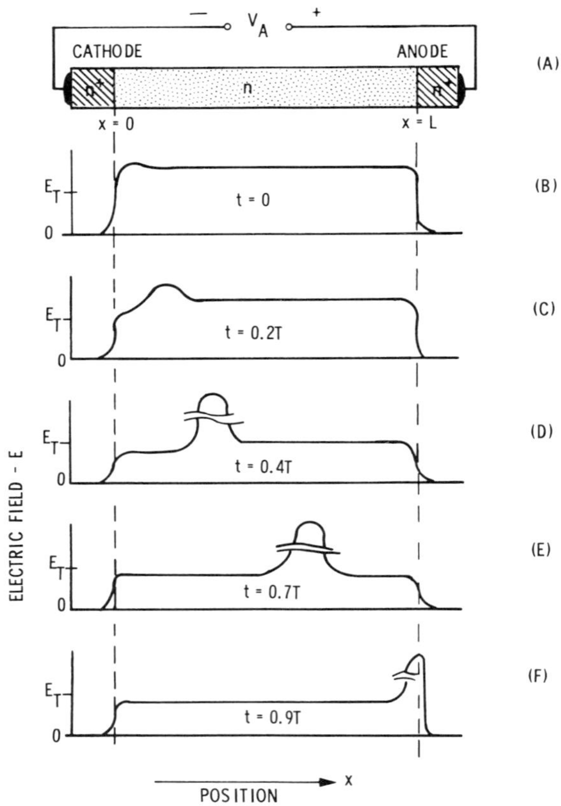

FIGURE 9.4
Domain propagation in a Gunn oscillator.

regions are used as contacts). It is assumed that a terminal voltage, V_A, is applied, at $t = 0$, to the device. Let us also assume that at $t = 0$ the value of E is above E_T. In this case a dipole domain like the one shown in Figure 9.3 is started (usually at the cathode end) and quickly grows to maturity, i.e., the field outside the domain falls below threshold (see Figure 9.4D). This domain then continues to drift across the sample, almost at the saturated drift velocity [11]. After a time $T \cong L/v_s$, the domain exits at the anode (Figure 9.4F). Then the field throughout the device rises above threshold again, and the whole process repeats. The terminal current waveform resulting from these internal activities is depicted in Figure 9.5, from which it can be seen that the terminal current waveform contains considerable energy at a frequency f_{Gunn}, where

$$\frac{1}{f_{\text{Gunn}}} = T \cong \frac{L}{v_s} \tag{9.10}$$

and where L = device length and v_s = saturated electron drift velocity. It is important to note that the Gunn oscillator does not require a cavity or resonant circuit to operate. It would oscillate in a purely resistive circuit

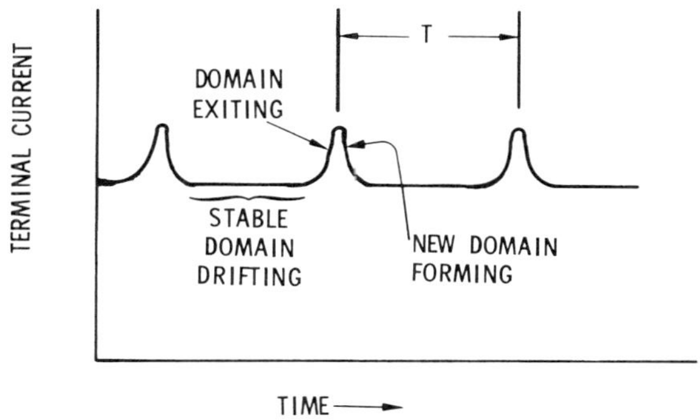

FIGURE 9.5
Terminal current waveform for a Gunn oscillator.

provided that the series equivalent load resistance is not too large. The frequency is determined primarily by the transit time, although a small amount of bias tuning is possible. The operating frequency of a Gunn device is determined by the length, L. Another way of stating this may be found by rewriting Eq. (9.10) as

$$(f_{\text{Gunn}})\cdot(L) = v_s \cong 10^7 \text{ cm/sec}. \tag{9.11}$$

Hence, once a desired frequency is known, the device length is determined by Eq. (9.11). If the growth of charge instabilities (domains) are to have a large effect on the terminal (I-V) characteristics of the diode, it is necessary that

$$|T_d| \ll |T| \tag{9.12}$$

where T_d is given by Eq. (9.9) (for the negative conductivity case) and T is given by Eq. (9.10), i.e., a domain must have several exponential time constants to grow before it drifts out of the device. Eq. (9.12) may be rewritten as

$$\frac{\epsilon}{n_0 e|\mu^-|} \ll \frac{L}{v_s} \tag{9.13}$$

where μ^- = negative differential mobility (taken here as $-2400\ \text{cm}^2/\text{V}\cdot\text{sec}$ [7]). Substituting with numerical values and rearranging Eq. (9.13), we find that the condition stated in Eq. (9.12) requires that

$$n_0 \cdot L \gg 2.5 \times 10^{10}\ \text{cm}^{-2}\,. \tag{9.14}$$

In practice it is found [12] that $n_0 \cdot L \geq 3 \times 10^{11}\ \text{cm}^{-2}$ is required to produce Gunn oscillations. Therefore, the principal design equations for low-power Gunn oscillators are found to be

$$\begin{gathered} f_0 \cdot L = 10^7\ \text{cm/sec}\,; \\ n_0 \cdot L \geq 3 \times 10^{11}\ \text{cm}^{-2}\,. \end{gathered} \tag{9.15}$$

By enclosing the device within a resonant cavity, two major variations on the Gunn mode are possible. These are the quenched-domain mode and the delayed-domain mode. In the quenched-domain mode, the rf voltage swing is large enough to quench or extinguish a transiting domain at the voltage minimum of the rf cycle. Thus, the domains never transit fully across the device width, and the frequency is larger than f_{Gunn}. In the delayed-domain mode the rf swing is not large enough to extinguish a domain once it is formed, but it is large enough that the field remains below threshold (during the negative rf swing) after the old domain has exited. Therefore, the formation of a new domain is "delayed," and the operating frequency is below f_{Gunn}. In both the quenched- and delayed-domain modes, the frequency is determined by the cavity rather than the transit time.

Most practical Gunn oscillators utilize one of the preceding cavity-controlled modes because the cavity improves efficiency and reduces the carrier noise and allows control of the frequency.

9.4 The LSA Mode

While the domain mechanism discussed in the previous section does produce microwave oscillations, the dc-rf conversion efficiency is quite low (1% to 5%). One reason for this low efficiency can be seen from Figure 9.4 which shows that, while a domain is propagating, most of the device is below threshold ($E > E_T$) and so acts like a parasitic, lossy resistance. If a way could be found to keep the entire device above threshold ($E < E_T$) and prevent domain formation, then the entire n-region could act to generate power and the efficiency would be much higher. Copeland [4] discovered a means to accomplish this result called the LSA (*L*imited *S*pace-charge *A*ccumulation) mode. The operation of the LSA mode will be explained with the aid of Figure 9.6.

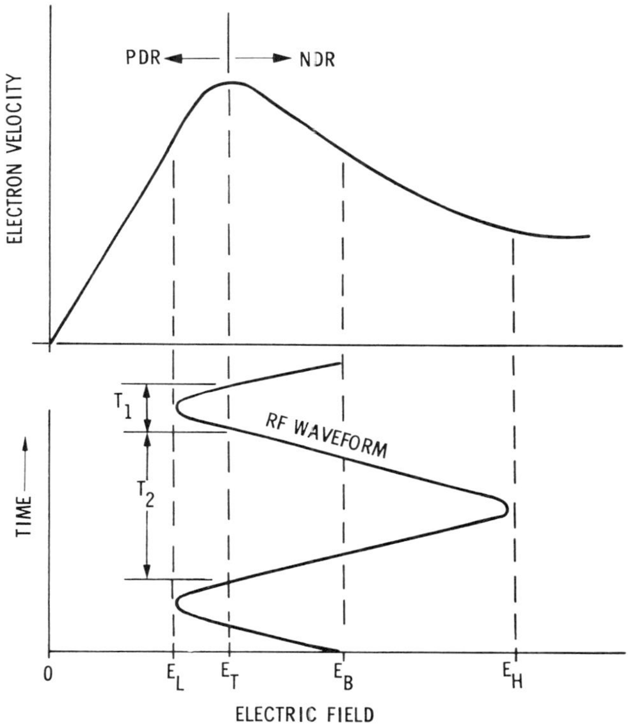

FIGURE 9.6
LSA mode waveforms.

In the LSA mode, a large voltage swing is used in conjunction with a frequency, whose period is short compared to the dielectric growth time [Eq. (9.7)], to limit the amount of space charge that can grow during the negative conductivity part of the rf cycle (T_2 on Figure 9.6). The peak-to-peak rf voltage swing is required to be large enough that the device's field swings below threshold for a part of each rf cycle (T_1 on Fig. 9.6). While the field is below threshold, the space charge growth that occurred during the NDR part of the cycle is dispersed by the dielectric relaxation effect. In order to have stable operation in this mode, two basic requirements must be met:

$$\frac{1}{f_0} < \frac{\epsilon}{\sigma^-} \cong \frac{\epsilon}{n_0 e|\mu^-|}, \tag{9.16A}$$

and

$$\frac{\epsilon}{\sigma^+} \cong \frac{\epsilon}{n_0 e \mu_1} \ll \frac{1}{f_0}. \tag{9.16B}$$

The condition in Eq. (9.16A) says that, effectively, an rf period must be much shorter than one growth time constant. The condition in Eq. (9.16B) says that, effectively, the decay time constant must be much, much less than one rf period. Equation (9.16) may be rewritten as

$$\frac{\epsilon}{e\mu_1} \ll \frac{n_0}{f_0} < \frac{\epsilon}{e|\mu^-|}. \tag{9.17}$$

Copeland [12] has predicted that the best range of n_0/f_0 values for LSA oscillations is given by

$$2 \times 10^4 \leq \frac{n_0}{f_0} \leq 2 \times 10^5 \text{ cm}^{-3}\text{ Hz}^{-1}. \tag{9.18}$$

Equation (9.18) is the basic design equation for an LSA device to operate a given frequency, f_0. Note, however, the lack of dependence of design frequency on device length Since transit-time effects are not used in the LSA mode, the device can have as long an active region as the technology will allow ($\sim$0.02 cm). This means that very large pulse powers can be achieved from a single large chip. Most high-powered LSA devices are quite long; they usually have $n_0 \cdot L \geq 10^{13}$ cm^{-2} and operate at several times the threshold voltage. Therefore, if the oscillations should be perturbed, or slow to start, a large domain will form, and the internal field inside the domain will cause avalanche breakdown and possible device burnout.

The theory derived by Copeland [4] and discussed with the aid of Figure 9.6 is simple to analyze, however it does not accurately describe the operation of most LSA oscillators. The basic problem is that in most cases the voltage waveform is not sinusoidal as shown in Figure 9.6 but looks more like the waveform shown in Figure 9.7. The waveform in Figure 9.7 is for the relaxation mode LSA oscillator [13,14]. In the relaxation mode, the field grows with an L/R time constant towards the bias value E_B (line A-B in Figure 9.7). When the voltage gets to threshold, the current drops rapidly (due to the negative resistance effects), and an inductive voltage (or field) overshoot rings for $\frac{1}{2}$ cycle (line B-C in Figure 9.7). As soon as the field drops below threshold, the cycle again repeats. Figure 9.7 also shows a simplified circuit diagram of a relaxation mode LSA oscillator. L, C, and R_L represent cavity parameters, and the diode is represented by $-R_n$ (most of C is due to the device capacitance). The operating frequency is determined by both the cavity and the L/R time constant. Hence

$$T = \frac{1}{f_0} \cong 2\pi\sqrt{LC} + \frac{L}{R_0 M} \tag{9.19}$$

where L, C = cavity parameters, R_0 = diode low-field ($E < E_T$) resistance, and $M = E_B/E_T$ = bias to threshold multiple. The condition for oscillation†

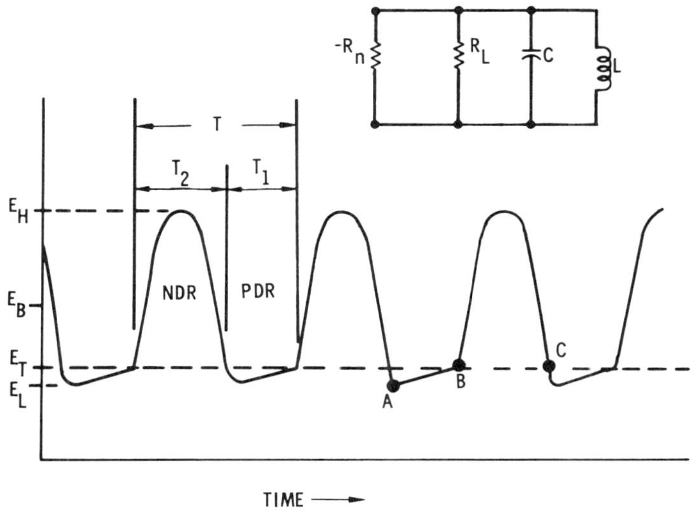

FIGURE 9.7
Electric field versus time and simplified circuit for a relaxation mode LSA oscillator.

requires that $|R_n| \leq R_L$. An upper limit on power output occurs due to the fact that $|R_n|$ increases with bias. This can be demonstrated as follows: Near breakdown $|R_n|$ has the form [15]

$$|R_n| \cong K \cdot R_0 \cdot M \tag{9.20}$$

where K is constant for large values of M. Therefore, it can be seen that as E_B increases, M increases, and if $|R_n| > R_L$, oscillation will cease and breakdown can occur.

† This is also a condition for survival; if the device does not oscillate, a domain will form and burnout may occur.

Pulsed LSA oscillators are available which can reliably produce a few hundred watts of rf power in the 3 GHz to 15 GHz frequency range with duty cycles of 0.01 or less [16].

9.5 Permanent Radiation Effects in Gunn Devices

From the discussions in Section 9.3 it can be seen that critical parameters for Gunn Oscillator operation are the following: transit time, threshold voltage, $n_0 \cdot L$ product, negative and positive mobilities, and the saturated drift velocity. The saturated drift velocity and the transit time are assumed not to change below neutron fluences of 10^{16} n/cm^2. The reasons for this have been outlined in Section 8.4. The $n_0 \cdot L$ product will change with radiation due to carrier removal effects. If the threshold voltage remains constant, the negative mobility will decrease if the positive mobility does. This is due to a decrease in the peak velocity (see Figure 9.2) with a decrease in low-field mobility. As discussed in Section 5.3, the room-temperature mobility is not expected to change significantly for radiation-induced trap concentrations below 10^{15} cm^{-3}. The discussions in this section will be broken into two parts: First an examination of radiation-induced changes in the device itself, and then a discussion of radiation-induced changes in oscillator performance.

DC and Low-Frequency Effects. Permanent radiation effects in Gunn devices have been examined by several authors [17–24]. Both neutron and Co^{60} gamma irradiation studies have been carried out. In general, it has been found that X-band Gunn oscillators quit operating at a neutron fluence of approximately 10^{14} n/cm^2 ($\mathscr{E} > 10$ keV) or at a gamma fluence of 10^{17} photons/cm^2 (Co^{60}). Researchers [20,22] have measured the mobility of Hall samples (see Section 4.4) made from the same material as the Gunn oscillators. They find (see Figures 5.41 and 5.43) that the room-temperature mobility is only degraded 5% to 10% at 10^{14} n/cm^2, and as a result this effect can be ignored.

The most important low-frequency effect seen in irradiated Gunn devices is a large increase in the low-field resistance.

$$R_0 = \frac{L}{n_0 e \mu_1 A} \tag{9.21}$$

where L and A are the device dimensions and n_0, μ_1 have been discussed earlier in this chapter. Figure 9.8 shows a typical result [23] for an X-band Gunn diode which has $L = 12\ \mu$m, $A = 4 \times 10^{-4}$ cm^2, and $n_0 = 8 \times 10^{14}$ cm^{-3}. Since we may assume μ_1 does not change with irradiation, Eq. (9.21) and Figure 9.8 show that n_0 must decrease with irradiation. This is the carrier removal effect discussed in Chapter 5. Carrier removal is known to occur in both neutron and gamma-irradiated GaAs (see Figures 5.39 and 5.42). This is important in that it changes the $n \cdot L$ product (to be discussed later). The I-V characteristic of Gunn diodes (as measured with a curve tracer) also changes with irradiation, as is shown in Figure 9.9. There are two important regions of interest in each photograph in Figure 9.9. First, below the threshold voltage (about 3 volts), the low field resistance (which is the inverse of the initial slope of the I-V characteristic) increases with irradiation, as is also shown in Figure 9.8. The second region of interest on each I-V characteristic shown in Figure 9.9 is the region above threshold voltage. For $V > 3$ volts,

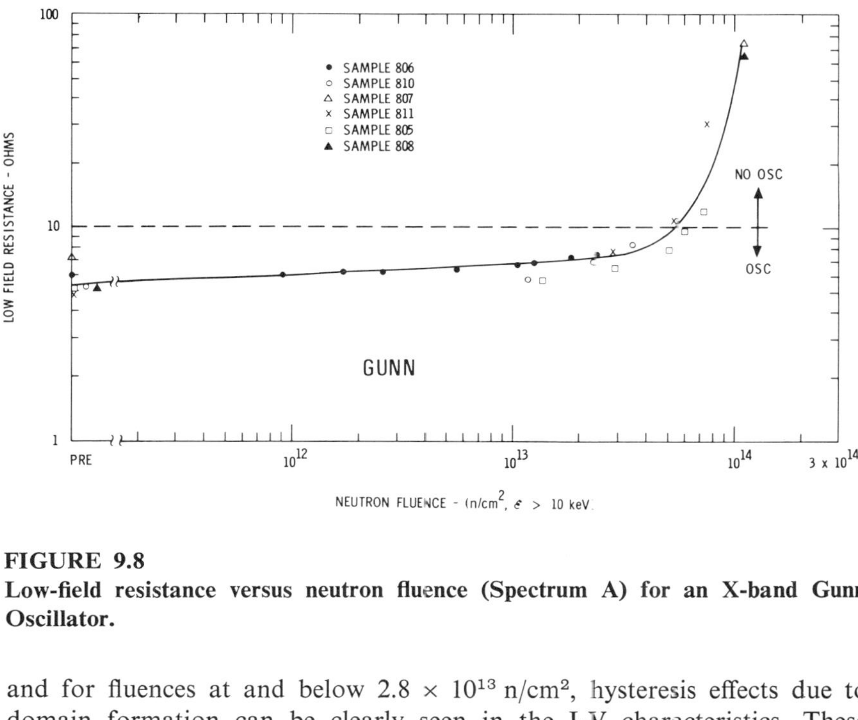

FIGURE 9.8
Low-field resistance versus neutron fluence (Spectrum A) for an X-band Gunn Oscillator.

and for fluences at and below 2.8×10^{13} n/cm^2, hysteresis effects due to domain formation can be clearly seen in the I-V characteristics. These hysteresis effects are due to the NDR that occurs above threshold, i.e., domain formation. After 7.5×10^{13} n/cm^2, the device represented by Figure 9.9 appears as a positive resistance at all voltages and shows no hysteresis effects. The device also no longer functions as a Gunn oscillator at this fluence level.

Radiation Effects on Microwave Behavior. In the microwave behavior of irradiated Gunn oscillators, the most significant effect seen is a reduction in

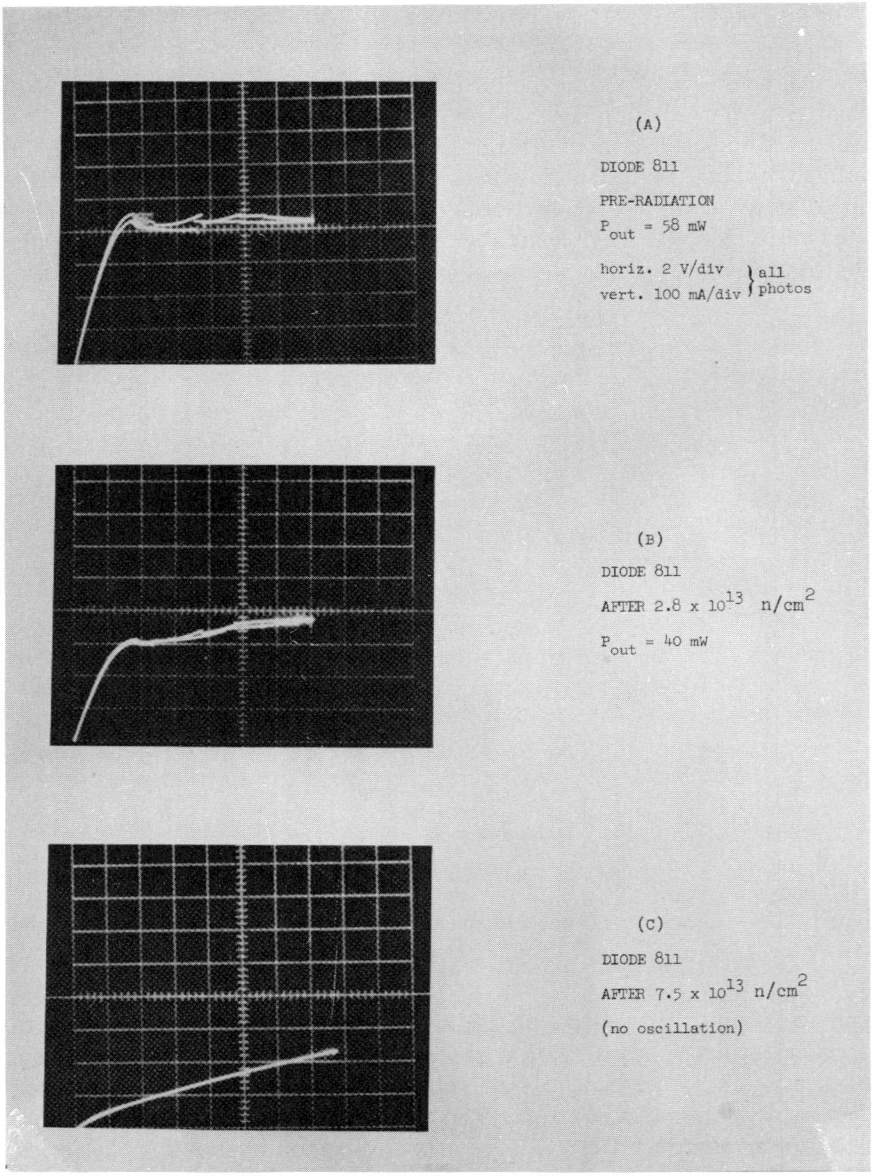

FIGURE 9.9
Measured V-I Characteristic for a Gunn diode versus neutron fluence.

power output and eventual cessation of oscillations with increasing irradiation fluence. A typical result is presented in Figure 9.10 which shows the behavior of the particular Gunn diodes described earlier in this section (see also [23]). From Figure 9.10 it can be seen that the output power remains essentially constant up to 2×10^{13} n/cm² and then falls off rapidly above this point, typically failing at 10^{14} n/cm². The reason for this behavior can be

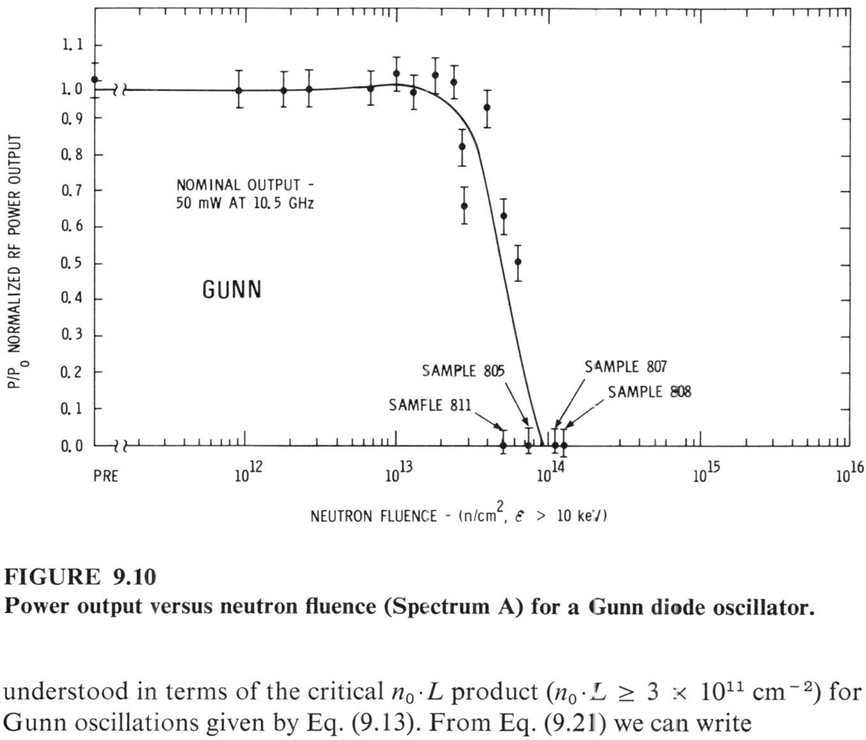

FIGURE 9.10
Power output versus neutron fluence (Spectrum A) for a Gunn diode oscillator.

understood in terms of the critical $n_0 \cdot L$ product ($n_0 \cdot L \geq 3 \times 10^{11}$ cm^{-2}) for Gunn oscillations given by Eq. (9.13). From Eq. (9.21) we can write

$$n \cdot L = \frac{R_0}{R} \cdot [n_0 \cdot L] \tag{9.22}$$

where n = electron density after irradiation and R = low-field resistance after radiation. For the device represented by Figures 9.8 to 9.10, the initial $n_0 \cdot L$ product is 10^{12} cm^{-2}. Figure 9.8 shows that the Gunn diodes stop oscillating for $R \geq 10\ \Omega$. Since R_0 is approximately 5 Ω (see Figure 9.8), this means that these diodes stop oscillating when their $n \cdot L$ product drops below 5×10^{11} cm^{-2}. This value is close to the critical value predicted by Eq. (9.15). Similar results are obtained for gamma-irradiated Gunn oscillators

[22]. The physical mechanism that is occurring here is that the dielectric growth time constant [Eq. (9.9)] increases due to a decrease in negative conductivity $|\sigma^-| \cong n_0 e|\mu^-|$; i.e., the domains do not have time to form before they drift out of the device. The falloff in power before failure is reached is due to two effects: First, since these devices are operated at constant voltage, the decrease in operating current that occurs with increased resistance (due to carrier removal) results in a reduced input power, as is shown in Figure 9.11. Second, for $n \cdot L$ products below the optimum value of 2×10^{12} cm^{-2} (determined by Copeland [25]), the efficiency decreases as $n \cdot L$

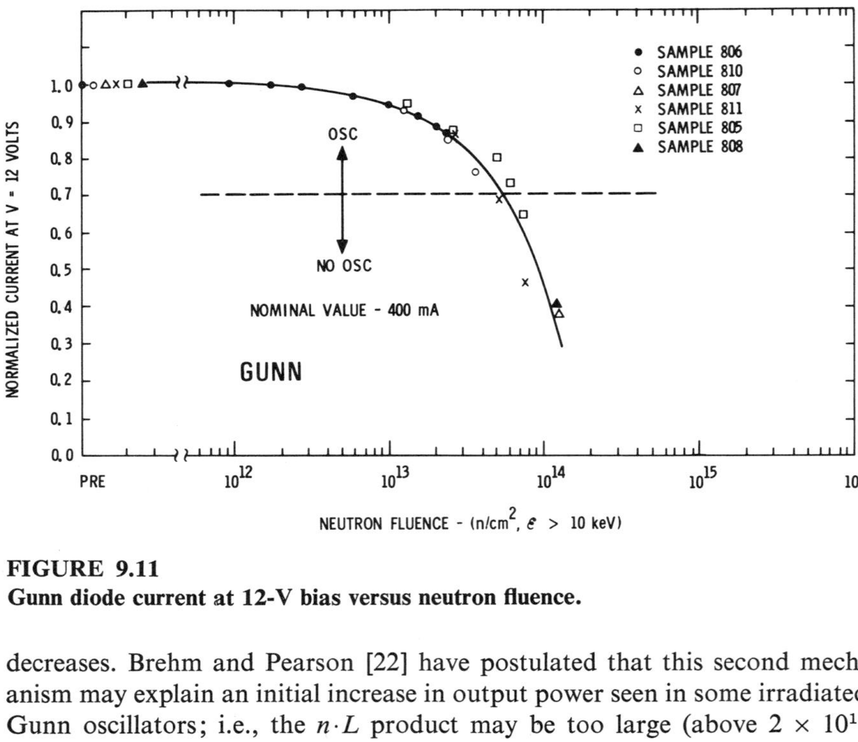

FIGURE 9.11
Gunn diode current at 12-V bias versus neutron fluence.

decreases. Brehm and Pearson [22] have postulated that this second mechanism may explain an initial increase in output power seen in some irradiated Gunn oscillators; i.e., the $n \cdot L$ product may be too large (above 2×10^{12} cm^{-2}) to begin with, and the radiation-induced carrier removal may reduce it to the optimum value. For $n_0 \cdot L = 2 \times 10^{12}$ cm^{-2} (the optimum value [25]) Eq. (9.15) may be solved to give

$$\frac{n_0}{f_{\text{Gunn}}} = 2 \times 10^5 \text{ sec cm}^{-3}. \tag{9.23}$$

From Eq. (9.23) it can be seen that higher-frequency Gunn oscillators will require heavier doping densities. Figure 5.42 shows that the carrier removal rate will also increase with doping density. The carrier removal rate may be expressed as [19]†

$$\frac{dn}{d\Phi} \simeq 9 \times 10^{-6}(n_0)^{0.4}\ \text{cm}^{-1}. \tag{9.24}$$

If we assume that the initial doping density is given by Eq. (9.23), that $(f_{\text{Gunn}})\cdot L = 10^7$ cm sec^{-1}, and that failure occurs when $n \cdot L$ is reduced to 5×10^{11} cm^{-2}, then we can show [using Eq. (9.23) and (9.24)] that

$$\Phi_f = 1.2 \times 10^8 (f_{\text{Gunn}})^{0.6}\ \text{n/cm}^2 \tag{9.25}$$

where Φ_f is the neutron failure fluence‡ due to increased domain growth time.

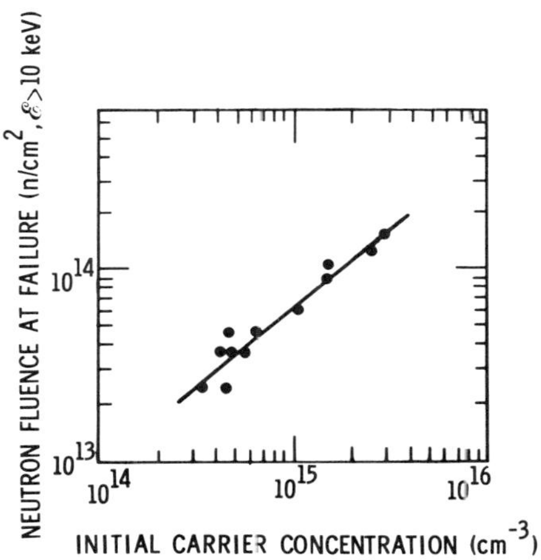

FIGURE 9.12
Failure neutron fluence versus initial doping concentration for Gunn oscillators (after Marcus and Bruemmer [19]).

For a device with $f_{\text{Gunn}} = 10$ GHz, Eq. (9.23) predicts failure at 1.2×10^{14} n/cm^2. A similar expression could be derived for γ-ray fluence. Marcus and Bruemmer [19] have also experimentally observed a similar relation [Eq. (9.25)] by plotting the failure fluence versus the initial doping concentration for Gunn oscillators. Their result is shown in Figure 9.12.

The frequency of a Gunn oscillator is determined by the transit time for a pure Gunn mode or by the cavity in the quenched- and delayed-domain

† Spectrum A.
‡ Spectrum A (see Appendix C).

modes. In any case, the frequency of oscillation is not expected to vary significantly with radiation fluence. Anderson [24] observed frequency shifts of less than 5 MHz in a cavity-controlled 9000-MHz Gunn oscillator irradiated nearly to failure at 2×10^{14} n/cm^2 ($\mathscr{E} > 10$ keV, Spectrum B).

Brehm and Pearson [22] report that the FM noise levels in a Gunn oscillator increased, at most, by a factor of 2 after a heavy irradiation with Co^{60} gamma rays. A sample of their data is shown in Figure 9.13. Their experiments were performed at X-band in a waveguide cavity. The gamma fluence (2×10^{17} photons/cm^2) is sufficient to reduce the output power by 3 dB.

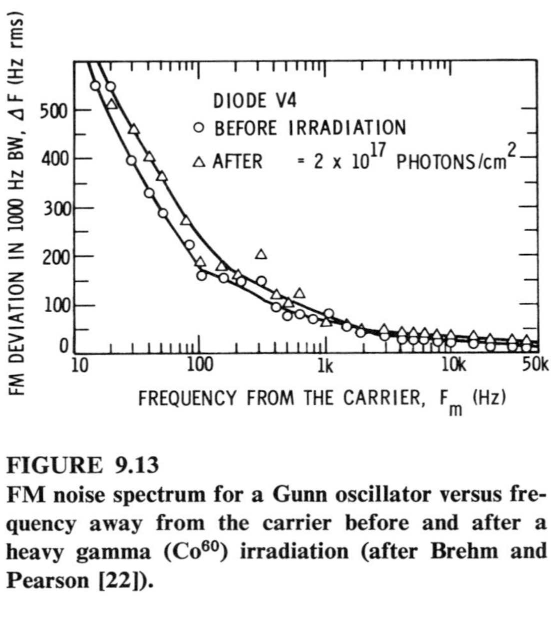

FIGURE 9.13
FM noise spectrum for a Gunn oscillator versus frequency away from the carrier before and after a heavy gamma (Co^{60}) irradiation (after Brehm and Pearson [22]).

Another failure mechanism that is seen in some [20] neutron-irradiated pulsed Gunn oscillators is the slow trapping effect. As the neutron fluence approaches the failure level [Eq. (9.25)], appreciable numbers of electrons become trapped with a long time constant in the neutron-created damage clusters. These trapped charges can give rise to a very slow domain propagation speed and produce very low frequency (100 kHz) oscillations which suppress the microwave oscillations. This effect has been studied by Law and Kao [26].

Most of the irradiation studies [17–24] report no observable changes in threshold field value E_T.

9.6 Permanent Radiation Effects in LSA Oscillators

The parameters critical to LSA operation have been discussed in Section 9.4. LSA oscillators are usually designed so that $n_0 \cdot L \gg 3 \times 10^{11}$ cm^{-2} and, thus, unlike Gunn devices, it would take a large radiation fluence to reduce the $n_0 \cdot L$ product to 3×10^{11} cm^{-2}. However, LSA devices still fail at neutron fluences near 10^{14} n/cm^2, but for a different reason. Radiation studies of LSA devices have been made by Wilson [14,15] and Berg [27]. The principal result they find is an increase in R_0 [low-field resistance, see Eq. (9.19)] and a slight decrease in mobility resulting in degraded oscillator performance.

Wilson [14,15] measured the effects of neutron bombardment (Spectrums A and B) on n^+-n-n^+ epitaxial GaAs LSA devices. After each irradiation, the low-field resistance [Eq. (9.21)] and the magneto-resistance mobility (see Section 4.5), as well as the microwave performance, were measured. The low-field resistance increased by 75%, and the low-field mobility decreased by 14% after a fluence of 1.1×10^{14} n/cm^2 ($\mathscr{E} > 10$ keV, Spectrum A). The devices exposed to Spectrum A irradiations had an active length of 0.008 cm and an active region doping density of 1.5×10^{15} cm^{-3}. The devices exposed to Spectrum B irradiations had an active length of 0.011 cm and active region doping density of 2.7×10^{15} cm^{-3}. Typical results from these irradiations are presented in Figures 9.14 and 9.15 which show that neutron irradiation causes an upward shift in oscillator frequency and a decrease in the maximum peak-power output of LSA oscillators. The reason for the increase in frequency with irradiation can be understood from an inspection of Eq. (9.19) and Figure 9.7. Part of the rf cycle time is determined by an L/R_0 time constant (line A-B in Figure 9.7). Therefore, an increase in R_0 causes a reduction in this time constant and results in an increase in operating frequency. This increase in frequency can be as large as 10% after 10^{14} n/cm^2 (see Figures 9.14 and 9.15). The decrease in maximum peak-power output (before breakdown occurs) can be understood from an inspection of the circuit shown in Figure 9.7 and Eq. (9.20). When $|R_n| > R_L$, LSA oscillations will cease, and breakdown will occur due to domain formation. Therefore, it can be seen from Eq. (9.20) that as R_0 increases with irradiation, the maximum allowable value of M [M = (bias/threshold) voltage ratio] must decrease if $|R_n| < R_L$ is to be satisfied. This in turn means that the maximum peak-power output decreases, since the maximum allowable bias voltage must be reduced. An approximate equation for the failure fluence due to breakdown (assuming the frequency shift can be ignored) can be derived. Assume that LSA oscillators are designed† with $n_0/f_0 = 2 \times 10^5$ cm^{-3} Hz^{-1}. Then using Eq. (9.22) and

† In actual practice, most LSA oscillators are designed with $1 \times 10^5 \leq n_0/f_0 \leq 3 \times 10^5$ cm^{-3} Hz^{-1}.

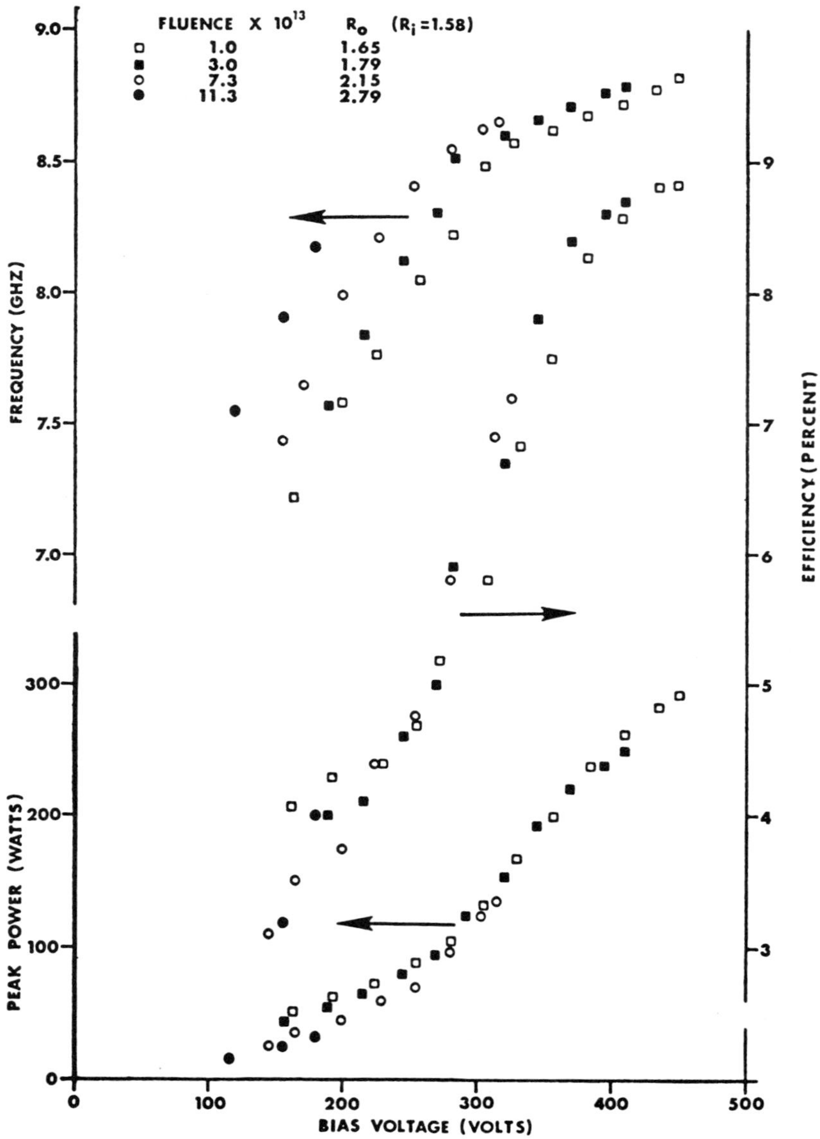

FIGURE 9.14
Effect of neutron irradiation (Spectrum A) on LSA oscillators (after Wilson [14]).

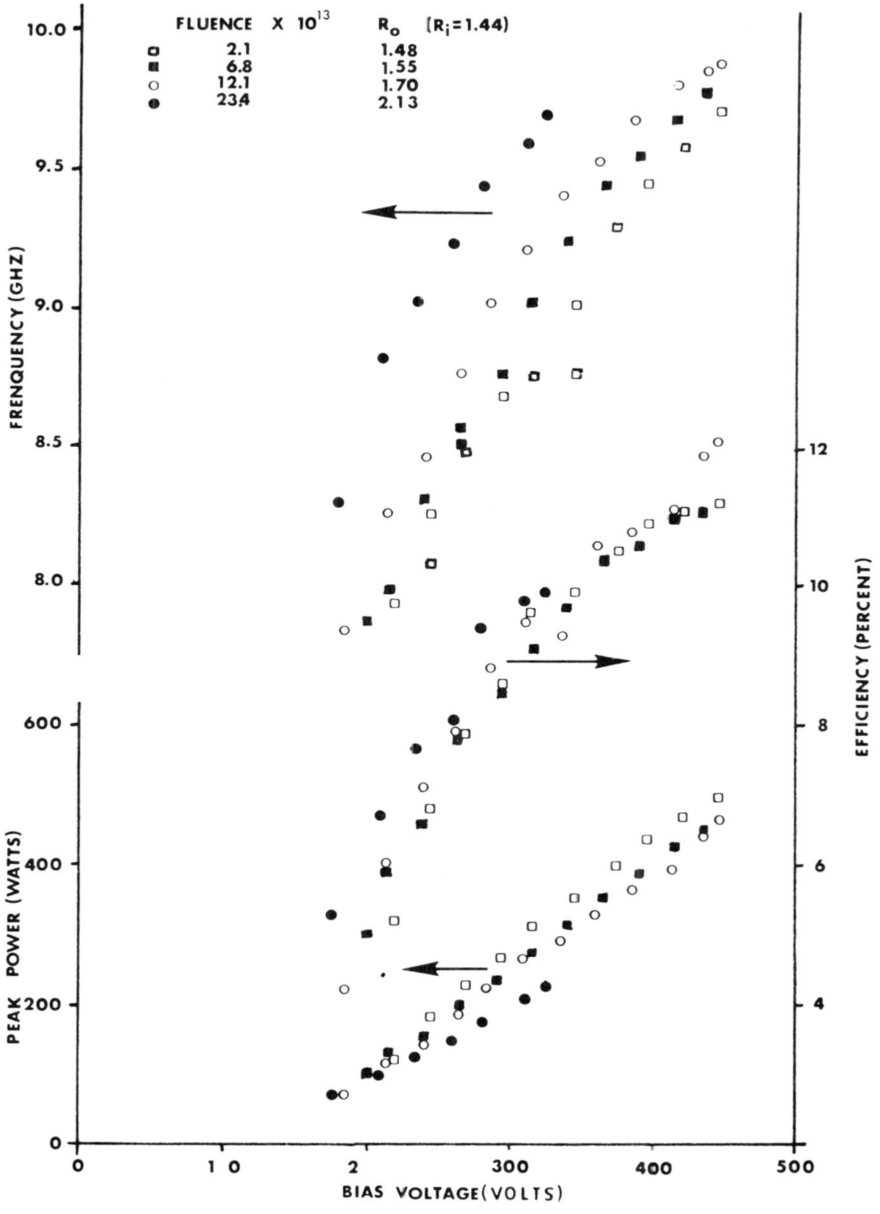

FIGURE 9.15
Effect of neutron irradiation (Spectrum B) on LSA oscillators (after Wilson [14]).

(9.24), and assuming failure of the type described by Eq. (9.20) occurs when R (postrad) $= 1.5R_0$, we may show† for the LSA mode

$$\Phi_f \cong 5.5 \times 10^7 (f_0)^{0.6} \text{ n/cm}^2 \tag{9.26}$$

where Φ_f = Spectrum A neutron fluence at failure. For a 10 GHz LSA oscillator, Eq. (9.26) predicts that failure would occur at approximately 5.5×10^{13} n/cm^2 ($\mathscr{E} > 10$ keV, Spectrum A). The failure fluence can be increased by derating the device, i.e., operating a 600 V device at 300 V; then from Eq. (9.20) we can see that an increase in R_0 of a factor of 2 to 2.5 could be tolerated before breakdown occurs.

Berg [27] has observed an increase in frequency and a decrease in maximum peak-power output for gamma (Co^{60}) irradiated LSA oscillators. His devices were similar to those described in this section. Berg observed a slight degradation in maximum power output and a 2% frequency increase after a fluence of 5×10^{16} photons/cm^2 of Co^{60} gamma rays; he observed much larger effects at fluences above this level.

9.7 Transient Ionizing Radiation Effects in Gunn Oscillators

The effects of transient ionizing radiation (x-ray, γ-ray pulses) on Gunn oscillators has been investigated by several researchers [18,23,28,29,30]. If the devices are protected by a current limiter, no permanent effects occur for pulsed ionizing radiation rates up to 10^{10} rad/sec. It has been found that ionizing radiation rates above 10^8 rad/sec cause an interruption of oscillations; however, the oscillations resume shortly ($\sim$200 nanoseconds) after the ionizing radiation ceases.

The effect of ionizing radiation on the Gunn device is to create extra holes and electrons by pair generation. From Section 5.6 we know that an ionizing radiation dose rate of 1 rad per second produces 7.25×10^{13} pairs cm^{-3} sec^{-1} in GaAs. If the ionizing radiation pulse lasts for a time much longer than one rf period (10^{-10} sec at 10 GHz), then we have a steady-state problem to solve. From Eq. (5.36) we know that for the steady state the excess photocurrent is given by

$$\Delta J = g \cdot e \cdot L \tag{9.27}$$

Here W_T has been replaced by L (the active device length), and L_p and L_n are negligible. If the velocity of all of the carriers is near to the saturated drift velocity, then we can write

$$\Delta J = g \cdot e \cdot L = \Delta n \cdot e \cdot v_s \tag{9.28}$$

† Neglecting mobility variation with irradiation fluence.

or

$$\Delta n = g \cdot \frac{L}{v_s} = g \cdot T \tag{9.29}$$

where Δn = excess electron density and T is given by Eq. (9.10). For example, Eq. (9.29) predicts that a 10-GHz Gunn device exposed to an ionizing radiation dose rate of 7×10^7 rad sec^{-1} will contain $\Delta n = 5.1 \times 10^{11}$ excess electrons per cm^3 due to the ionizing radiation. From Eq. (9.23) we know that an optimally doped 10-GHz Gunn device would contain approximately 2×10^{15} electrons/cm^3. It is important to note that this density is approximately four orders of magnitude larger than the excess-electron density produced at an ionizing radiation dose rate of 7×10^7 rad sec^{-1}. This implies that a 10-GHz device should not be affected at or below this dose rate. Figure 9.16 shows the effect of a combined neutron and gamma pulse (from a fast burst reactor) on an operational Gunn oscillator. The device is a 10-GHz diode of the same construction as those described at the beginning of Section 9.5 and represented by Figures 9.8 to 9.11. Figure 9.16 shows essentially no transient effects occurring at or below 7×10^7 rad sec^{-1}. The middle photograph in Figure 9.16 does show some permanent damage effects due to the 10^{13} neutrons/cm^2 received during the pulse. Compare Figure 9.16 with Figure 8.23 for IMPATT diodes. Figure 8.23 shows no permanent damage effects due to the same dose, but the IMPATT device stops operating temporarily at this dose rate. For a larger dose/dose rate (10^{14} n/cm^2, 5.4×10^8 rad/sec), Figure 9.16 shows that the Gunn oscillator fails permanently due to carrier removal effects. From the discussions given below Eq. (9.29) it might appear that it would take a dose rate of nearly† 3×10^{11} rad sec^{-1} to affect oscillator output. In practice it is observed that most Gunn oscillators are interrupted temporarily at dose rates above 10^8 rads sec^{-1} (ionizing radiation doses only) [18,29]. The reason for this is that most Gunn devices are mounted in high-Q microwave cavities, and the effect of the ionizing radiation on the air in these cavities must be considered. It can be shown (see Appendix B) that for ionizing radiation dose rates above 10^8 rad sec^{-1} the air plasma that forms in the cavities is sufficiently lossy to perturb the oscillator output. If this problem is alleviated by evacuating the cavity (or operating without a cavity [29]), then some rf output can be detected during dose rates as high as 2×10^{10} rad sec^{-1} in a 5-GHz Gunn oscillator [29]. This is the point where the excess generated electron density [Eq. (9.29)] begins to approach the initial free-carrier concentration [Eq. (9.23)], and a perturbation of the $n \cdot L$ product occurs.

The effect of transient ionizing radiation pulses on Gunn diode amplifiers has also been investigated [30]. Ref. [30] reports that if a current limiter is

† This is the dose rate which makes $\Delta n = n_0 = 2 \times 10^{15}\ \text{cm}^{-3}$.

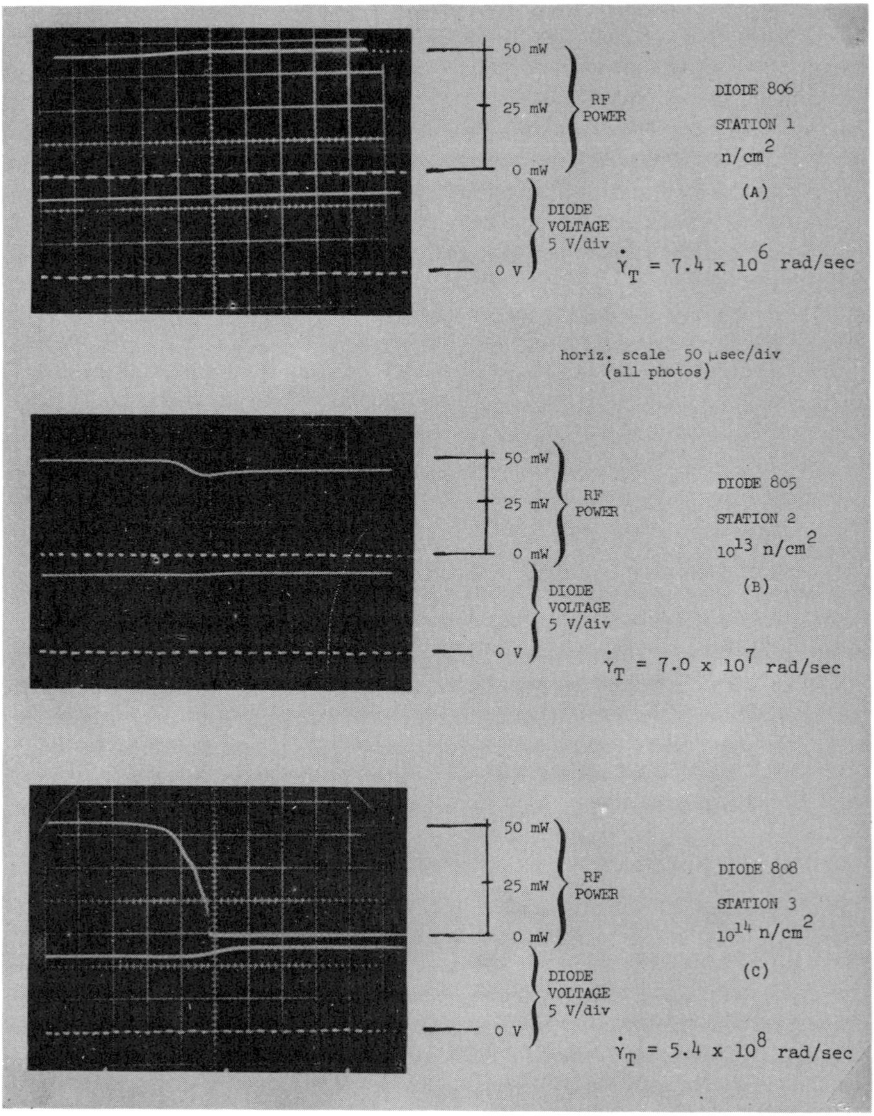

FIGURE 9.16
Effect of a combined neutron and gamma pulse on a Gunn oscillator.

used, the devices still function as normal after (but not during) an ionizing radiation pulse whose dose rate is 6×10^{10} rad sec^{-1}.

REFERENCES

[1] B. K. Ridley and T. B. Watkins, "The Possibility of Negative Resistance Effects in Semiconductors," *Proc. Phys. Soc.* (*London*), **78**, 293 (1961).

[2] C. Hilsum, "Transferred Electron Amplifiers," *Proc. IEEE*, **50**, 185 (1962).

[3] J. B. Gunn, "Microwave Oscillation of Current in III-V Semiconductors," *Sol. State Comm.*, **1**, 88 (1963).

[4] J. A. Copeland, "A New Mode of Operation for Bulk Negative Resistance Oscillators," *Proc. IEEE*, **54**, 1479 (1966).

[5] B. S. Perlman and C. L. Upadhyayula, "Transferred Electron Amps Challenge the TWT," *Microwaves*, **59** (December 1970).

[6] H. A. Watson, ed., *Microwave Semiconductor Devices and Their Circuit Applications*, McGraw-Hill, New York, 1969.

[7] S. M. Sze, *Physics of Semiconductor Devices*, Wiley-Interscience, New York, 1969.

[8] J. E. Carroll, *Hot Electron Microwave Generators*, Edward Arnold Pub., London, 1970.

[9] H. V. Shurmer, *Microwave Semiconductor Devices*, Wiley-Interscience, New York, 1971.

[10] W. H. Hayt, Jr., *Engineering Electromagnetics*, McGraw-Hill, New York, 1958.

[11] P. N. Butcher et al., "A Simple Analysis of Stable Domain Propagation in the Gunn Effect," *Brit. J. Appl. Phys.*, **17**, 841 (July 1966).

[12] J. A. Copeland, "Characterization of Bulk Negative Resistance Diode Behavior," *IEEE Trans. PGED*, **ED-14**, 461 (1967).

[13] W. O. Camp, Jr., *Optimum Circuits for GaAs Oscillators · A Theoretical and Experimental Study of Device-Circuit Interaction and Maximum Limits of Operation*, Report No. RADC-TR-71-95, Rome Air Development Center, Griffiss AFB, New York, 1971.

[14] W. E. Wilson, "Neutron Induced Degradation and Failure in High Power LSA Diode Oscillators," *Proceedings of the High Frequency Generation and Amplification Conference*, Ithaca, New York, 1971.

[15] W. E. Wilson, "Neutron Irradiation Effects in LSA Oscillators," *Proc. IEEE*, **59**, 1633 (1971).

[16] W. E. Wilson, *Pulsed Solid-State Sources for Microwave Application*, Sandia Corporation, Albuquerque, N.M., Report No. SC-DR-70-698 (1970).

[17] G. C. Novotny, Private Communication.

[18] R. H. Schnurr and H. D. Southward, "Radiation Effects upon Gallium Arsenide Devices," *IEEE Trans, Nuc. Sci.*, **NS-15**, 6, 306 (1968).

[19] G. H. Marcus and H. P. Bruemmer, "Radiation Damage in GaAs Gunn Diodes," *IEEE Trans. Nuc. Sci.*, **NS-17**, 6, 230 (1970).

[20] N. Berg and H. Dropkin, "Neutron Displacement Effects in Epitaxial Gunn Diodes," *IEEE Trans. Nuc. Sci.*, **NS-17**, 6, 233 (1970).

[21] H. P. Bruemmer, "Neutron Damage in Gunn Effect Devices," U.S. Army Report ECOM-3238 (AD-868893) (February 1970).

[22] G. E. Brehm and G. L. Pearson, "Effects of Gamma Radiation on Gunn Diodes," *IEEE Trans. PGED*, **ED-17**, 475 (1970).

[23] R. J. Chaffin, *Transient and Permanent Radiation Effects in Gunn and IMPATT Oscillators*, Sandia Corporation, Albuquerque, N.M., Report No. SC-RR-71 0378 (June 1971).

[24] W. A. Anderson, "Comparing Radiation Effects in Gunn and IMPATT Oscillators," *IEEE Trans. Nuc. Sci.*, **NS-18**, 5, 404 (1971).

[25] J. A. Copeland, "Theoretical Study of Gunn Diode in a Resonant Circuit," *IEEE Trans. PGED*, **ED-14**, 55 (1967).

[26] H. C. Law and K. C. Kao, "On the Gunn Effect in Two Valley Semiconductors with Traps," *Sol. State Elec.*, **13**, 1119 (1970).

[27] N. Berg, private communication.

[28] W. A. Anderson, "The Performance of a Gunn Diode under the Influence of Gamma Radiation," *IEEE Proc.*, **57**, 1198 (1969).

[29] N. Berg and H. Dropkin, "The Effects of Ionizing Radiation on Gunn Diode Oscillators," *IEEE Trans. Nuc. Sci.*, **NS-18**, 295 (1971).

[30] H. Dropkin and N. Berg, "Effect of Ionizing Radiation on Gunn Diode Amplifiers," presented at IEEE Conference on Nuclear and Space Radiation Effects, Seattle, July 1972.

10

Radiation Damage in BARITT Devices

10.1 Introduction

The use of punch-through injection and transit time as the mechanisms for a two-terminal semiconductor microwave negative-resistance device has been suggested by several authors [1–4]. However, only recently, further refinements in the device theory and fabrication techniques have resulted in the demonstration of operational devices [5,6]. The first operational devices were demonstrated by Coleman and Sze [5] and were given the name BARITT (for BARrier Injection Transit Time).

The practical significance of the BARITT device is that it is very easy to make (it can be made using silicon technology), and it is a low-noise device (much lower than an IMPATT oscillator). Thus, it is possible that this device will replace Gunn oscillators as local oscillators for radars, small doppler intrusion alarms, etc. The device is inherently less efficient than an IMPATT oscillator (as will be discussed later), and for this reason it is limited to low-power applications like those just listed.

The purpose of this chapter is to study permanent radiation effects, as well as transient ionizing radiation effects, on the BARITT device. A brief discussion of the basic mechanisms utilized by the BARITT will be given in Sections 10.2 and 10.3. Sections 10.4 and 10.5 examine permanent radiation effects in the device and the circuit, respectively, and Section 10.6 discusses transient ionizing radiation effects.

10.2 BARITT Device Physics

The device fabricated by Coleman and Sze [5] was a two-sided Schottky-barrier device with a 12-micron drift region. The basic principles can be derived however by considering a p^+-n-p^+ structure such as is shown at the top of Figure 10.1. The device is assumed to be a pair of abrupt junctions back to back, with a lightly doped n-type region in between them. Figure 10.1 shows that, at zero bias, one junction is slightly forward-biased (left side) and

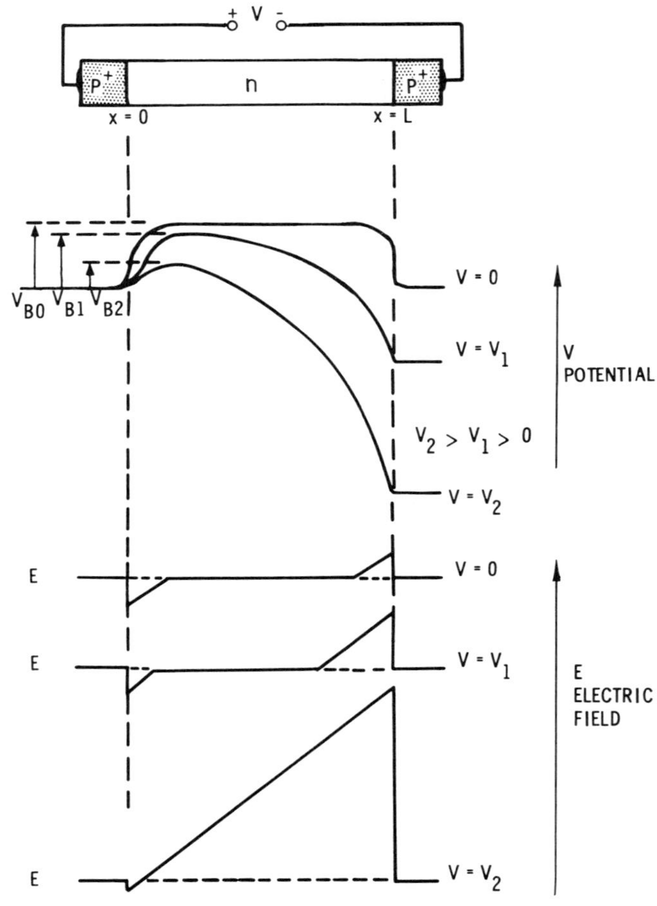

FIGURE 10.1
Potential diagram of a BARITT diode.

the other (right side) is slightly reverse-biased. Current continuity requires that the largest current that can flow is limited by the reverse-biased junction. This is because there is only a very small probability of carriers injected at the forward-biased junction diffusing through to the p^+ region at the right side.

As the bias is increased to $V = V_1$, the reverse-bias depletion width increases, and the depletion width of the forward-biased junction decreases (see Figure 10.1). The current at $V = V_1$ is still essentially limited only to reverse-bias current. If the bias voltage is increased further to $V = V_2$, then the depletion region "reaches through" to the forward-biased junction (see Figure 10.1). Now all of the injected carriers (holes) at the forward-biased junction are swept across the device (moving left to right) by the high field in the depletion

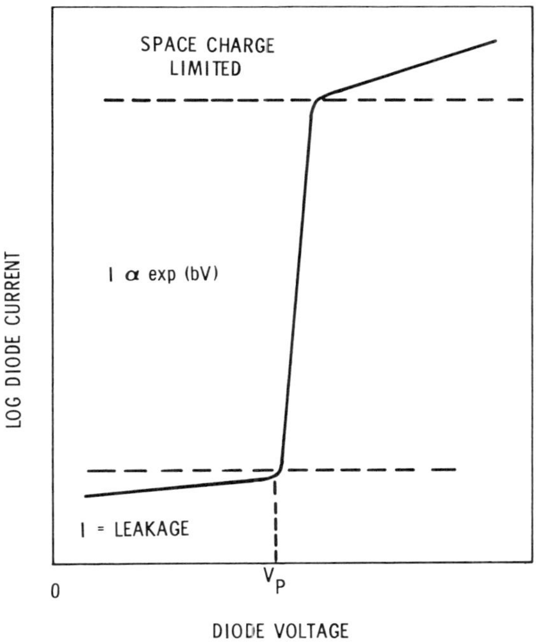

FIGURE 10.2
Typical BARITT V-I curve.

region, and the current now increases quite rapidly. The potential diagram in Figure 10.1 also shows that as the bias is increased further, a dramatic lowering of the hole barrier (V_B) occurs, which also increases the injection and therefore the current. Eventually a point is reached (when the injected hole density approaches the background doping density) where space-charge limiting occurs, and the current does not increase as rapidly with voltage. The overall shape of the dc V-I characteristic is shown in Figure 10.2 where three distinct regions of operation can be seen: (1) voltage below punch-through $I \approx I_{\text{leakage}}$; (2) voltage near punch-through, I increasing exponentially; and

(3) space-charge limited, I increasing slowly. If the voltage is increased further, a fourth region (not shown in Figure 10.2) will occur due to avalanche breakdown, and the current will again increase rapidly. The area of interest here will be the voltage near the punch-through region.

Sze et al. [7], have shown, through a solution of Poisson's equation, that the hole† current density is given by:

$$J_p \cong A_p T^2 \exp\left\{\frac{-e}{kT}\,[\phi_p + (V_f - V_a)^2/4V_f]\right\} \tag{10.1}$$

where
$$\phi_p = 0.2 \text{ volts for Pt-Si Schottky barrier}$$
$$A_p = 30 \text{ A/cm}^2/^\circ\text{K for Pt-Si Schottky barrier}$$
$$T = \text{temperature } ^\circ\text{K}$$
$$e/kT = 38.6 \text{ volts}^{-1} \text{ at } 300^\circ\text{K}$$
$$V_a = \text{applied voltage}\,.$$

The voltage at which punch-through occurs, V_f, is approximately [5]:

$$V_f = \frac{eN_D}{2\epsilon} L^2 \tag{10.2}$$

where L = n-region width. The value of V_f may be taken as the approximate operating voltage of the device. It should be noted at this point that the punch-through voltage can be changed by varying either N_D or L. The electric field in the device (above punch-through) is given approximately as [7]

$$E(x) = \frac{eN_D}{\epsilon} x + \frac{V_a - (eN_D/2\epsilon)L^2}{L}\,, \tag{10.3}$$

and a typical plot of $E(x)$ versus x is shown by the bottom line in Figure 10.1.

There are two important points to be noted here: (1) the field can be large enough so that the injected holes (moving left to right in Figure 10.1) move across almost the entire depletion region at saturated drift velocity; (2) if the electric field at the right-hand edge is too large, avalanche breakdown can occur.

10.3 BARITT Microwave Oscillators

The microwave behavior of the BARITT will be described with the aid of Figure 10.3. Assume that at operating voltage ($V_a \cong$ punch-through, V_f) the electric field is high enough so that the holes (moving left to right on

† The injected electron current (moving right to left on Figure 10.1) can be ignored because of minority carrier suppression in the p^+ region.

Figure 10.1) transit the entire depletion region ($x = 0$ to $x = L$, Figure 10.1) at saturated drift velocity, v_s. Then the transit angle is given by

$$\theta = \frac{\omega L}{v_s} \tag{10.4}$$

where ω = operating frequency and $v_s = 8 \times 10^6$ cm·sec^{-1}, saturated hole velocity. Also assume a frequency ω_0 where $\theta = 3\pi/2$ ($\omega_0 = 1.5\pi v_s/L$).

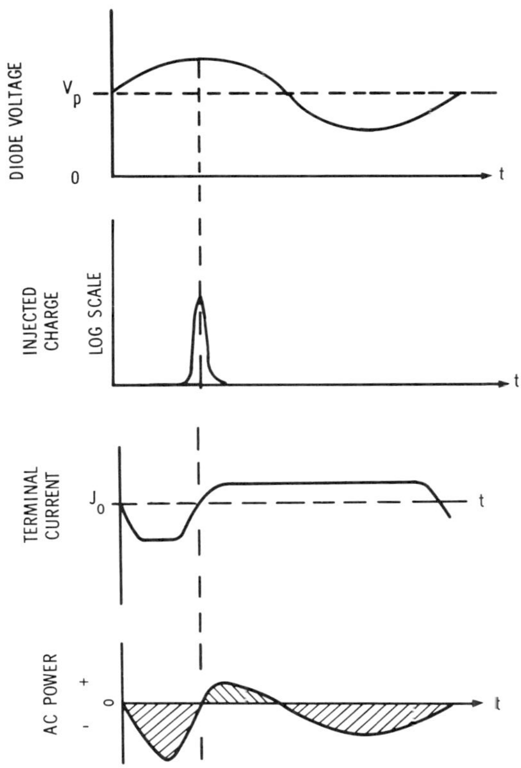

FIGURE 10.3
BARITT waveforms.

Further assume that the diode is dc biased to punch-through and a small sinusoidal voltage source is driving the diode at $\omega = \omega_0$. Then the resulting waveforms that will occur are shown in Figure 10.3. The resulting injected hole density (at the left edge of the device, Figure 10.1) as a function of time is shown in the second trace in Figure 10.3. The injected hole density peaks when the driving voltage does. It can be shown that the component of terminal current due to drifting carriers is given by†

$$I(t) = A \cdot J(t) = A \cdot \int_0^L \frac{\rho(x, t)\, dx}{T_t} \tag{10.5}$$

where A = device area, $T_t = L/v_s$, $\rho(x, t)$ = injected charge density. The terminal current therefore rises to a higher value right after the charge is injected, and remains at this value‡ until the injected charge exits the right-hand edge of the n-region (Figure 10.1). At this time the terminal current drops to its original value until more charge is injected, and then the cycle repeats. This is shown in the third trace in Figure 10.3. The resulting ac power is the product of the ac voltage times the ac current. This multiplication has been performed graphically in Figure 10.3 (bottom trace), and it can be seen that the average ac power is *negative*. This means that the device can generate power at the frequency assumed ($\omega_0 = 1.5\pi v_s/L$), and once the device gets started it can continue to oscillate. If the usual assumption is made that thermal noise starts the oscillator, it can be seen that self-starting, coherent oscillations are possible from this device. The reason for low efficiency can also be seen from the bottom trace in Figure 10.3. AC power flows into the diode during part of the cycle and out during the rest. By comparison, the extra 90° delay picked up in an IMPATT (due to avalanche delay) and the shorter transit angle used ($\theta = \pi$) makes the ac power flow out of an IMPATT for the full cycle, making an IMPATT inherently more efficient.

Coleman [9] has derived the small-signal impedance of the BARITT by using a small-signal expansion of the differential equations governing the operation of the device. The result is

$$Z = \frac{1}{j\omega C_D}\left[1 - \left(\frac{g}{j\omega C_D + g}\right) \cdot \left(\frac{1 - e^{-j\theta}}{j\theta}\right)\right], \tag{10.6}$$

where

$$C_D = \frac{\epsilon}{L} \quad \text{depletion capacitance/area},$$

and

$$g = \frac{e}{kT} \cdot \frac{J_p(V_f - V_a)}{2v_f} \quad \text{conductance/area}.$$

Equation (10.6) has been evaluated for a device with $N_D = 5.6 \times 10^{14}/\text{cm}^3$ and $L = 12\ \mu$ at a current density $J_p = 8\ \text{amp/cm}^2$. The resulting impedance is plotted in Figure 10.4 where it can be seen that the device has negative

† Following an argument similar to Read [11].

‡ Neglecting recombination effects.

resistance between approximately 4.25 GHz and 6.5 GHz. The peak in the negative resistance occurs close to the $3\pi/2$ frequency (assumed in Figure 10.3) which is

$$\omega_0 = \frac{3\pi v_s}{2L} = 3.1 \times 10^{10}/s(4.9 \text{ GHz}) . \tag{10.7}$$

The Q of the device is given by

$$Q = \frac{|X|}{R} .$$

From Figure 10.4 it can be seen that the minimum Q is approximately -40; thus care must be taken to reduce the series resistance losses in this fairly high-Q device.

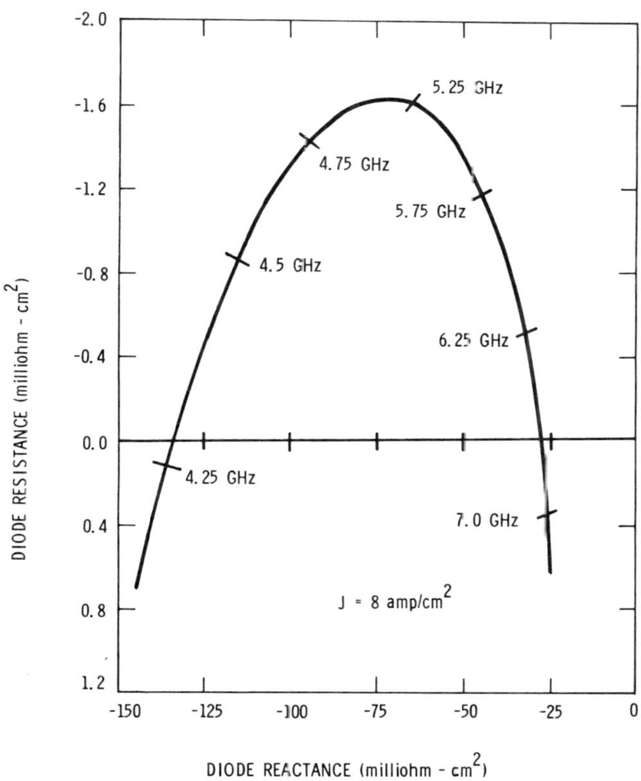

FIGURE 10.4
Small-signal diode impedance.

At a single frequency, the impedance values shown in Figure 10.4 may be modeled as a simple negative resistance in series with a capacitor. To use the BARITT as an oscillator, it is necessary to match the load impedance to the device impedance with a suitable matching network such as is shown in Figure 10.5A. In practice this can be accomplished simply by using sliding

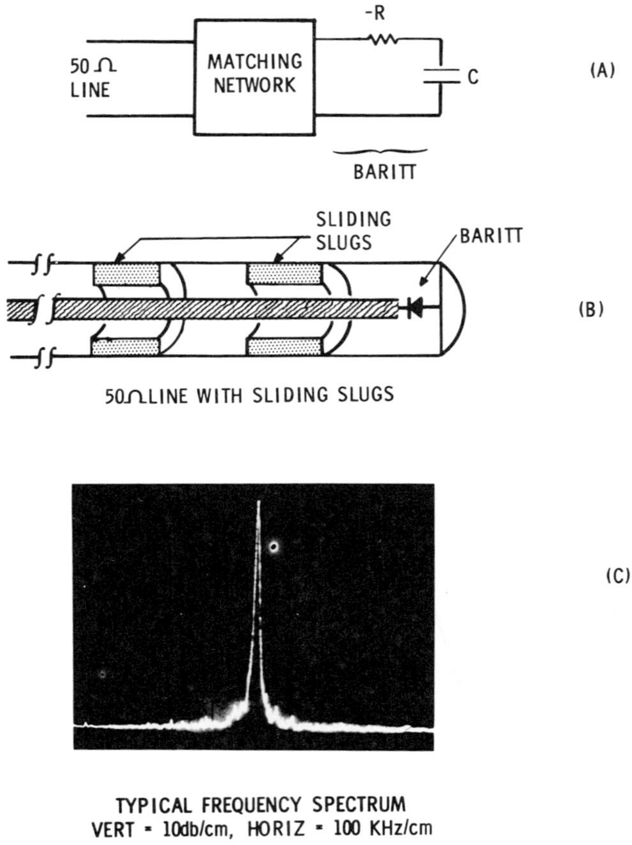

FIGURE 10.5
Diode circuits and typical frequency spectrum.

$\lambda/8$ slugs in a 50-Ω transmission line such as is shown in Figure 10.5B. A typical output frequency spectrum of a BARITT oscillator is shown in Figure 10.5C. The center frequency is 5 GHz and the power output is 10 mW. The previously mentioned low noise operation of the BARITT can be seen quite clearly from Figure 10.5C (the vertical scale is 10 dB/division). The reason for the low noise of the BARITT in comparison with the IMPATT is as follows: In the IMPATT the moving charge carriers are produced by avalanche multiplication of a small amount of randomly injected charge. This process produces multiplied shot noise, while in the BARITT the noise is strictly shot noise (reduced somewhat by space-charge smoothing). The noise performance of a BARITT is typically 20 dB better than an IMPATT.

It was mentioned earlier that for a specific frequency the optimum length for the device is given by Eq. (10.7). It has been shown [8] that the optimum doping density, N_D, increases with increasing frequency. In particular, the device should be doped to the point just below where avalanche breakdown occurs on the right-hand edge (Figure 10.1) at punch-through. This will insure that the carriers will transit most of the depletion region at saturated drift velocity. Figure 3.8 shows that the critical peak field at breakdown of an abrupt junction is approximately 2.8×10^5 V/cm. The maximum field in the BARITT is approximately given by

$$\begin{aligned} E_{\max} &= \frac{eN_D}{\epsilon} L \qquad (10.8) \\ &= 2.8 \times 10^5 \text{ V/cm to be just at breakdown} \\ &= 2.0 \times 10^5 \text{ V/cm, to stay safely below breakdown}\,. \end{aligned}$$

By combining Eq. (10.7) and (10.8) we may write

$$N_{D\text{opt}} \cong 2 \times 10^5 \frac{2\omega_0 \epsilon}{3\pi v_s e}, \qquad (10.9\text{A})$$

and we already know

$$L_{\text{opt}} \cong \frac{3}{2} \pi \frac{v_s}{\omega_0}. \qquad (10.7)$$

These are the approximate design equations for the BARITT doping density and length. It is also known [8] that the optimum operating current density increases nearly linearly with increasing frequency and with an optimum value of 30 A/cm^2 at 5 GHz or:

$$J_{\text{opt}} \cong 30\left(\frac{f_0}{5 \times 10^9}\right). \qquad (10.9\text{B})$$

These design relationships will be utilized later to predict the failure fluence of these devices in general.

10.4 Permanent Radiation Effects in BARITT Devices

This section analyzes the effects of neutron irradiation on the BARITT device. The actual device to be examined is a metal-semiconductor-metal device made by forming a Pt-Si Schottky barrier on each side of a 12-μ-thick silicon n-region. The n-region is doped with $N_D = 5.6 \times 10^{14}/\text{cm}^3$ and has an area of 3.8×10^{-4} cm^2. Neutron damage in silicon was discussed in Chapter 5, where it was found that the most lightly doped regions are affected first (the n-region in this case). Calculations show that, for the fluences of interest, recombination to first order may be neglected (since carrier transit time is short in comparison to lifetime). Changes in low-field mobility may also be neglected since we are concerned with carriers moving at saturated drift velocity (see Section 8.4). The effects of the traps introduced by the neutron irradiation will be shown to be the primary damage mechanism of interest. The trapping phenomena is assumed to be described by the divacancy model (discussed in Sections 2.7 and 5.3).

The metal-semiconductor-metal device was subjected to the following experimental sequence:

1. Measure C-V characteristic.
2. Measure static V-I characteristic.
3. Measure microwave power and frequency at fixed slug settings.
4. Measure tuning curve.
5. Irradiate the device.
6. Repeat 1 through 5.

The devices were irradiated at the Sandia Annular Core Pulsed Reactor (ACPR) and the fluences given are for $\mathscr{E} > 10$ keV (Spectrum B).

Figure 10.6 shows the capacitance-voltage characteristic (measured at 1 MHz) for a typical BARITT diode pre- and postrad. From Figure 10.6 it can be seen that the zero bias capacitance decreases dramatically with radiation while the punch-through capacitance ($V_a \cong 35$ volts) remains essentially unchanged. This is to be expected since at low bias voltages one finds that the traps are occupied in a negative state (trapping electrons from the undepleted n-region). This negative trapped charge compensates the material (by canceling the effect of the positively charged donor atoms) and the result is an increased depletion width W (and a lower depletion capacitance, $C = \epsilon A/W$) at low bias voltages. The lower limit on depletion capacitance occurs at punch-through and is fixed by the spacing between the heavily doped p^+-regions (Schottky barriers) and so the punch-through capacitance does not change with radiation.

Figure 10.7 shows the measured V-I characteristic pre- and post-rad for a typical BARITT diode. From Figure 10.7 it can be seen that at low currents the voltage required for a given current *decreases* with increasing neutron fluence. This occurs for the same reason as mentioned above in the capacitance-voltage discussion, i.e., the material is compensated by trapped electrons which lowers the punch-through voltage. This is equivalent to reducing

N_D in Eq. (10.2). At higher currents the voltage for a given current is seen to *increase* with increasing neutron fluence. This occurs because now the material is fully punched through (depleted of free electrons) and the only carriers available to be trapped are the holes flowing through the device (recall from Section 10.2 that the electron current is negligible). In this case the traps are charged in a positive state. This positive trapped charge now adds to the ionized donor charge density and effectively increases N_D (as far as

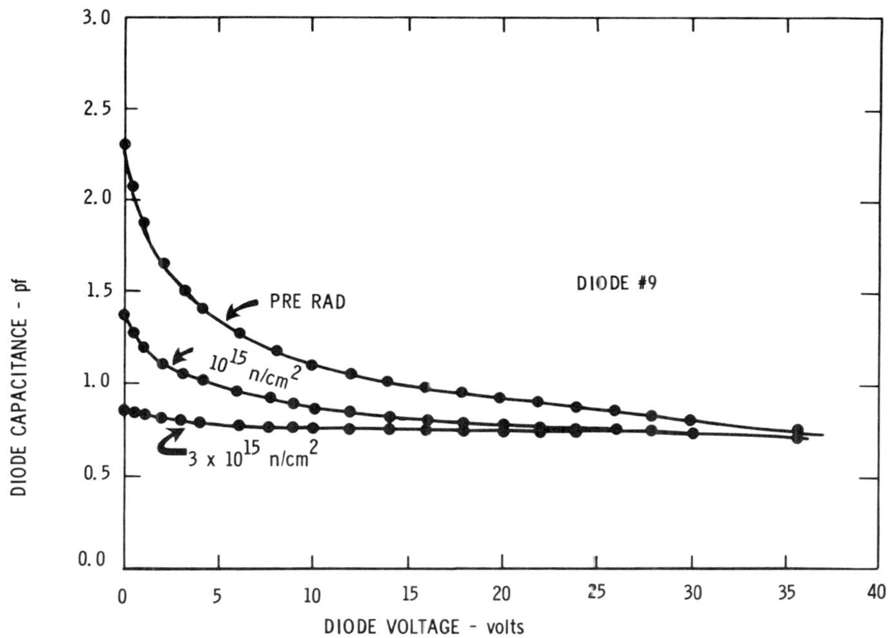

FIGURE 10.6
Capacitance-voltage plot, pre- and post-rad.

the solution to Poisson's equation is concerned), and from Eq. (10.2) it can be seen that this increases the operating voltage.

At higher currents postrad it can be seen that the V-I curve bends back to the left (a dc differential negative resistance). This is due to heating effects in the device at high currents. For each point in Figure 10.7 the diode was allowed to reach thermal equilibrium. At higher temperatures one finds an increase in the rate at which the holes are reemitted from the trap, and thus a net reduction in the trapped positive charge [10] is seen. This reduces the effective N_D and voltage [see Eq. (10.2)]. This reduction in voltage with heating can occur quickly, and the device in this experiment actually burned out before the power supply's current limiter could respond. This is one potential radiation-induced failure mechanism of BARITT devices. The occupation probabilities of the two-level trap are defined by Eq. (2.63) to (2.65). The divacancy capture cross-sections are given in Section 5.3. Assume that an

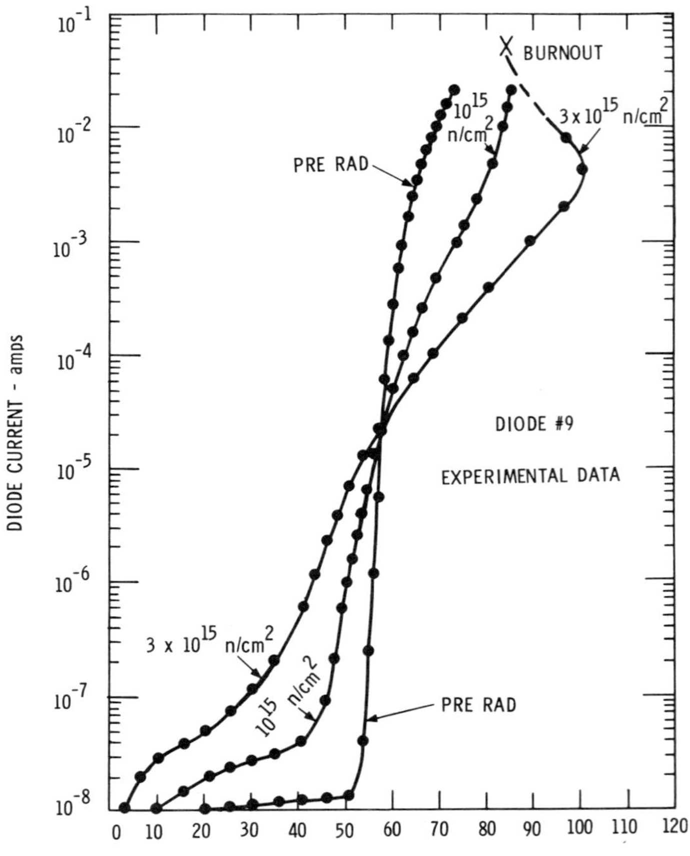

FIGURE 10.7
Diode V-I curve as a function of neutron fluence.

acceptor level at $\mathscr{E}_i + 0.085$ eV and a donor level at $\mathscr{E}_i - 0.22$ eV are produced by neutron irradiation. If it is assumed that the device is just at punch-through, then n (free electron density) will be very small and p (free hole density) will be given by

$$p \cong \frac{J_p}{ev_s}, \tag{10.10}$$

i.e., the free holes are those due to the injected hole current. The occupation probabilities using the above assumptions have been calculated and the results are tabulated in Table 10.1 for $T = 293°K$. The damage in the device due to changes in N'_D (effective ionized donor concentration) may be modeled through Poisson's equation, i.e., assume

$$N'_D = N_D + \eta^+ \cdot N_T - \eta^- \cdot N_T \tag{10.11}$$

where N_T = trap density. In particular, if it is assumed that three traps/cm³ per n/cm² (a nominal value) are introduced by radiation, Eq. (10.11) becomes

$$N'_D = N_D + 3 \cdot (\eta^+ - \eta^-) \times \text{fluence}\,. \tag{10.12}$$

If this value is used for N_D in Eq. (10.1) to calculate the V-I characteristic (remembering that η^+ and η^- are current dependent, see Table 10.1), it is

TABLE 10.1

Diode Current Density (A/cm²)	Diode Current (Amp)	Hole Density $p = \frac{J}{ev_s}\Big/\text{cm}^3$	$\frac{N_T^+}{N_T} = \eta^+$	$\frac{N_T^0}{N_T} = \eta^0$	$\frac{N_T^-}{N_T} = \eta^-$
2.6×10^{-5}	10^{-8}	2.4×10^{7}	7×10^{-7}	0.983	1.7×10^{-2}
2.6×10^{-4}	10^{-7}	2.4×10^{8}	5×10^{-6}	0.983	1.7×10^{-2}
2.6×10^{-3}	10^{-6}	2.4×10^{9}	5×10^{-5}	0.985	1.5×10^{-2}
2.6×10^{-2}	10^{-5}	2.4×10^{10}	5×10^{-4}	0.993	7×10^{-3}
2.6×10^{-1}	10^{-4}	2.4×10^{11}	5×10^{-3}	0.994	1×10^{-3}
2.6×10^{0}	10^{-3}	2.4×10^{12}	4.6×10^{-2}	0.954	1×10^{-4}
2.6×10^{1}	10^{-2}	2.4×10^{13}	2.53×10^{-1}	0.747	2×10^{-5}
2.6×10^{2}	10^{-1}	2.4×10^{14}	7.22×10^{-1}	0.278	1×10^{-6}

possible to obtain a theoretical V-I characteristic pre- and postrad. This has been done and is shown in Figure 10.8. The theoretical curves in Figure 10.8 show essentially the same behavior as the experimental curves shown in Figure 10.7. The dotted region of Figure 10.8 is the region in which space-charge limiting and temperature effects invalidate Eq. (10.1). The agreement between the theoretical model and the experimental measurements reinforce the arguments made earlier about the charge states of the two-level trap first compensating the material at low currents then effectively adding to the ionized donor atom charge at higher currents.

The theoretical electric field distribution in the device as a function of neutron fluence can also be calculated using N'_D from Eq. (10.12) and Eq.

(10.3). This has been done for the BARITT under discussion for prerad and an assumed fluence of 10^{15} n/cm². The result is plotted in Figure 10.9. The effect of neutron fluence is to increase the electric field inside the device (at the operating current density). This makes sense since the terminal voltage is

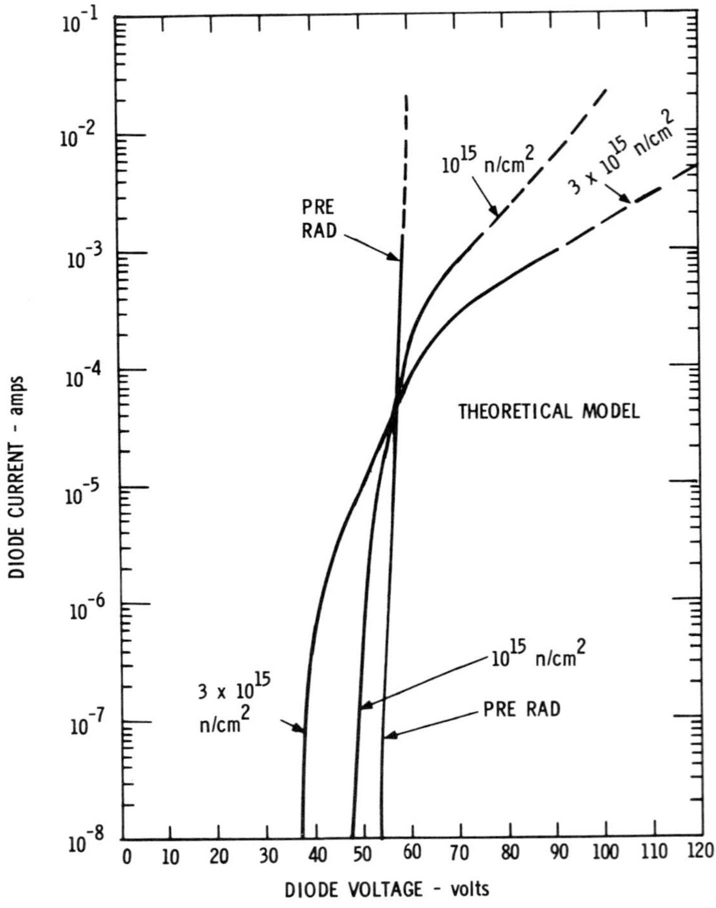

FIGURE 10.8
Theoretical V-I curves for irradiated BARITT.

simply the integral of the field across the device and the terminal voltage increases with radiation at the operating current (see Figures 10.7 and 10.8). This will have two effects: (1) the fraction of the depletion region which the carriers transit at below saturated velocity is reduced postrad ($d_2 < d_1$, see

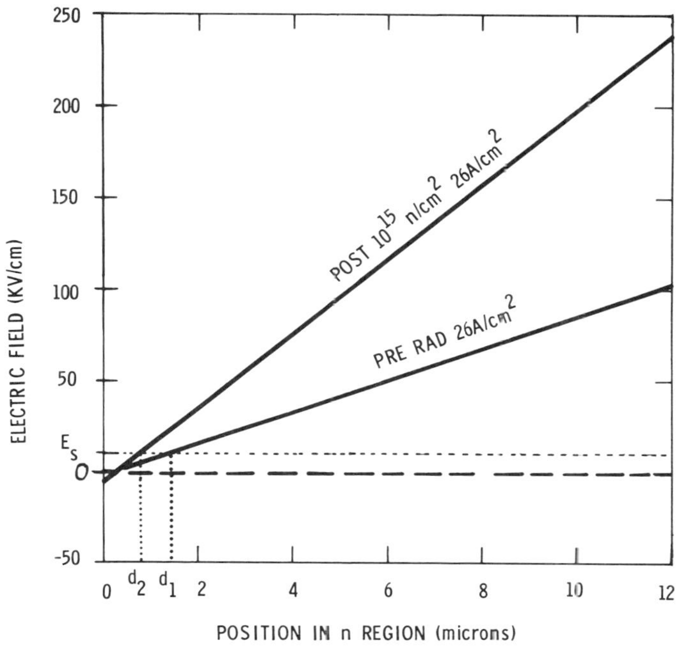

FIGURE 10.9
Electric field distribution for BARITT.

Figure 10.9); (2) the peak electric field on the right-hand edge is beginning to approach the 280 kV/cm avalanche breakdown value, and breakdown could occur.

10.5 Permanent Radiation Damage in BARITT Oscillators

The microwave characteristics of the irradiated devices were measured in a sliding slug fixture like the one shown in Figure 10.5B. The tuning characteristic (the range over which oscillations could be obtained using all possible slug settings) was measured for a constant current of 15 mA. Typical results of this measurement for several values of neutron fluence are shown in Figure 10.10. The tuning range of the device does not change appreciably even up to the point of failure at 3×10^{15} n/cm^2; however, the power output changes drastically—first increasing by nearly 3 dB at 10^{15} n/cm^2 and then decreasing 10 dB at 3×10^{15} n/cm^2. The efficiency varies similarly (0.4% prerad, 0.52% after 10^{15} n/cm^2, and 0.05% after 3×10^{15} n/cm^2). The increase in power output and efficiency is attributed to the increase in internal electric field with radiation which reduces the fraction of the depletion region transited at less than saturated drift velocity and also increases the applied voltage. The falloff in power output that occurs at a higher neutron fluence is attributed to avalanche multiplication occurring within the device (to be discussed in detail later).

The data in Figure 10.10 were obtained for all possible slug settings. Another experiment was performed for fixed slug settings of the fixture shown in Figure 10.5B. To accomplish this the device and oscillator fixture were tuned up, and the slugs were locked into place. Then the device and

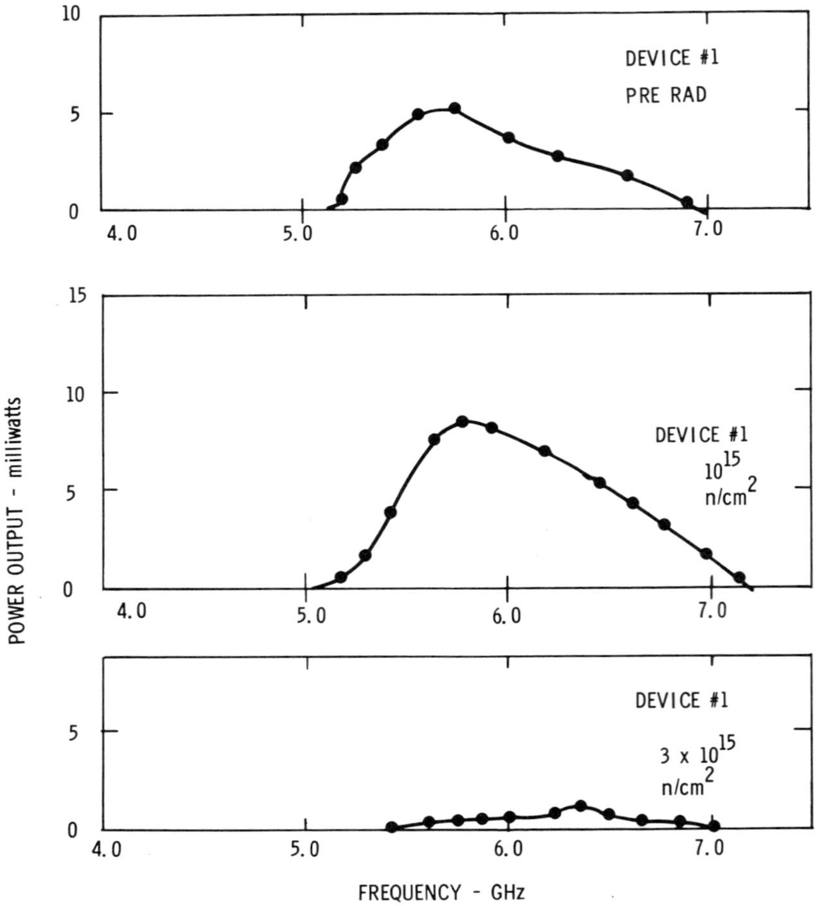

FIGURE 10.10
BARITT tuning curves for several values of neutron fluence.

fixture were lowered into the core of the Sandia ACPR reactor, and the device was connected through cables to the outside world. The experimental procedure used was as follows:

1. Allow heatsink temperature to return to 30°C (measured with a thermocouple).
2. Turn on power and measure operating voltage at a current of 15 mA.
3. Measure power output and frequency.
4. Shut down voltage.
5. Do a reactor power run.
6. Repeat 1 through 5.

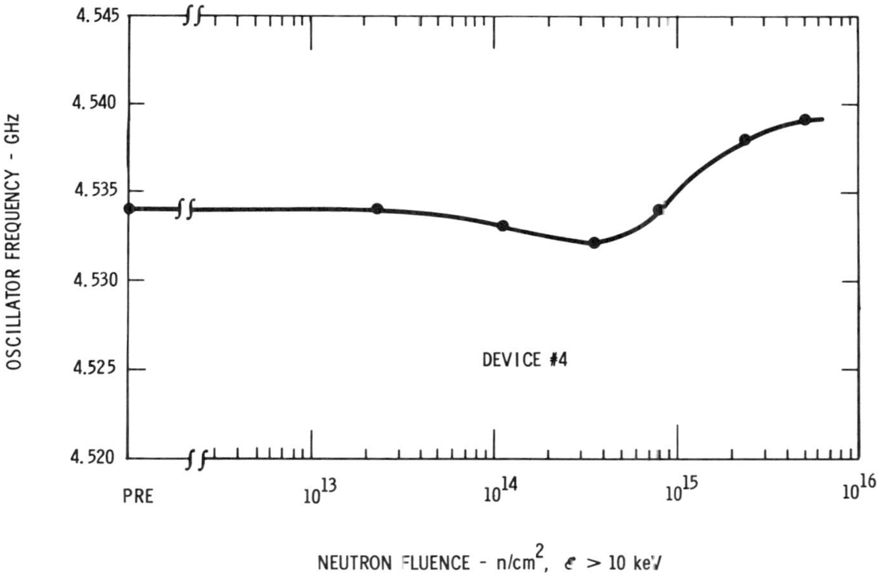

FIGURE 10.11
Frequency shift versus fluence for BARITT oscillator.

By using the above procedure it was possible to eliminate variations in the data due to mechanical variations, i.e., diode socketing problems, slug movement, etc. The data obtained in this way is shown in Figures 10.11 to 10.13. Figure 10.11 shows that frequency shift was less than ± 5 MHz at 5×10^{15} n/cm^2 (where the power output was severely degraded). Recall from Figure 10.4 that this is a high-Q device (i.e., displacement current $\gg$ conduction current), and thus the operating frequency is determined primarily by the resonance of the punch-through capacitance (which is independent of radiation, see Figure 10.6) with the inductance provided by the oscillator fixture (also independent of radiation).

The power output, Figure 10.12, is seen to peak at 2×10^{15} n/cm^2 and then fall off dramatically. The reasons for this have been discussed previously (an increase in electric field with radiation until avalanche is reached) and will be discussed in more detail later.

The operating voltage (at $I = 15$ mA) is shown in Figure 10.13. The operating voltage increases from 85 volts prerad to 140 volts at 5×10^{15} n/cm^2. The reason for this is that the trapped holes add to the ionized donor concentration, as has been discussed previously.

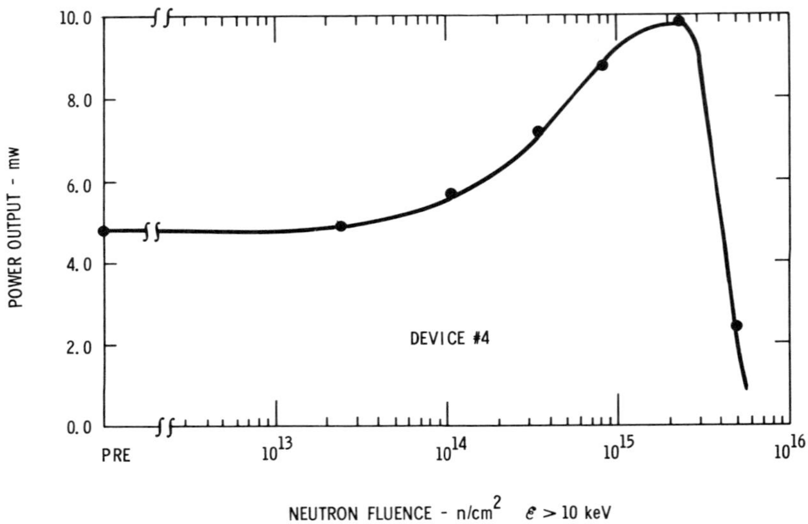

FIGURE 10.12
Power output versus fluence for BARITT oscillator.

The severe degradation in power output that occurs at and above fluences of 3×10^{15} n/cm^2 in the BARITT are attributed to avalanche multiplication within the device. The increase in internal electric field that occurs with neutron irradiation (see Figure 10.9) eventually causes avalanche multiplication to be a problem. At fields near breakdown it is possible to have significant avalanche multiplication without having avalanche breakdown. Since hole current is the only concern, the multiplication factor may be obtained from Section 3.5 as

$$M = M_p = \frac{J_p(L)}{J_p(0)} = \frac{1}{1 - \int_0^L \alpha_p \, dx}, \tag{10.13}$$

where $J_p(L)$ = hole current at $x = L$ (Figure 10.1)

$J_p(0)$ = hole current at $x = 0$ (Figure 10.1)

$$\alpha_p = 2.25 \times 10^7 \exp\left(-\frac{3.26 \times 10^6}{E}\right) \quad \text{(Table 3.1)}$$

E = electric field, V/cm.

For the device of interest it is possible to evaluate Eq. (10.13) as a function of fluence using Eq. (10.12), (10.3), and (10.1). The calculated value of $\int_0^L \alpha_p\, dx$

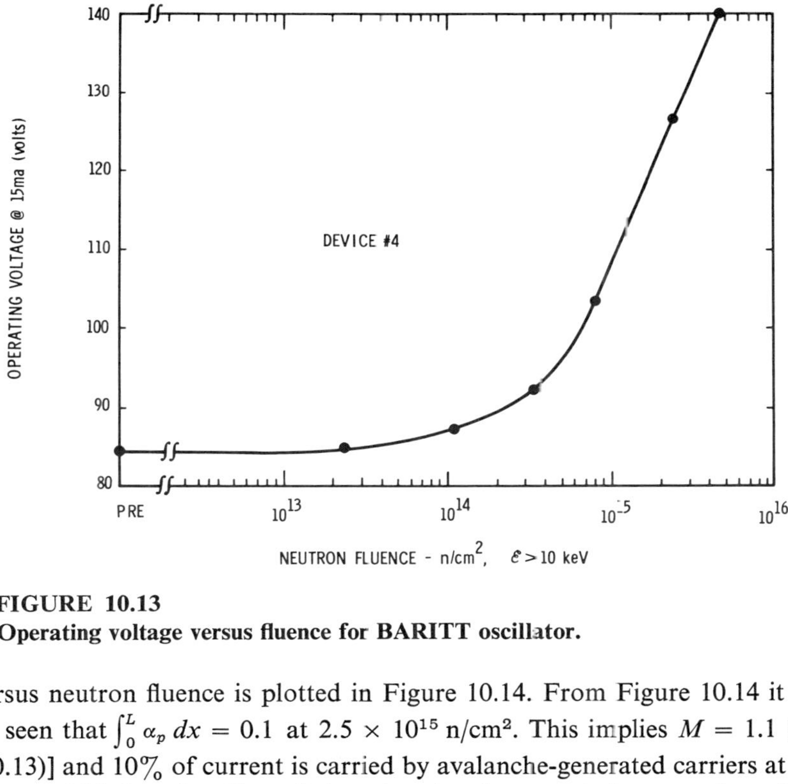

FIGURE 10.13
Operating voltage versus fluence for BARITT oscillator.

versus neutron fluence is plotted in Figure 10.14. From Figure 10.14 it can be seen that $\int_0^L \alpha_p\, dx = 0.1$ at 2.5×10^{15} n/cm^2. This implies $M = 1.1$ [Eq. (10.13)] and 10% of current is carried by avalanche-generated carriers at this fluence. This has the effect of changing the phase angle between the voltage and conduction current components and reducing the average power output. For an explanation of this, refer to Figures 10.3 and 10.9. The charge pulse is injected at $t = 0$ (Figure 10.3) and at $x = 0$ (Figure 10.9). If this is assumed to be a small-signal case, then M [the multiplication ratio, Eq. (10.13)] is essentially independent of time. Also, it is obvious that most of the avalanche multiplication will take place near the peak in electric field (right edge, Figure 10.9). This means that the peak in avalanche generation will occur when the charge pulse (Figure 10.3) has drifted nearly out of the device. Normally the terminal current would decrease at this time (Figure 10.3, third trace), but due to avalanche generation of an electron-hole pair an electron starts drifting back (right to left on Figure 10.9) in the opposite direction across the device and exits $\frac{3}{4}$ cycle later. The avalanche-generated hole exits at

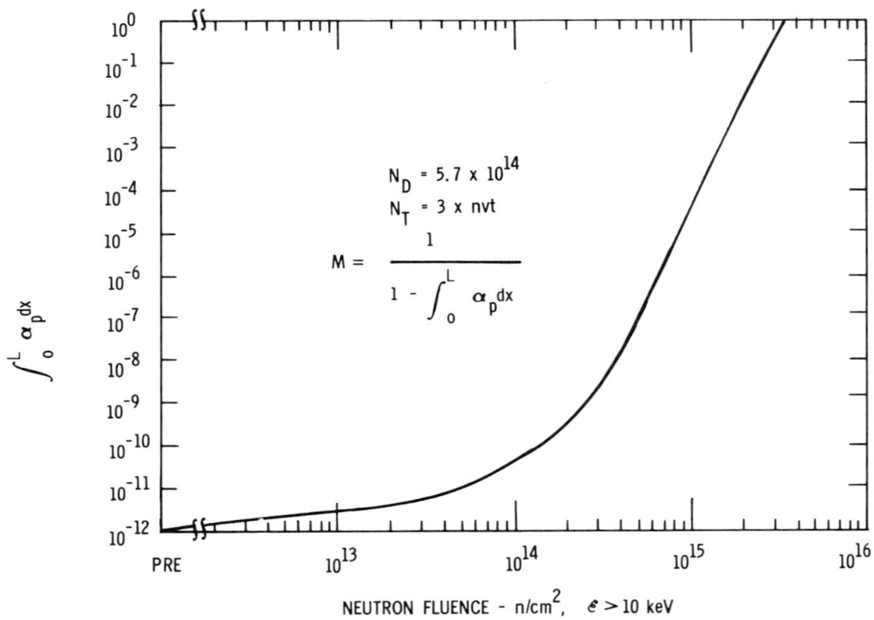

FIGURE 10.14
Ionization integral versus neutron fluence.

the right-hand edge almost immediately after being generated and is of no consequence. The reverse drifting electrons keep the terminal current up [see Eq. (10.5)] until they exit the left-hand edge (Figure 10.9). This extra current component (the electrons) changes the phase between the voltage and current, and it reduces the average negative resistance per cycle. It can be shown that when one electron is generated for each hole (or $M = 2$) the negative resistance vanishes completely. Consequently, as the internal electric field increases with neutron fluence, a point will be reached where avalanche multiplication causes a drop-off in device performance.

An equation for the failure fluence, in general, for these devices will now be developed. If it is assumed that the problem created by the dc differential negative resistance due to thermal effects can be overcome with a suitable bias network, then the primary failure mechanism in these devices will be failure due to avalanche multiplication. The failure fluence can be obtained from the device design equations, given in Section 10.3, and the trap occupation probabilities. Assume that the trap occupation probability for the positive state is 10%.† This is accurate at 5 GHz but is too small at higher frequencies where the operating current densities are higher. From Eq. (10.8) we have

$$E_{\max} = \frac{eN_D'}{\epsilon} L , \tag{10.8}$$

and from Eq. (10.12) we have

$$N_D' = N_{D\,\mathrm{opt}} + \frac{3\Phi}{10} , \tag{10.14}$$

where Φ = neutron fluence. From Eq. (10.9) we have

$$N_{D\,\mathrm{opt}} = 2 \times 10^5 \left(\frac{2\omega_0 \epsilon}{3\pi v_s e}\right) . \tag{10.9}$$

Combining Eq. (10.8), (10.9), and (10.14), we can show that

$$E_{\max}(\omega, \Phi) = \frac{eL}{\epsilon} \left(\frac{4 \times 10^5 \omega_0 \epsilon}{3\pi v_s e} + 0.3\Phi\right) . \tag{10.15}$$

If we set $E_{\max} = 2.8 \times 10^5$ V/cm (the avalanche value), we may express [also using Eq. (10.7)] the neutron fluence at which failure occurs as

$$\Phi_{\mathrm{failure}} \cong 4.8 \times 10^4 \cdot \omega_0 \tag{10.16}$$

where ω_0 = design frequency. For example, the device studied in this chapter operated at $\omega_0 = 3.14 \times 10^{10}/s$ (i.e., 5 GHz) and Eq. (10.16) predicts neutron-induced failure at 1.5×10^{15} n/cm^2, fairly close to the experimentally observed value.

10.6 Transient Ionizing Radiation Effects

This section describes experimental and theoretical studies of the effects of transient ionizing radiation on BARITT oscillators. The device utilized in the experiments was described in Section 10.4. The transient ionizing radiation was obtained in the form of short, intense x-ray pulses from a febetron

† This will depend on the current density (see Table 10.1).

flash x-ray source. This source is capable of generating pulses of nearly† Gaussian shape with a half width of approximately 60 nsec. The ionizing radiation rate (rads/sec) was calculated by the use of TLD-200 dosimeters (which measure the integrated x-ray dose). The integrated dose,‡ divided by 60 nsec, was taken to be the peak ionizing radiation rate.

The experimental setup used is shown in Figure 10.15. The BARITT oscillator runs CW, and its output power is monitored by a crystal detector (see Figure 10.15). The BARITT is biased through a 500-Ω resistor and operates

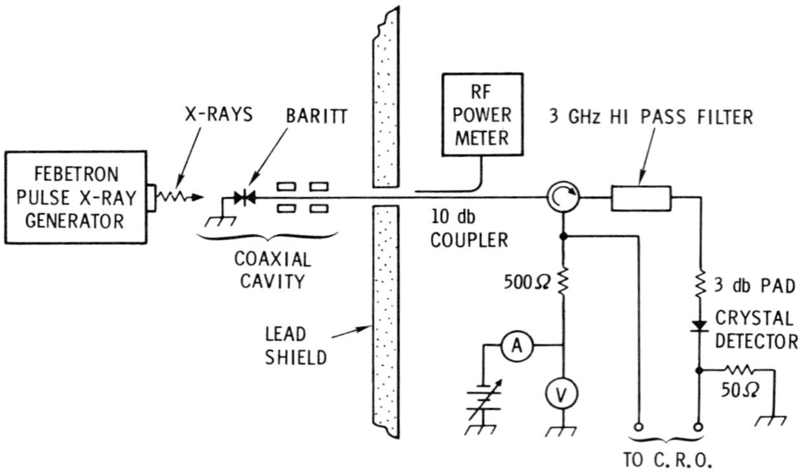

FIGURE 10.15
Experimental setup for ionizing radiation test.

at a nominal 100 V and 15 mA. The diode voltage and crystal detector voltage are brought out to the vertical channels of a dual beam C.R.O. The horizontal deflection of the C.R.O. is set to single-sweep and triggered at the beginning of the x-ray pulse. Typical oscilloscope traces obtained in this way are shown in Figure 10.16. The top trace in each photograph in Figure 10.16 is the diode voltage (offset to the 110-V nominal operating voltage). A decrease in diode voltage (with the power supply held at a constant voltage) corresponds to increased current through the diode (see Figure 10.15). The increase in current may be simply calculated from

$$\Delta I = \frac{-\Delta V}{R} = \frac{-\Delta V}{500} \tag{10.17}$$

† The pulse is not symmetrical in that the rise time is considerably shorter than the fall time.

‡ Corrected to account for the absorption of x-rays by the aluminum microwave cavity walls.

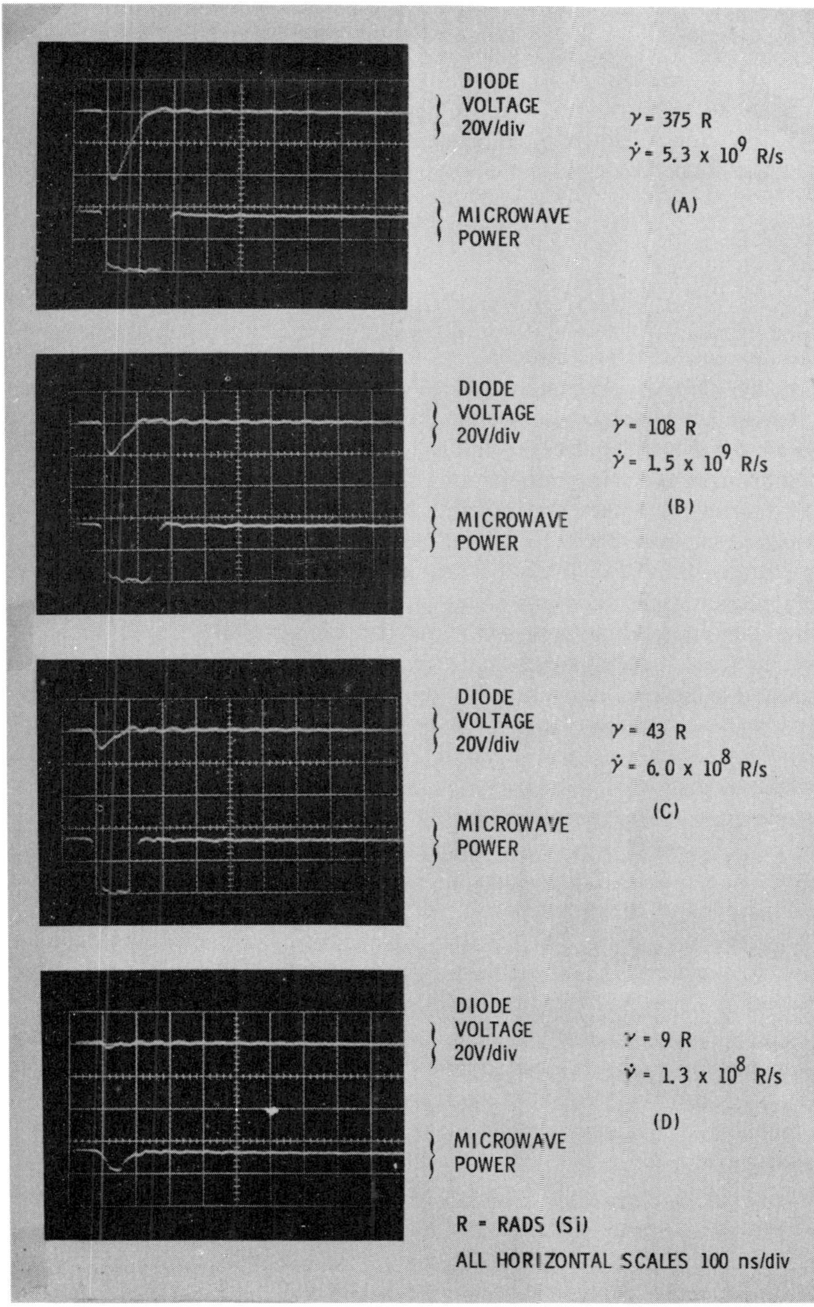

FIGURE 10.16
Experimental data for ionizing radiation test.

where ΔV = decrease in diode voltage. The lower trace in each photograph in Figure 10.16 corresponds to the rf power output from the BARITT oscillator. The bottom line on the graticule corresponds to zero power output while the steady-state power output level corresponds to 15 mW. The deflection between zero and 15 mW was calibrated and recorded. For example, the dip in power output shown in Figure 10.16D corresponds to a 3 dB reduction in power output. Figure 10.16A–C, shows that for large ionizing radiation dose rates (greater than or equal to 6×10^8 rad/sec) the oscillations cease during the pulse and restart shortly (in approximately 60 nsec) after the pulse has ended. It can also be seen that diode current increases during the pulse (by 80 mA in Figure 10.16A). At lower levels (1.3×10^8 rad/sec), the oscillation does not cease during the pulse, but the power output drops slightly during the pulse. At levels lower than 10^8 rad/sec no changes in output during the x-ray pulse can be seen.

The principal effect of ionizing radiation on semiconductor junction devices (such as a BARITT) is the generation of hole-electron pairs in the depletion regions. The BARITT device operates punched-through, and thus the depletion region consists of the entire n-region. If the ionizing radiation pulse lasts longer than a few rf cycles (which is usually true), the problem may be treated as a steady-state problem similar to the IMPATT case discussed in Section 8.6. From Eq. (8.19) we may write

$$\Delta I = g \cdot e \cdot L \cdot A \tag{10.18}$$

where $g = 4 \times 10^{13}$ pairs $\text{cm}^{-3}\ \text{sec}^{-1}$ for each rad/sec^{-1} in Si (see Section 5.6), and L and A are device dimensions. For the devices represented by Figure 10.17, $L = 12\ \mu\text{m}$ and $A = 5 \times 10^{-4}\ \text{cm}^2$. An evaluation of Eq. (10.18) at 5.3×10^9 rad/sec predicts $\Delta I = 20$ mA. An inspection of Figure 10.16A and use of Eq. (10.17) shows a measured value of $\Delta I = 80$ mA at 5.3×10^9 rads/sec. Therefore, Eq. (10.18) agrees with experiment within a factor of 4. The same statement holds true for all of the measurements shown in Figure 10.16.

The BARITT depends on punch-through injection and transit time to generate a time-varying terminal current sufficiently out of phase with rf terminal voltage to produce the negative resistance required for oscillations to exist. The x-ray generated carriers are created uniformly throughout the n-region for a time much longer than an rf period, and thus, for these two reasons, they cannot contribute to the transit-time negative resistance. The presence of this excess current by itself would not cause the oscillations to cease. The reason the oscillations cease is that the increased current caused by the x-ray pulse causes a drop in the diode's voltage to below the punch-through value, and the needed punch-through injection current essentially vanishes† (this voltage drop is due to the increased $I \cdot R$ drop across the bias resistor, see Figure 10.16).

From Figure 10.16 it can also be seen that the oscillator requires about 60 nsec to reach full rf output after the diode current returns to normal. This is attributed to two effects: (1) the oscillations need time to grow exponentially to the final value once the oscillator restarts; and (2) the air in the microwave cavity is ionized and forms a plasma which perturbs the resonance (see Appendix B). The cavity returns to normal as soon as the plasma decays. Either of these two effects could easily account for delays of 60 nsec.

An analysis of the ionizing effects reported here can be extended to other devices of this general type through use of Eq. (10.18). In the practical case, the device would be operated from a current source so that when the excess current [ΔI, Eq. (10.18)] was equal to the normal operating current, the component of current due to punch-through injection would be nil, and oscillations would cease. Eq. (10.18) may be rewritten as

$$\Delta J = \frac{\Delta I}{A} \tag{10.19}$$

where ΔJ = increased current density due to the ionizing radiation pulse. It is known that the optimum operating current density in the BARITT device increases approximately linearly with frequency. Equation (10.9B) may be rewritten as:

$$J_{\text{opt}} \cong 6 \times 10^{-9} \cdot f_0 \tag{10.20}$$

where f_0 = design frequency. By combining Eq. (10.18), (10.19), (10.20), and (10.7), and arbitrarily saying that (temporary) failure‡ occurs when $\Delta J = J_{\text{opt}}/2$, it can be shown that failure (cessation of oscillations) for BARITT oscillators will occur at or above ionizing radiation dose rates of

$$\dot{\text{R}} = 4 \times 10^{-11}(f_0^2) \text{ rad/sec}\,. \tag{10.21}$$

where f_0 = the design frequency. For the device utilized in this experiment, $f_0 = 4 \times 10^9$ Hz, and Eq. (10.21) predicts oscillations would be interrupted at dose rates above 6.4×10^8 rad/sec which is in reasonable agreement with the data shown in Figure 10.16.

It should be pointed out that Eq. (10.21) is only approximate, due to the many simplifying assumptions that were made. However, the important fact to note is the f_0^2 dependence of the ionizing radiation sensitivity.

† This implies that the x-ray generated current (Figure 10.16A) is really 80 mA + 15mA = 95 mA.

‡ This neglects effects due to the ionization of the air in the cavity (see Appendix B).

REFERENCES

[1] W. Shockley, "Negative Resistance Arising from Transit Time in Semiconductor Diodes," *Bell System Tech. J.*, **33**, 799 (1954).

[2] H. Yoshimura, "Space Charge Limited and Emitter Current Limited Injections in Space Charge Region of Semiconductors," *IEEE Trans. PGED*, **ED-11**, 414 (1964).

[3] H. Rüegg, "A Proposed Punch-Through Microwave Negative Resistance Diode," *IEEE Trans. PGED*, **ED-15**, 577 (1968).

[4] U. Sheorey, I. Lundström, and E. Ash, "Analysis of Punch-Through-Injection for a Transit-Time Negative Resistance Diode," *Int. J. Electronics*, **30**, 19 (1971).

[5] D. Coleman and S. Sze, "A Low Noise Metal-Semiconductor-Metal (MSM) Microwave Oscillator," *Bell System Tech. J.*, **50**, 1695 (1971).

[6] C. A. Lee and G. C. Dalman, "Local Oscillator Noise in a Silicon p^+ n-p^+ Microwave Diode Source," *Electronics Lett.*, **7**, 565.

[7] S. M. Sze, D. J. Coleman, and A. Loya, "Current Transport in Metal-Semiconductor-Metal Structures," *Sol. St. Elec.*, **14**, 1209 (1971).

[8] E. P. EerNisse, "Small-Signal Negative Conductance in BARITT Devices," *Appl. Phys. Lett.*, **20**, 301 (1972).

[9] D. J. Coleman, Jr., "Transit-Time Oscillations in BARITT Diodes," *J. Appl. Phys.*, **43**, 1812 (1972).

[10] E. P. EerNisse, "Temperature Coefficients of Avalanching p^+-n-n^+ Junctions with Carrier Trapping," *Appl. Phys. Lett.*, **16**, 12, 506 (1970).

[11] W. T. Read, Jr., "A Proposed High-Frequency Negative-Resistance Diode," *Bell System Tech. J.*, **47**, 400 (1958).

11

Radiation Damage in Step-Recovery Diode Frequency Multipliers

11.1 Introduction

The Step-Recovery Diode (or SRD) frequency multiplier [1–4] is a fairly new device which provides a simple means of obtaining relatively small amounts of coherent microwave energy at frequencies beyond the operating range of microwave transistors. The output of such a multiplier is typically a few tens of milliwatts, and it is frequently used as the local oscillator signal in radar sets or as a pump signal for parametric amplifiers. The SRD multiplier has almost completely replaced the older Varactor Frequency Multiplier (VFM). The VFM utilized the nonlinear capacitance-voltage relationship found in *p-n* junctions at reverse bias ($C \propto V^{-1/2}$, etc., see Chapter 3) to produce harmonics. The efficiency of the VFM was typically $1/n^2$ (n = harmonic number). The SRD multiplier uses a fast current step to produce harmonics and has typical efficiencies of $1/n$. The best result reported for an SRD is an astounding 30% efficiency operating as a times-20 multiplier [5].

This chapter is directed at determining the radiation-induced failure mechanism of these devices in sufficient detail so that, given the parameters of any SRD multiplier, its failure level can be accurately predicted. This chapter also contains suggestions for increasing the radiation tolerance of the SRD multiplier. Sections 11.2 and 11.3 are devoted to a discussion of the device physics and circuit operation, respectively, of SRD multipliers. Section 11.4 examines permanent radiation effects on the SRD device, and Section 11.5 discusses radiation-induced changes in multiplier circuit operation. Section 11.6 is devoted to a brief discussion of transient ionizing radiation effects.

11.2 SRD Device Physics

The mechanism utilized in the SRD multiplier is *charge storage*. Figure 11.1 depicts idealized voltage and current waveforms for three types of diodes: (B) a Schottky-barrier diode; (C) a simple abrupt *p-n* junction; (D) a step-recovery diode. From Figure 11.1 it can be seen that the effect of minority-carrier charge storage is to cause finite reverse currents to flow until charge

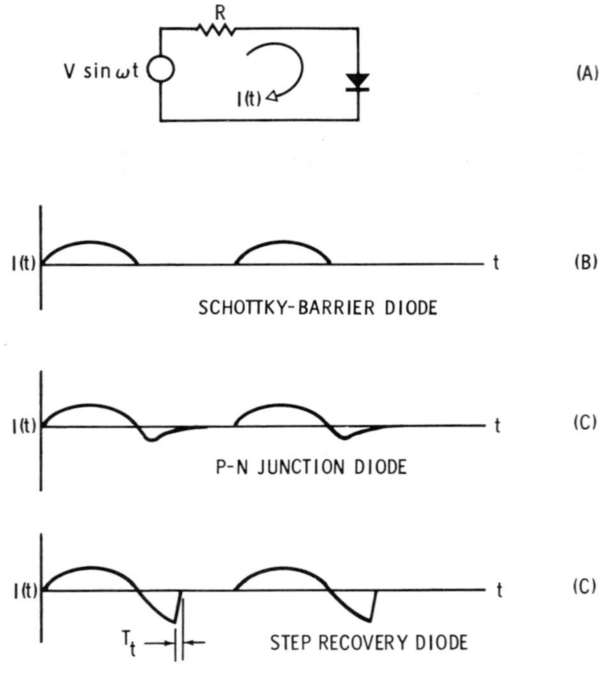

FIGURE 11.1
Effect of charge storage on the current waveform.

injected at forward bias is either extracted or has recombined. Figure 11.1B shows a Schottky-barrier diode, which is a majority-carrier device and as such has no minority-carrier charge storage effects. Figure 11.1C shows a typical *p-n* junction which exhibits charge storage. However, no effort to control the location of injected minority carriers is made; for this reason the reverse conduction does not end abruptly. The minority carriers are widely dispersed in the diode and return at different times. Figure 11.1D shows the current waveform of an SRD. In the SRD an effort is made to control the location of the stored charge through the use of "built-in" fields. As a result, the reverse conduction can end quite abruptly in the SRD ($T_t \sim 40$ psec, see Figure 11.1D). This fast-current step is rich in harmonics, and this effect is the one used in the SRD multiplier to produce the high-frequency rf output.

The injected minority carriers are kept localized by the use of a "built-in" field which is a direct result of the large impurity gradients deliberately put into these devices. Figure 11.2 shows a simplified mathematical model of a doping profile which is typical of SRDs. The device is a silicon p^+-n-n^+ structure with a fairly steep impurity profile.

The simplified doping profile shown in Figure 11.2 may be expressed as:

$$N_A = N_0 \exp\left(-\frac{x}{A}\right) \tag{11.1}$$

$$N_D = N_0 \exp\left(\frac{x - x_0}{A}\right) + (5 \times 10^{14}) \tag{11.2}$$

where $N_0 = 5 \times 10^{19}/\text{cm}^3$, $A = 0.2605\ \mu\text{m}$, and $x_0 = 10\ \mu\text{m}$. The "built-in" fields are always present when an impurity gradient is present; they are necessary to maintain equilibrium within the device. If we assume complete ionization of the donor and acceptor atoms, and assume the quasi-neutral approximation is valid, then Figure 11.2 also represents a rough plot of free-hole and electron densities. The large gradient of free carriers will cause significant diffusion currents to flow. To remain at thermal equilibrium (zero net current) an electric field must be set up which counteracts the diffusion currents. The net electron and hole current densities must vanish separately at thermal equilibrium. This may be expressed in the one-dimensional case as [see Eq. (2.43A,B)]:

$$J_n = ne\mu_n E + eD_n \frac{dn}{dx} = 0\,, \tag{11.3}$$

$$J_p = pe\mu_p E - eD_p \frac{dp}{dx} = 0\,. \tag{11.4}$$

Equation (11.3) may be rewritten (using the Einstein relation, $D/\mu = kT/e$)

$$E = \frac{kT}{e} \frac{1}{n} \frac{dn}{dx} \tag{11.5}$$

[a similar expression may also be written using Eq. (11.4)]. An approximation to the "built-in" field may be found if $n(x)$ and $p(x)$ are known. With $n(x)$ and $p(x)$ approximated by Eq. (11.1) and (11.2), it is possible to estimate roughly the size of the "built-in" fields by inserting Eq. (11.2) into Eq. (11.5) (to find E in the n^+ region). If this is done, the result is

$$E \cong \frac{kT}{e} \cdot \frac{1}{A} = \frac{0.026 \text{ volts}}{2.6 \times 10^{-5} \text{ cm}} = 1000 \text{ V/cm}\,. \tag{11.6}$$

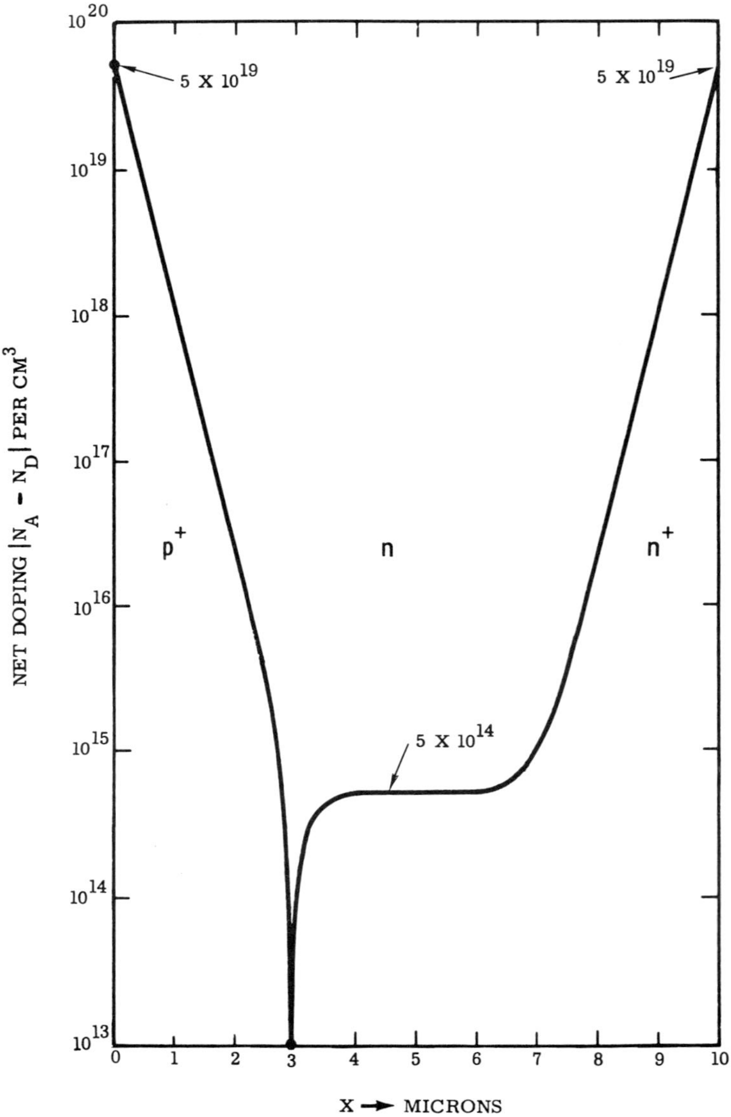

FIGURE 11.2
SRD doping profile.

Equation (11.6) is, at best, a crude approximation to the "built-in" fields within the device. This is due to the fact that in the presence of steep impurity profiles the quasi-neutral approximation no longer holds [6]. In this case, the two-way interaction between Boltzman's equation and Poisson's equation must be considered. An exact analytical solution of these coupled equations is virtually impossible in the general case. A nearly exact solution for the electric fields and excess carrier densities can, however, be found by using computer techniques [7]. The doping profile shown in Figure 11.2 was run through such a computer program [8], and the results are shown in Figures 11.3 and 11.4.

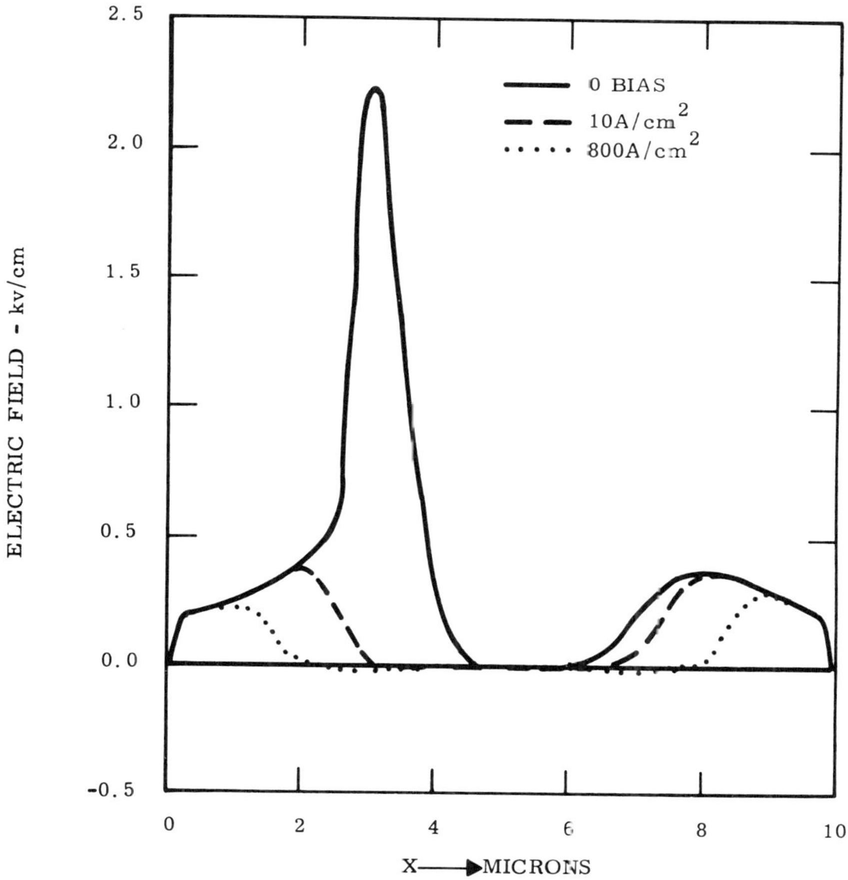

FIGURE 11.3
Computer solution for the electric fields in the SRD.

An inspection of Figure 11.3 very clearly shows the "built-in" fields for several bias current densities. At zero bias a large peak in the field at $x = 3$ μm can be seen. This is due to the depletion region which forms at the zero-biased *p-n* junction (see Figure 11.2). At reasonable forward biases, this depletion region collapses, and all that remains are the "built-in" fields on the right and left sides of the *n*-region which are due to the p^+ and n^+ impurity gradients. The direction of these "built-in" fields is such that they force holes to the left and electrons to the right (in Figure 11.3). This means that at forward bias the injected electrons and holes will be constrained to stay inside the *n*-region as may be seen in Figure 11.4. Therefore, on the reverse half of the applied rf voltage cycle, the conduction ends quite abruptly, due to the well-defined location of minority carriers (as is shown in Figure 11.1D). If the retarding fields were not present, such as in a perfectly abrupt *p-n* junction which has no impurity gradient except right at the junction, then the minority carriers would rapidly diffuse into the adjacent high conductivity, field-free regions where they could not be rapidly recovered on the reverse half of the voltage cycle (see Figure 11.1C).

Figure 11.4 also shows that for increasing forward currents the charge storage region widens; this is primarily due to field suppression brought on by the conductivity modulation caused by the injected carriers. The transition time T_t (Figure 11.1D) can be approximated as the time it takes a carrier to move across the charge storage region at saturated drift velocity, i.e.:

$$T_t = \frac{W}{10^7 \text{ cm/sec}} \tag{11.7}$$

where W = width of charge storage region. From Figure 11.4 it can be seen that for $J = 10$ A/cm^2, $W \cong 6$ μm, which gives $T_t \cong 60$ psec. The multiplier's harmonic efficiency is inversely proportional to transition time. The detrimental effect of increasing W (and T_t) with current density (seen in Figure 11.4) places an upper limit on the value of current density used. This suggests that as large an area as possible ought to be used to keep the current density small. This however would lead to large capacitances which cause an RC time constant problem (the cutoff frequency is $1/2\pi R_s C$) where R_s is the diode's series resistance. Therefore, a compromise area usually has to be selected. If the width of the lightly doped region is increased, the breakdown voltage and power handling ability can increase at the expense of the transition time and efficiency. For this reason, it is also necessary to compromise on the width of the lightly doped region. Therefore, the SRD design is usually optimized for a particular frequency range of interest. The *n*-region widths vary from 0.5 μm to 7 μm and the area from 10^{-2} to 10^{-4} cm^2.

Another parameter of interest in this device is the minority-carrier lifetime. In normal diode operation an attempt is made to recover, on the reverse-bias

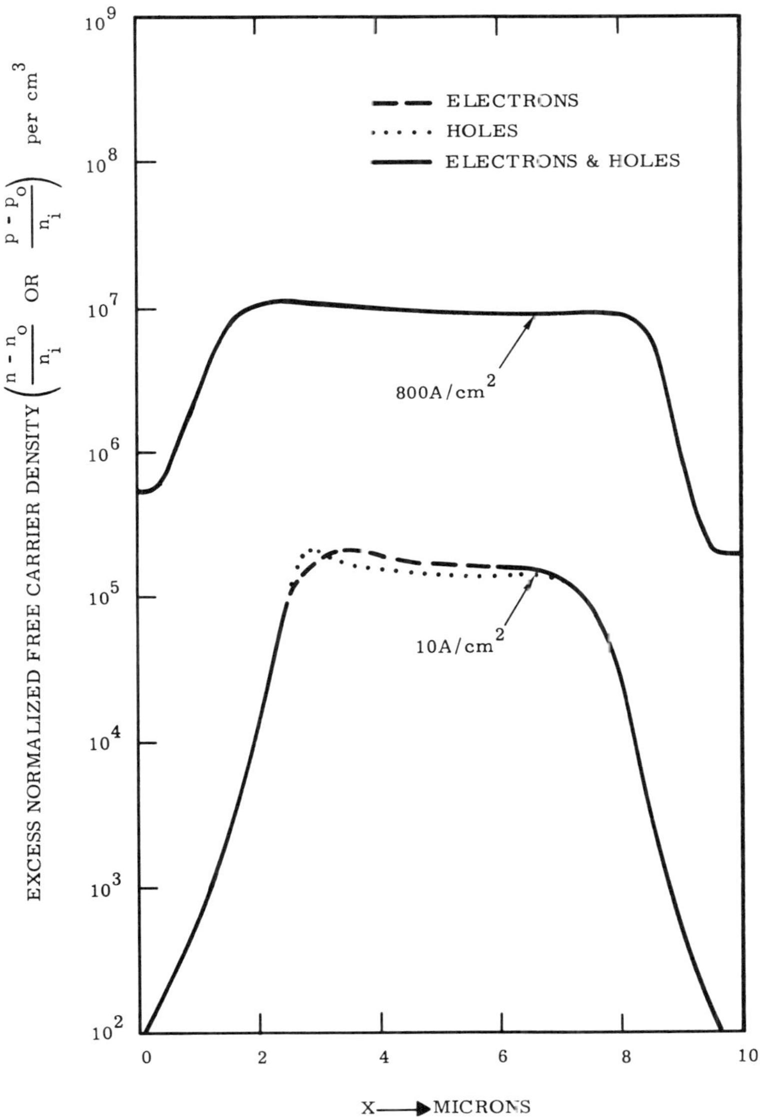

FIGURE 11.4
Computer solution for the excess-carrier densities in the SRD.

half cycle, all of the charge which was injected during the forward-bias half cycle. This implies that the minority-carrier lifetime should be at least one rf cycle long for satisfactory operation

$$\tau_{\min} \geq \frac{1}{f_i} \cdot \tag{11.8}$$

It will be shown, however, that too long a lifetime tends to make the multiplier more susceptible to neutron damage.

The capacitance-voltage relation for the SRD will be a large, nearly infinite, capacitance at forward bias (due to diffusion capacitance) and a small, nearly

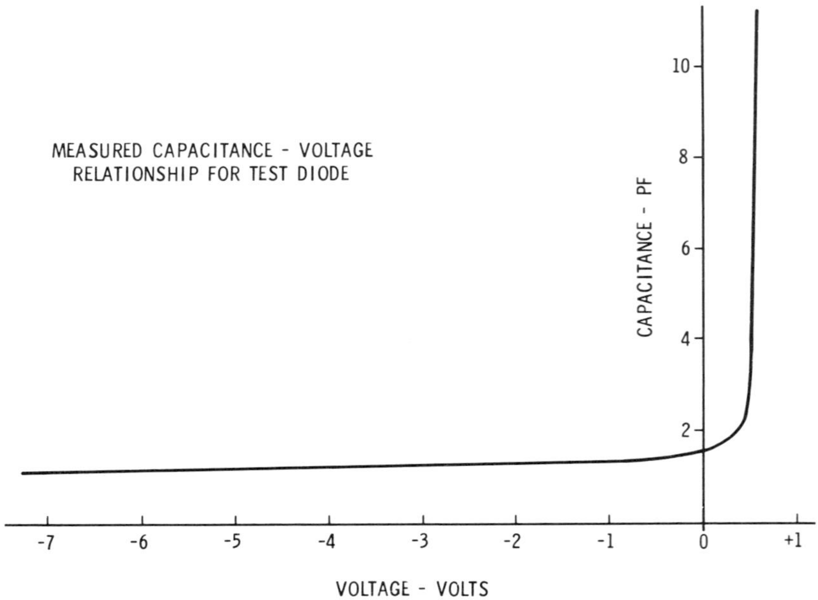

FIGURE 11.5
Measured C-V relation for the SRD.

constant, capacitance at reverse bias. The diode punches through at low reverse voltages, and once the heavily doped regions are reached, there is very little change in depletion width with increasing reverse bias. Basically this can be thought of as a two-state capacitance, and the difference between this multiplier and the older Varactor Frequency Multiplier described in the introduction is quite apparent. Figure 11.5 shows the measured C-V relation obtained from an SRD diode (the actual diode used will be described in Section 11.4). From Figure 11.5 it can be seen that the measured C-V curve is very close to the description given above.

11.3 SRD Multiplier Operation

Production of microwave harmonics via the fast-fall current pulse shown in Figure 11.1D requires a fair amount of engineering skill. Correct bias for the diode, input and output impedance matching, impulse generating circuitry, etc., must be considered. Since many papers have been written on this subject (see Ref. [1–5] and [9]), this problem will not be discussed in

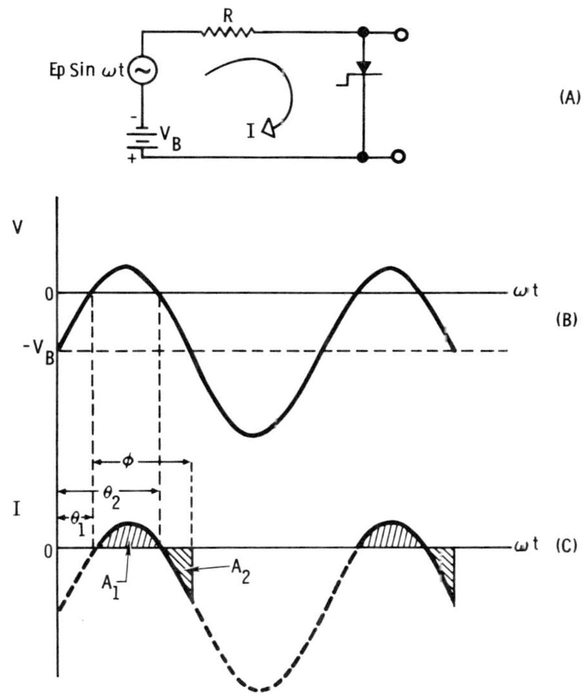

FIGURE 11.6
SRD in a resistive circuit.

depth here. Two subjects will be covered briefly: (1) bias methods and (2) general impedance matching. It is felt that these are the most important aspects of multiplier operation.

Bias Methods. Figure 11.6A shows a simple resistive circuit with an SRD in it. Figures 11.6B and 11.6C show approximate voltage and current waveforms that result from the combination of dc and rf drive voltages.

$$V(t) = E_p \cdot \sin \omega t - V_B . \tag{11.9}$$

The current waveform shown in Figure 11.6C is for an SRD that has zero transition time. At $\omega t = \theta_1$ (see Figure 11.6C), the diode becomes forward biased and charge injection begins. At $\omega t = \theta_2$, the diode becomes reverse biased, but conduction does not cease because there is still charge stored in the diode. Finally, at $\omega t = \theta_1 + \phi$, all of the stored charge has been extracted or recombined, and the conduction ceases abruptly ($T_t = 0$). The motion of charge in and out of the diode is governed by the equation:

$$\frac{dQ}{dt} = I(t) - \frac{Q}{\tau} \tag{11.10}$$

where Q = charge stored, $I(t)$ = current, and τ = carrier lifetime. The areas A_1 and A_2 (which correspond to the injected and extracted charge) in Figure 11.6C are not quite equal due to the effects of recombination, and consequently a small dc current must flow which is given by:

$$I_{dc} \cong \frac{Q_0}{\tau} \tag{11.11}$$

where Q_0 = charge stored at $\omega t = \theta_2$. A rough estimate can be made of the value of Q_0 by assuming a sinusoidal current waveform, $I_p \cdot \sin \omega t$ and also by assuming $\theta_1 = 0$ (i.e., $V_B = 0$). The solution proceeds as follows: Eq. (11.10) is a linear first-order differential equation of the general form

$$\frac{da}{dt} + Pa = R. \tag{11.12}$$

The solution is known [10] to be of the form:

$$a = e^{-Pt} \int Re^{Pt}\, dt + Ke^{-Pt} \tag{11.13}$$

where K is supplied through a boundary condition. The boundary condition used is $Q(t) = 0$ at $\omega t = 0$. By making the appropriate substitutions into Eq. (11.13), we find $Q(t)$ at $\omega t = \theta_2 = \pi$ to be: ($\theta_2 \equiv \pi$, since $\theta_1 = 0$)

$$Q\left(t = \frac{\pi}{\omega}\right) = Q_0 = \frac{\omega \tau^2 I_p}{1 + (\omega\tau)^2}\left[1 + \exp\left(-\frac{\pi}{\omega\tau}\right)\right]. \tag{11.14}$$

This value of Q will determine I_{dc} [Eq. (11.11)], the recombination current. In normal operation, the bias battery shown in Figure 11.6A is replaced by a resistor which produces a voltage drop across it due to the recombination current, I_{dc}, flowing through it. The value of this bias voltage is fairly critical ($\pm 5\%$) since it affects the phase angles θ_1, θ_2, and ϕ (see Figures 11.6B and 11.6C). It would be permissible to obtain the bias voltage with a battery, but in general this is more complicated than simple resistor bias.

General Impedance Matching. In order to convert power efficiently from one frequency, f_i (input frequency) to another frequency, f_n (nth harmonic of f_i), the input must be impedance matched to the generator and the output to the load. Figure 11.7 shows a simplified schematic of such an SRD multiplier. The required matching networks shown in Figure 11.7 are, at best, difficult to calculate. This is because, in general, the waveforms are nonsinusoidal; also, the circuit is no longer a simple resistive one such as shown in Figure 11.6 The matching networks in Figure 11.7 contain inductors and capacitors.

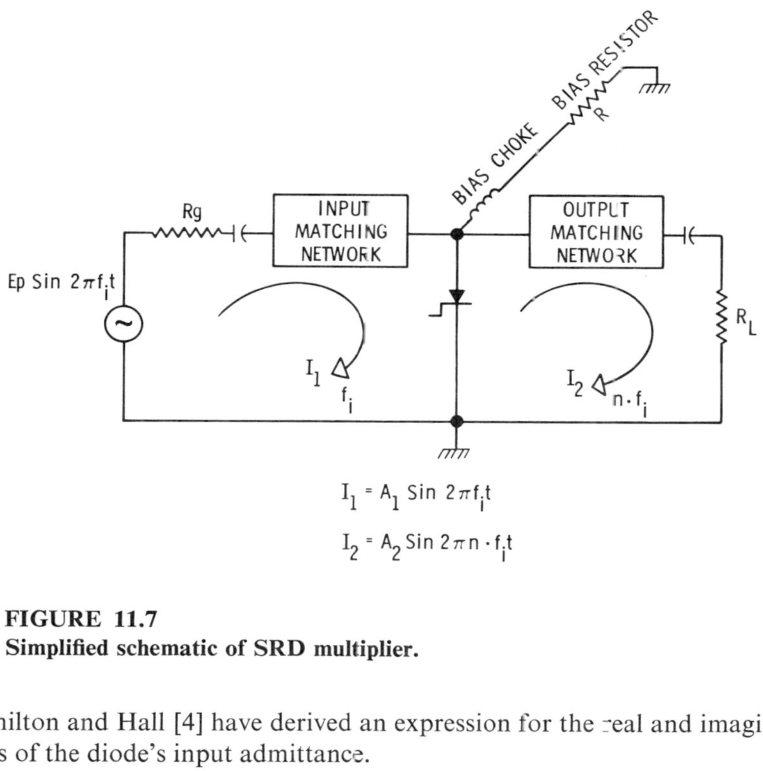

FIGURE 11.7
Simplified schematic of SRD multiplier.

Hamilton and Hall [4] have derived an expression for the real and imaginary parts of the diode's input admittance.

$$G_i \cong \frac{2 \cos \theta_1 \sin (\theta_1 - \pi/n)}{\omega L} \tag{11.15}$$

$$B_i \cong \frac{1 + 2 \sin (\theta_1 - \pi/n) \sin \theta_1}{\omega L} \tag{11.16}$$

where L = series inductance of the matching network, n = harmonic number, and θ_1 is defined in Figure 11.6C. It is not the intent here to calculate

G_i and B_i but merely to show that input admittance is a function of θ_1. Care must be taken in this design to make sure that θ_1 is not a function of rf input power; for if it is, then this would require retuning of the matching network each time the power is changed. An inspection of Figure 11.6 shows that for battery bias, θ_1 is indeed a function of the rf voltage. This problem is alleviated to a large extent by the use of resistor bias (or so-called self-bias). In the case of self-bias, when the rf voltage increases, the injected charge Q_0 will increase and I_{dc} will increase [see Eq. (11.11)]. As a result the increased $I \cdot R$ drop will bias the diode further in the reverse direction, thus reducing Q_0. This mechanism is seen to stabilize the value of θ_1 against input power variations. It will be seen later that this helpful mechanism causes the multiplier to fail under neutron irradiation.

11.4 Permanent Radiation Effects on the SRD Device

This section discusses the experimental results of neutron irradiation on the SRD device. The actual device utilized in the experiments was a p^+-n-n^+ silicon structure consisting of a shallow p^+ diffusion into an epitaxial n-layer which was grown on an n^+ substrate. The p^+ diffusion is approximately 3 μm deep, leaving an n-layer thickness of 4 μm. The area of the device was 1.25×10^{-4} cm^2. An approximate doping profile is shown in Figure 11.2. Neutron damage to silicon has been discussed in Chapter 5. The most lightly doped region will be affected first, while damage in the p^+ and n^+ regions will require much larger fluences. In the n-region, changes in mobility, thermal-equilibrium carrier concentrations, and lifetime are expected. In the SRD, high-field saturated drift velocities are the primary concern, and these do not change with radiation (see Section 8.4); therefore, degradation of the low-field mobility is only of secondary importance. Reduction of thermal equilibrium carrier densities is not significant since the lightly doped region is heavily conductivity-modulated by injected carriers in normal operation. Lifetime of the injected carriers is a very important parameter since the device utilizes charge storage for its primary operational mechanism. From Eq. (5.23) it can be seen that the carrier lifetime as a function of neutron fluence can be expressed

$$\tau = \frac{\tau_0}{1 + K_\tau \Phi \tau_0} \tag{11.17}$$

where τ_0, τ = lifetime before and after a radiation fluence Φ. It can be seen from Eq. (11.17) that as Φ increases, τ decreases; this implies that there will be a change in stored charge in the lightly doped region of the diode. The effects of this upon multiplier behavior will be discussed later.

The experimental measurements made on the diode described here were as follows: (1) reverse bias punch-through capacitance, (2) forward V-I characteristic, and (3) minority-carrier lifetime. These measurements were repeated on all diodes (irradiated and controls) after each irradiation. The irradiations were performed using the Sandia ACPR (Annular Core Pulsed Reactor), and all fluences given are for $\mathscr{E} > 10$ keV, Spectrum B.

The capacitance measurements were made at -10 V bias (punch-through) using a 1 MHz digital capacitance bridge. The results of these measurements are summarized in Figure 11.8. It can be seen that the punch-through capacitance does not change for fluences up to 2×10^{15} n/cm^2. This is to be

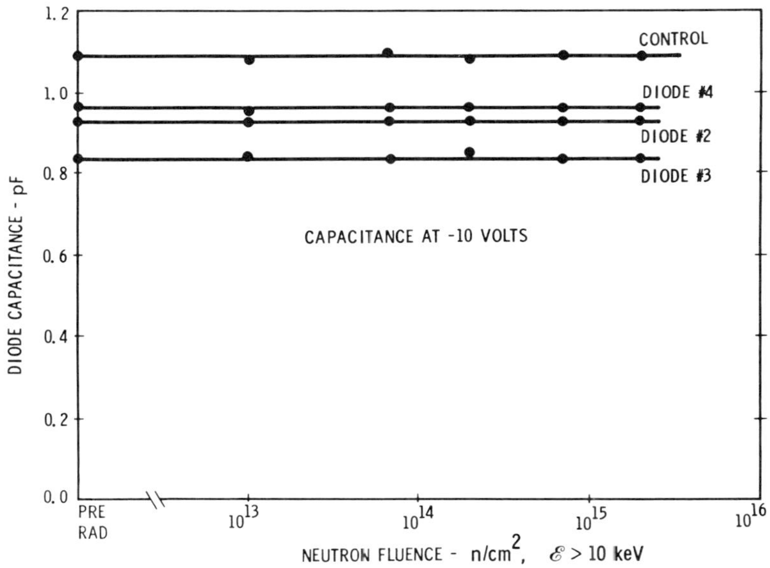

FIGURE 11.8
Reverse-bias punch-through capacitance versus neutron fluence.

expected since the depletion width is essentially fixed by p^+-n and n-n^+ boundaries, and the p^+ and n^+ regions are not significantly affected by 2×10^{15} n/cm^2. This invariance of capacitance with fluence is significant since the values chosen for the impedance matching networks (Figure 11.7) depend partly on this capacitance.

The forward V-I characteristic of the diodes was measured, and the voltage across the diode at a given current was found to decrease slightly after irradiation. Typical results are shown in Figure 11.9. The behavior seen in Figure 11.9 has been discussed in Section 7.4 [see the text by Eq. (7.30)]. This

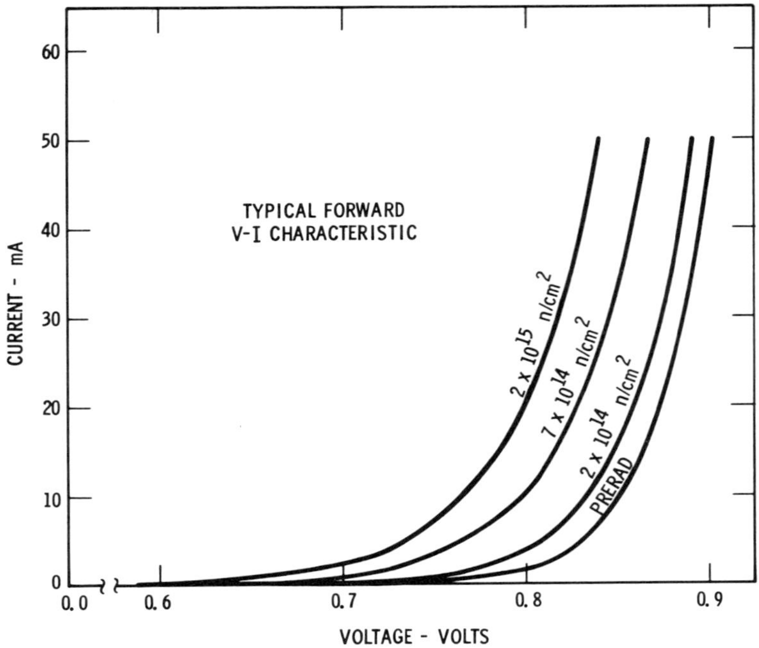

FIGURE 11.9
Typical V-I characteristic as a function of neutron fluence.

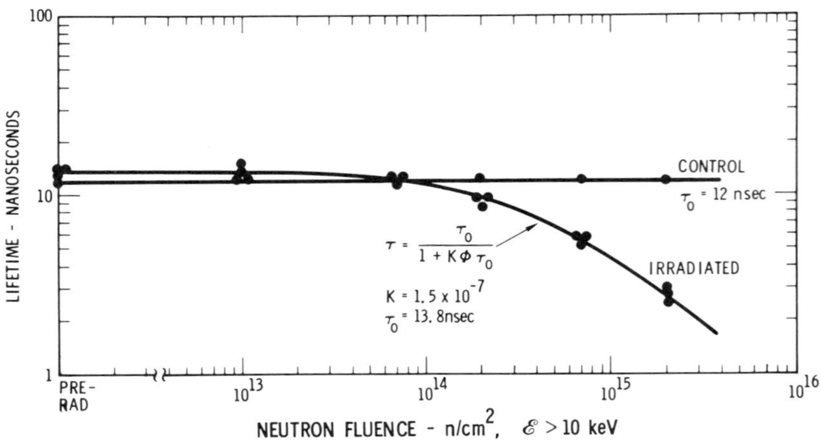

FIGURE 11.10
Minority-carrier lifetime versus neutron fluence.

behavior is typical for an irradiated, narrow-based diode. The slight variation in forward voltage drop seen in Figure 11.9 should not cause any real problem in the mutliplier since typically the diode is dc biased at -5 V with a peak-to-peak rf voltage swing of about 10 V.

The minority-carrier lifetime was measured using the charge removal method discussed in Section 4.5. A plot of the results versus neutron fluence is shown in Figure 11.10. Figure 11.10 also shows that the theoretical curve [Eq. (11.17)], with $K_\tau = 1.5 \times 10^{-7}$ cm^2/n·sec, fits the experimental data very well. An inspection of Figure 5.25 shows this value of K_τ to be quite reasonable for the high injection level case.

11.5 Permanent Radiation Effects on SRD Multiplier Performance

This section analyzes the effects of neutron fluence on the complete SRD multiplier. The diode used is the one described in the preceding section. The circuit used is of the type shown in Figure 11.7. The multiplication factor was $\times 10$ with an input frequency of 250 MHz and an output frequency of 2500 MHz. The input power was a constant 250 mW, and the output was nominally 15 mW. The multiplier adjustments were optimized prerad and then fixed. The only variable allowed was the bias. The bias terminal was arranged so that either self-bias (with an external resistor) or voltage bias (with an external supply) could be used. Figure 11.11 shows the multiplier power output versus neutron fluence. It can be seen that the power output falls off rapidly with fluence if a fixed bias resistance is used (i.e., the bias resistor remains at its optimum prerad value). However, it can also be seen from Figure 11.11 that the power output can be almost completely recovered (even after 2×10^{15} n/cm^2) if the bias resistor is reoptimized after each irradiation. The value of the optimum bias resistance versus n/cm^2 is plotted in Figure 11.12. The optimum value of the bias resistance decreases from approximately 4.2 kΩ prerad to 300 Ω after 2×10^{15} n/cm^2. This effect may be understood if the optimum self-bias current, I_{dc}, flowing through the bias resistor is plotted, as is shown in Figure 11.13. From this graph it can be seen that the optimum self-bias current increases from approximately 1 mA to 14 mA in going from prerad to 2×10^{15} n/cm^2. The self-bias voltage is simply the product of the self-bias resistor times the self-bias current. The optimized self-bias voltage is therefore obtained by multiplying the self-bias resistance plotted in Figure 11.12 by the self-bias current plotted in Figure 11.13. This has been done in Figure 11.14 where it can be seen that the optimum self-bias voltage decreases slightly ($\sim 10\%$) going from prerad to 2×10^{15} n/cm^2. It is interesting to note that the optimum self-bias voltage changes by only 10% over the fluence range, while I increases by $\sim 1400\%$, and R decreases by $\sim 1400\%$. If it is assumed that the optimum bias voltage is constant, then the

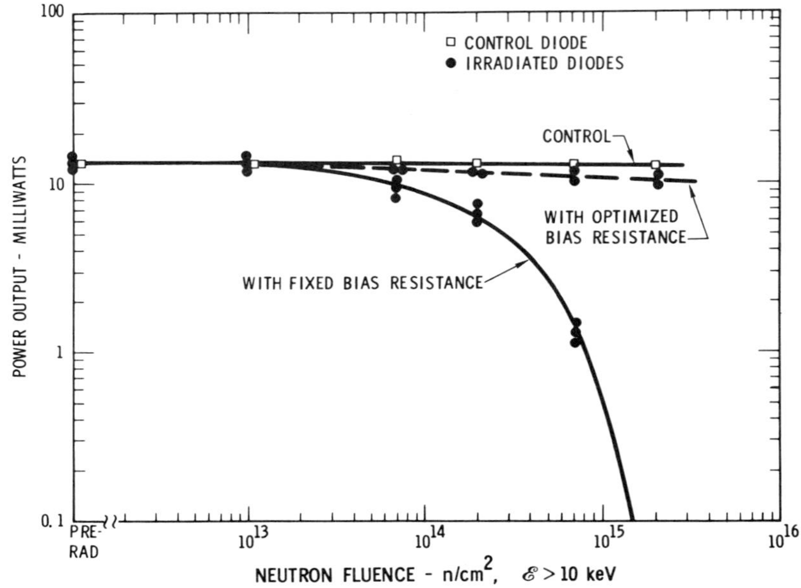

FIGURE 11.11
Power output versus neutron fluence.

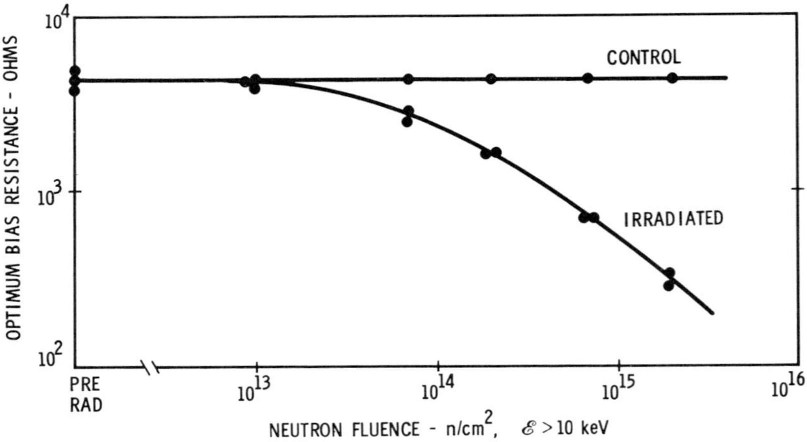

FIGURE 11.12
Optimum bias resistance versus neutron fluence.

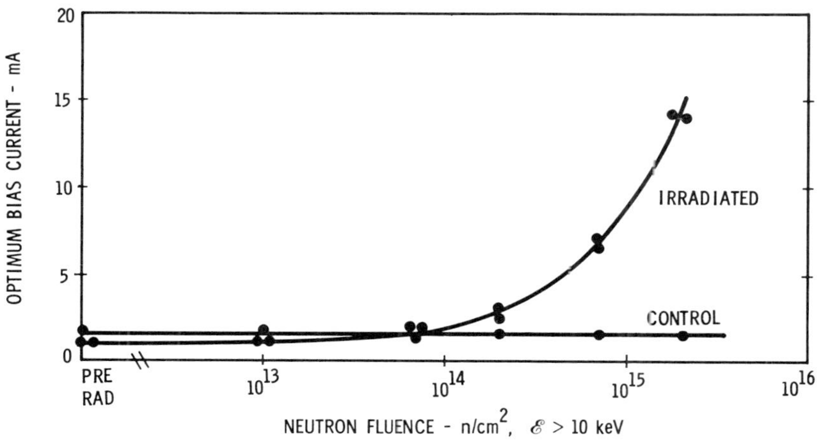

FIGURE 11.13
Optimum bias current versus neutron fluence.

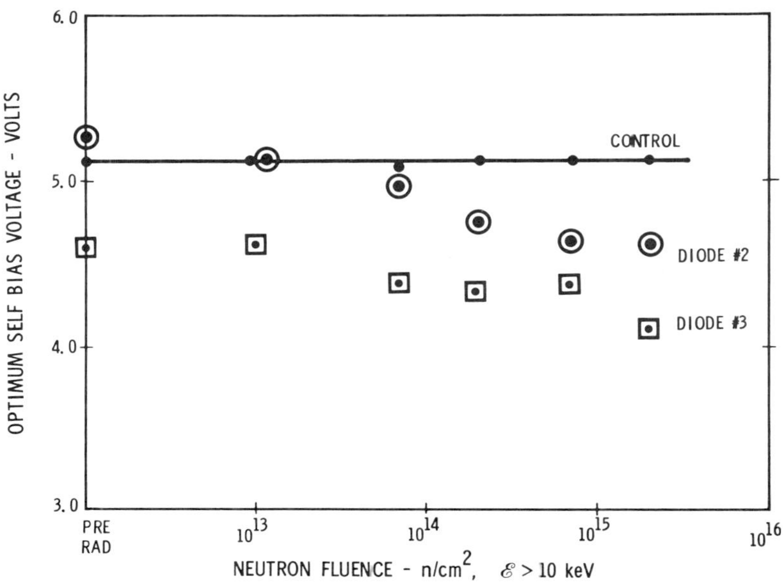

FIGURE 11.14
Optimum self-bias voltage versus neutron fluence.

failure mechanism with self-bias can be explained as follows: To first order, the V-I characteristic does not change. This means that a constant input power develops a constant amplitude rf current in the diode. This in turn injects a given amount of charge Q_0. Since $1/f < \tau$, Q_0 remains constant, to first order, while as τ decreases, I_{dc} increases [see Eq. (11.11)]. As I_{dc} increases, increased reverse bias is developed across the self-bias resistor (Figure 11.6A and 11.6B), and this tries to bias the diode off. Consequently, the power

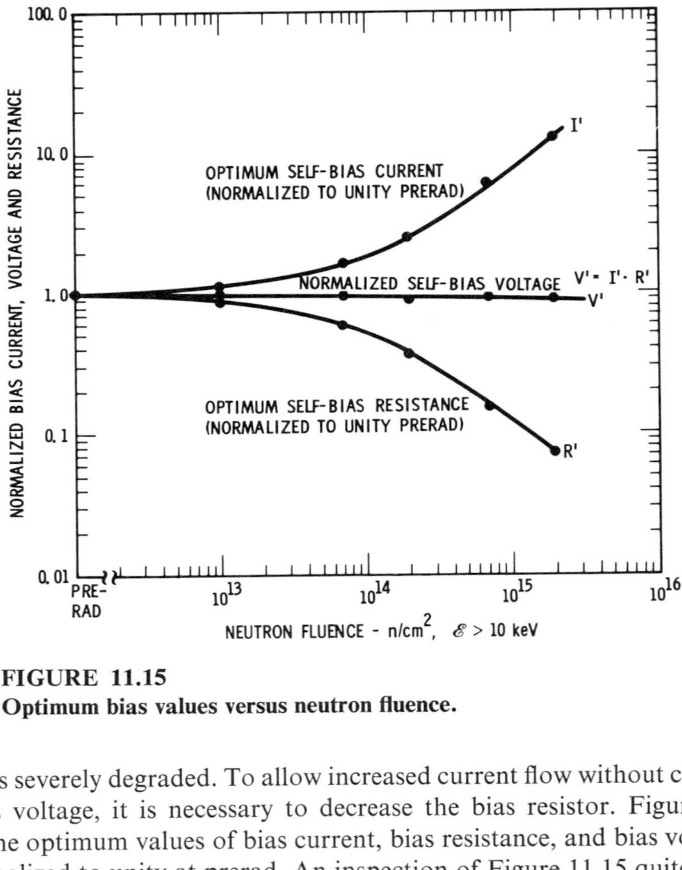

FIGURE 11.15
Optimum bias values versus neutron fluence.

output is severely degraded. To allow increased current flow without changing the bias voltage, it is necessary to decrease the bias resistor. Figure 11.15 shows the optimum values of bias current, bias resistance, and bias voltage—all normalized to unity at prerad. An inspection of Figure 11.15 quite clearly shows this mechanism. Therefore, it can be seen that self bias may not be desirable in a radiation environment since an unfavorable shift in the bias point would occur if a fixed bias resistor were used. Since the bias voltage remains essentially constant (see Figure 11.14), it might be advantageous to supply it from a battery (this would, however, mean giving up the benefits of self bias mentioned in Section 11.3). The slight decrease in reverse bias (see Figure 11.14) together with the shift in the V-I characteristic (Figure 11.9) serve to increase the injected charge (recall that the input current swing was a constant corresponding to 250 mW). The reason for this can be seen from Eq. (11.14) which shows that since τ decreases, it is necessary to increase the peak current, I_p, by about 20% (at 2×10^{15} n/cm^2) to keep Q_0 constant. It is essential that Q_0 remain nearly constant so the phase angles, θ_1, θ_2 and ϕ in Figure 11.6 do not change appreciably; otherwise, the rf tuning would be off [see Eq. (11.5) and (11.16)].

The slight power falloff that occurs even under optimum bias conditions (see Figure 11.11) is thought to be due to one or all of the following: (1) rf detuning effects, (2) excessive recombination, and (3) slight reduction in carrier mobility causing an increase in T_t [see Eq. (11.7)]. The predominant loss is thought to be due to (2) above.

The failure fluence due to an increase in self-bias current can be calculated for SRD multipliers in general. Assume that failure occurs when I_{dc} [Eq. (11.11)] increases by 10%. Eq. (11.11) and (11.17) may be combined to show that

$$I_{dc}(\Phi) = \frac{Q_0}{\tau_0}(1 + K_\tau \Phi \tau_0) . \tag{11.18}$$

From Eq. (11.18) it can be seen that the dc bias current increases by 10% at a fluence

$$\Phi = \frac{0.1}{K_\tau \tau_0} \cdot \tag{11.19}$$

From Eq. (11.19) it can be seen that in order to make Φ as large as possible one should make τ_0 as small as possible [commensurate with Eq. (11.8)]. This will reduce the sensitivity of the self-bias circuit to neutron fluence. For example, the multiplier discussed in this section has $f_{in} = 250$ MHz, and from Eq. (11.8) we find $\tau_{min} = 4$ nsec. From Eq. (11.19), using $K_\tau = 1.5 \times 10^{-7}$ cm^2/n·sec, it is seen that the self-bias circuit could be used out to 1.6×10^{14} n/cm^2 without degradation if the diode lifetime were shortened to 4 nsec prerad. Another possibility would be to substitute voltage bias in place of self bias, but this has its drawbacks as discussed in Section 11.3.

In general the absolute minimum value of τ allowable will be

$$\tau \cong \frac{1}{4f_i} \cdot \tag{11.20}$$

At this point the peak current, I_p, must be increased by a factor of 3 to make up for the charge lost through recombination [see Eq. (11.14)]. It is possible to combine† Eq. (11.17) and (11.20) and predict the absolute upper limit for

† Neglecting 1 in comparison with $K_\tau \Phi \tau_0$, i.e., $\tau \cong 1/K_\tau \Phi$.

neutron fluence (at which $\tau = 1/4f_i$). The failure due to this excessive recombination is found to occur at a fluence of†

$$\Phi_f(\text{n/cm}^2) \cong \frac{4f_i}{K_\tau} = 2.67 \times 10^7 \cdot f_i(\text{Hz})\ \text{n/cm}^2 \qquad (11.21)$$

At this fluence the recombination is so great that the multiplier cannot work efficiently. For example, the multiplier discussed in this section has $f_i = 250$ MHz, therefore $\Phi_f = 6.7 \times 10^{15}$ n/cm^2, Spectrum B.

Finally, while this work has concentrated on permanent neutron damage effects in silicon devices, it can easily be extended to other types of radiation and different materials using the appropriate lifetime damage constant from Chapter 5.

11.6 Ionizing Effects in SRD Multipliers

Since the SRD multiplier usually operates in a self-bias mode, no permanent damage is likely to be caused by pulsed ionizing radiation bombardment. The effect that will occur at sufficiently high fluences is a temporary interruption of multiplier output during the radiation pulse. The level at which the effect occurs may be estimated as follows: If the ionizing radiation pulse lasts longer than a few rf cycles, consider it to be a steady-state problem. The effect of the ionizing radiation will be greatest during that part of the rf cycle when the carrier concentration is at its lowest value in the SRD, i.e., when $\theta_1 + \phi < \omega t < \theta_1 + 2\pi$ (see Figure 11.6). During this interval, the diode is reverse biased, and the effect of the radiation generated carriers will be to increase the reverse current, resulting in a decrease in the reverse-bias voltage, V_B. The effect of changes in the bias voltage is very detrimental, as shown in the previous section. A serious degradation in output power can be expected for 10% changes in the bias current or voltage. Eq. (5.36) shows‡ that interruption of operation of a silicon diode is likely to occur at a dose rate of

$$\dot{\text{R}}(\text{rads/sec}) = \frac{0.1\ I_{\text{dc}}}{(4 \times 10^{13}) \cdot e \cdot W \cdot A} \qquad (11.22)$$

where A, W are the diode dimensions. For the device being discussed in this chapter, $A = 1.25 \times 10^{-4}$ cm^2, $W = r \times 10^{-4}$ cm, and $I_{\text{dc}} = 1$ mA. For this case Eq. (11.22) predicts interruption will occur at an ionizing radiation dose rate of 3×10^8 rad/sec (Si). For dose rates near and above this rate, the

† This neglects the effect of mobility variations and Eq. (11.21) is probably not valid for $\Phi_f > 10^{16}$ n/cm^2.

‡ Neglecting L_p and L_n in comparison with W.

effects of the ionizing radiation on the microwave circuit must also be considered (see Appendix B).

REFERENCES

[1] D. Leenov and A. Uhlir, "Generation of Harmonics and Subharmonics at Microwave Frequencies with *p-n* Junction Diodes," *Proc. IRE*, **47**, 1724 (1959).

[2] S. Krakauer, "Harmonic Generation, Rectification, and Lifetime Evaluation with the Step Recovery Diode," *Proc. IRE*, **49**, 1665 (1962).

[3] J. Moll and S. Krakauer, "*p-n* Junction Charge Storage Diodes," *Proc. IRE*, **49**, 43 (1962).

[4] S. Hamilton and R. Hall, "Shunt Mode Harmonic Generation Using Step Recovery Diodes," *The Microwave Journal*, **10**, 69 (April 1967).

[5] R. Hall, *Harmonic Generation with Step Recovery Diodes*, Hewlett Packard Application Note No. 2, HP Associates, Palo Alto, Calif.

[6] W. Shockley, "The Theory of *p-n* Junctions in Semiconductors and *p-n* Junction Transistors," *BSTJ*, **28**, 435 (1949).

[7] D. Scharfetter and H. Gummel, "Large-Signal Analysis of a Silicon Read Diode Oscillator," *IEEE Trans. PGED*, **ED-16**, 72 (1969).

[8] E. P. EerNisse, private communication.

[9] *Harmonic Generation using Step Recovery Diodes and SRD Modules*, Hewlett Packard Application Note No. 920, Hewlett Packard, Palo Alto, Calif.

[10] M. E. Van Valkenburg, *Network Analysis*, Prentice-Hall, Englewood Cliffs, N.J., 1955, Section 4-3.

12

Radiation Damage in Microwave Transistors

12.1 Introduction

Recent technological advances have extended the operating frequency of transistors into the microwave range. Microwave transistors find applications as local oscillators for radars, as low-noise microwave amplifiers, and as rf sources for low-power transmitters. The majority of this chapter will be devoted to a discussion of radiation effects on small-signal, bipolar microwave transistors.

Section 12.2 discusses those aspects of microwave transistor physics that are important to an understanding of radiation effects. Section 12.3 examines permanent radiation effects on the dc and low-frequency behavior of microwave transistors. Section 12.4 discusses changes in the microwave properties of these devices due to irradiation. Sections 12.2 to 12.4 are devoted to small-signal bipolar devices. Sections 12.5 and 12.6 discuss large-signal bipolar devices and field-effect devices, respectively. Section 12.7 analyzes transient ionizing radiation effects in bipolar microwave transistors.

12.2 Device Physics

The principles behind the operation of small-signal, bipolar microwave transistors have been discussed in detail by many authors [1–13]. Therefore transistor operation will be only briefly discussed here, with emphasis on those points important to an understanding of radiation effects. Microwave transistors are essentially scaled-down (size-wise) versions of low-frequency transistors and therefore operate on the basic principles discussed in Sections 3.6 and 3.7. Most bipolar microwave transistors are silicon *n-p-n* devices, and the basic transistor physics may be inferred from Figure 3.9. Equation (3.68) shows that if ω_α (the common-base cutoff frequency) is to be in the microwave range, the base width, W_B, must be in the submicron range. The "current crowding" effect that occurs at microwave frequencies is explained with the

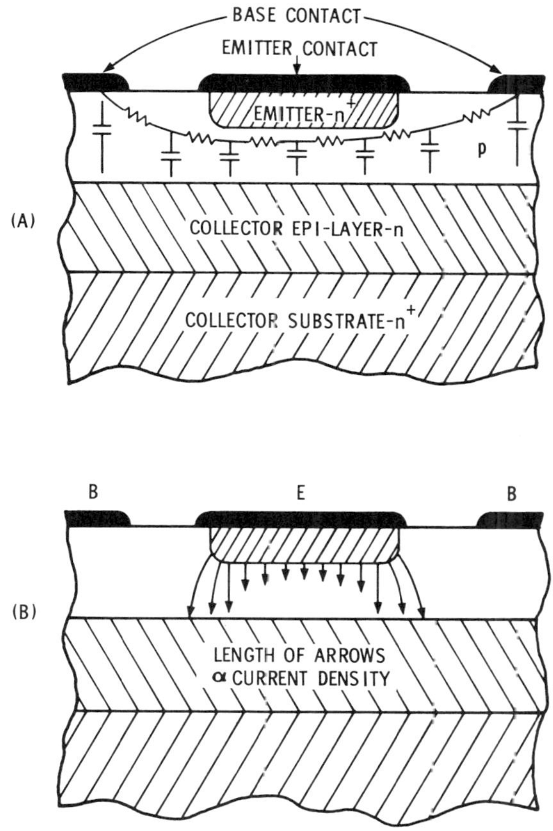

FIGURE 12.1
High-frequency "current crowding" effects in a double-diffused, epitaxial transistor.

aid of Figure 12.1 which shows a cross section of a double-diffused, epitaxial transistor. The base region is usually fairly lightly doped to improve the emitter efficiency [see Eq. (3.60)], and therefore the base-emitter current produces a lateral voltage drop which results in the edges of the emitter being more strongly forward biased than the center. This effect is referred to as "current crowding" in low-frequency transistors. At high frequencies this effect is aggravated by the *r-c* low-pass filter action of the base layer [11], as is demonstrated in Figure 12.1A,B. This *R-C* filter action results in the "current crowding" type of effects even at low currents. This explains the increase in saturation voltage usually seen at high frequencies. The region under the center of the emitter in Figure 12.1B does not contribute to current gain, but it does contribute to parasitic capacitance. Therefore this "current crowding" effect is very undesirable. A reduction in the importance of "current crowding" can be obtained by using an overlay or an interdigitated structure. In these emitter structures a large emitter-periphery-to-emitter-area ratio can be obtained. In the case of the overlay structure the emitter consists of many small emitter dots and the periphery-to-area ratio is $2\pi r/\pi r^2 = 2/r$, and as $r \to 0$, this ratio can become quite large. For the interdigitated structure the emitter takes the form of several long thin fingers (see Figure 12.2); if the fingers have a length L and a width S, the periphery-to-area ratios is $2/S$ (if $S \ll L$). As $S \to 0$, this ratio can also become quite large. Figure 12.2 shows the basic interdigitated, double-diffused, epitaxial, microwave transistor. The base contacts are interdigitated with emitter contacts in order to minimize the series base resistance. This is necessary in order to keep the $r_b \cdot C_c$ time-constant small. In some microwave transistors, a localized p^+ region is diffused into the base contact region before the base metallization is deposited. This has the advantage of decreasing r_b without decreasing the emitter efficiency.

Another feature usually found in microwave transistors is the epitaxial collector region (see Figure 12.2). This lightly doped region allows the fabrication of transistors with fairly high collector-base breakdown voltages (see Figure 3.8) and fairly small capacitances [due to the wide depletion width, Eq. (3.7)] without having to use a thick high-resistivity layer which would introduce a large series parasitic resistance. Under normal operating bias, the epitaxial layer is usually punched through to the n^+ substrate, which has a very low series resistance. From these discussions it can be seen that a very close tolerance on emitter geometry and base width is necessary. For this reason commercial microwave transistors are fabricated almost exclusively using the planar, diffusion process (see Chapter 4). Recently [14], good laboratory results have been obtained using ion-implantation techniques.

The equivalent circuit shown in Figure 3.11B is not adequate at microwave frequencies. In the microwave range, the effects of the emitter-base capacitance as well as the package parasitics must be included. This has been done in Figure 12.3 which shows a "T" equivalent [4] network (enclosed by the dashed lines) imbedded inside a network composed of the package parasitics [16]. The network shown in Figure 12.3 is found to be very satisfactory for modeling the small-signal behavior of microwave bipolar transistors.

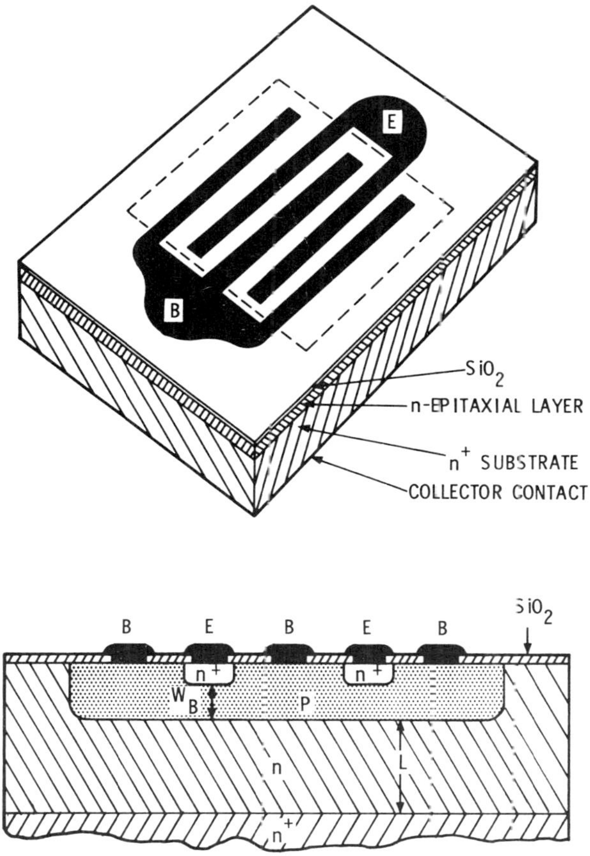

FIGURE 12.2
Interdigitated, double-diffused, epitaxial, microwave bipolar transistor.

For the moment consider an unpackaged device (inside the dashed lines on Figure 12.3). Several different figures of merit for microwave transistors have been established [1,2]. The first one is f_α (or ω_α), from Eq. (3.68) it can be seen that

$$2\pi f_\alpha = \omega_\alpha = \frac{\eta D_n}{W_B^2} \, . \tag{12.1}$$

The factor η has been added to include the effects of any drift field that may be present in the base [4]. f_α is the frequency at which the common-base current

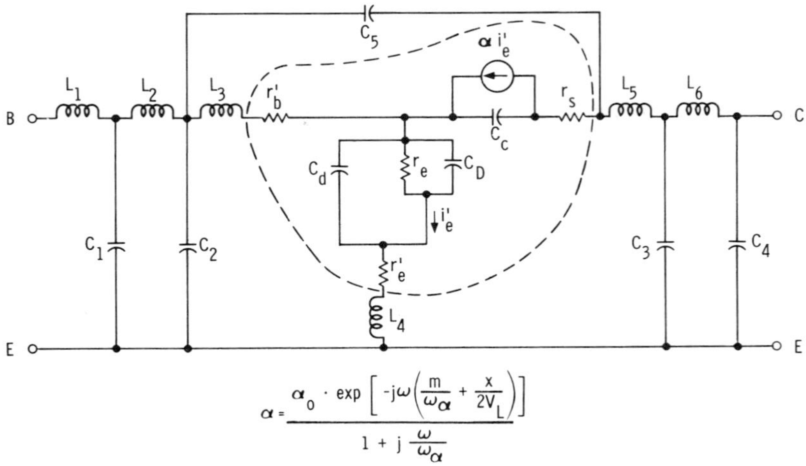

FIGURE 12.3
Equivalent circuit of a packaged microwave bipolar transistor.

gain, [see Eq. (3.64)] decreases to a value of 0.5. From Eq. (3.67) it can be seen that

$$\alpha(f) = \frac{\alpha_0}{1 + j(f/f_\alpha)} \tag{12.2}$$

where α_0 = low-frequency common-base current gain. For the common-emitter case we are interested in the common-emitter current gain, β. From Eq. (3.66) we can write

$$\beta(f) = \frac{\alpha(f)}{1 - \alpha(f)} \cdot \tag{12.3}$$

Using Eq. (12.2) and (12.3), we can show that

$$\beta(f) = \frac{\beta_0}{1 + j(f/f_\beta)} \tag{12.4}$$

where $\beta_0 = \alpha_0/(1 - \alpha_0)$ and $f_\beta = (1 - \alpha_0)f_\alpha$. The frequency at which $|\beta| = 1$ (with the collector shorted) is called the common-emitter unity gain frequency, f_T. For the simple model shown in Figure 3.11B, $f_T = f_\alpha$. However, for the model shown in Figure 12.3 this is no longer true. The common-emitter cut-off frequency for the circuit inside the dashed lines on Figure 12.3 is given by

$$f_T = \frac{1}{2\pi\tau_{ec}} \tag{12.5}$$

where

$$\tau_{ec} \cong \tau_e + \tau_b + \tau_c \tag{12.6}$$

where τ_e = emitter-base $R \cdot C$ time constant

$$\tau_e = \frac{kT}{eI_e}(C_d + C_c) \qquad \text{See Ref. [4]} \tag{12.7}$$

and τ_b = base transit time

$$\tau_b = \frac{1}{\omega_\alpha} \qquad \text{See Eq. (12.1);} \tag{12.8}$$

and τ_c = collector transit time

$$\tau_c = \frac{L}{2v_s} \qquad \text{See Ref. [1] .} \tag{12.9}$$

From these discussions it can be seen that in order for Eq. (12.4) to be valid, the definition $f_\beta = (1 - \alpha_0)f_T$ must be used. Using this definition, and assuming $\beta_0 \gg 1$ and $f \rightarrow f_T$ (which is usually the case in microwave transistor operation), we can show that

$$\beta(f) \cong -j\left(\frac{f_T}{f}\right) \cdot \tag{12.10}$$

Hence, at high frequencies the current gain is independent of β_0 and depends only on the cutoff frequency of the transistor. The consequences of this will be seen later in the discussion of radiation effects.

Microwave transistors are conveniently characterized by their S (scattering) parameters [15]. The S parameters are based on a two-port approach to transistor measurements. For the common-emitter case the emitter terminal is grounded, and the base and collector ports are taken as ports 1 and 2, respectively. This is shown in Figure 12.4. In the S-parameter measurement technique, the measurements are taken in terms of incident and reflected waves on transmission lines of impedance Z_0. The transistor is fully characterized by the scattering matrix defined as

$$\begin{bmatrix} b_1 \\ b_2 \end{bmatrix} = \begin{bmatrix} S_{11} & S_{12} \\ S_{21} & S_{22} \end{bmatrix} \cdot \begin{bmatrix} a_1 \\ a_2 \end{bmatrix} \tag{12.11}$$

which may be written as

$$\begin{aligned} b_1 &= S_{11} \cdot a_1 + S_{12} \cdot a_2 \\ b_2 &= S_{21} \cdot a_1 + S_{22} \cdot a_2 \end{aligned} \tag{12.12}$$

where all of the above quantities are complex. If a wave is incident at port 1 and a matched load is connected to port 2, then $a_2 \equiv 0$, and therefore

$$b_1 = S_{11} \cdot a_1 \, . \tag{12.13}$$

Hence, it can be seen that S_{11} is simply the input reflection coefficient with the output matched. Also, with the output matched ($a_2 \equiv 0$), Eq. (12.12) shows that

$$b_2 = S_{21} \cdot a_1 \, . \tag{12.14}$$

Hence, S_{21} is the forward-transfer ratio (gain) for an output load of Z_0. Similar statements can be made for the case when the wave is incident at port

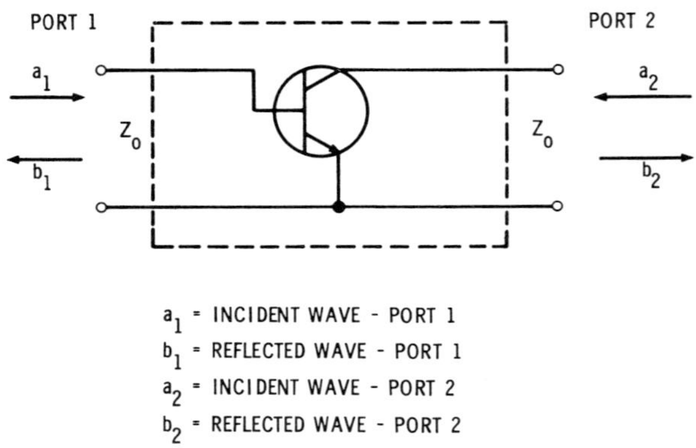

FIGURE 12.4
Two-port measurement of transistor *S* parameters.

2 and port 1 is terminated by a matched load. In this case S_{12} is called the reverse gain, and S_{22} is the output reflection coefficient. The four S parameters measured in this way can be used directly to determine amplifier designs, stability criteria, etc., or they may be converted to h parameters [15]. The S-parameter measurement technique has two main advantages over the direct measurement of h parameters. The first is that when a microwave transistor is terminated in a short or an open (required in direct h-parameter measurement) there is a strong chance that the transistor will oscillate. This does not usually occur when the transistor is terminated in a matched 50-Ω load. The second advantage for the S-parameter system is that since measurements are made in terms of traveling waves on transmission lines, accurate measurements can be made away from the location of the device chip if lossless lines are used. This is quite important in view of the high frequencies used.

The frequency at which $|S_{21}| = 1$ is called f_s. For the case when $2\pi C_c \cdot Z_0 \ll 1/f_T$,† it can be shown [4] that

$$f_s \cong f_T \,. \tag{12.15}$$

Equation (12.15) is usually valid for small-signal bipolar devices. Since most microwave transistors are characterized by their S parameters rather than their h parameters, f_s rather than f_T is usually specified.

12.3 DC and Low-Frequency Radiation Effects

Radiation damage to low-frequency bipolar transistors has been thoroughly studied [17–20]. The predominant effect seen is a decrease in current gain (β) due to radiation-induced recombination in the base region. The fluences at which this effect becomes important can be readily predicted from Eq. (5.23), (12.8), and (12.1). For microwave transistors, the base width is sufficiently narrow so that this effect can be neglected at neutron fluences up to 10^{17} n/cm^2. Changes in the low-frequency parameters of microwave transistors have been investigated by Gwyn and Gregory [21]. They found that eventual degradation of the current gain occurs due to excessive recombination in the emitter-base space region and emitter bulk regions.

Graham et al. [22] have reported extensive experimental and theoretical radiation studies on a commercially available microwave transistor, the Hewlett-Packard 35821E. The devices were mounted in a stripline package. The 35821E has an f_s of approximately 4.5 GHz. This is a silicon device which utilizes an interdigitated structure and has four emitter fingers and five base fingers (see Figure 12.2). The finger widths were 2.5 μm and their length was approximately 70 μm [9]. The device was formed by phosphorous and boron diffusions into a thin, 3.2 μm, n-type‡ epitaxial layer grown on an n^+ silicon substrate. An approximate doping profile for the device is shown in Figure 12.5. DC and low-frequency measurements as well as high frequency S-parameter measurements were made on the transistors. The transistors were then exposed to neutron fluences in the Sandia ACPR reactor (all fluences given are $\mathscr{E} > 10$ keV, Spectrum B). The measurements were repeated after each irradiation. Typical results are shown in Figures 12.6 and 12.7.

Figure 12.6 shows that the C-V characteristic of the collector-base junction changes with irradiation fluence. At zero bias, Figure 12.6 shows a 70% drop in capacitance after 7×10^{15} n/cm^2. This drop in capacitance is attributed to a widening of the zero-bias depletion width due to carrier removal effects. This was discussed in Chapter 5. Figure 12.6 shows that, at large values of reverse bias, capacitance decreases only slightly with irradiation (6%). This is

† $2\pi C_c \cdot Z_0$ is the transmission line impedance-collector capacitance time constant.

‡ $N_D \cong 2 \times 10^{15}$ per cm^3.

due to the fact that the collector depletion region has punched-through to the n^+ substrate at this bias, and much larger radiation fluences are required to produce significant carrier removal effects in the n^+ layer.

Figure 12.7 shows that the low-frequency common-emitter current gain falls off with fluence, dropping by 90% after 7×10^{15} n/cm^2. The base width

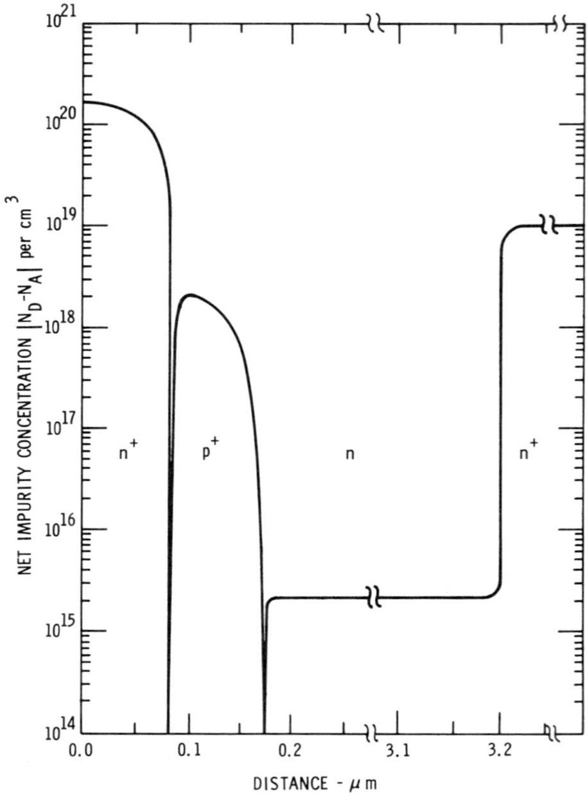

FIGURE 12.5
Approximate doping profile of the 38521E transistor (after Graham, Chaffin, and Gwyn [22]).

of the 35821E is sufficiently narrow so that this β falloff cannot be explained in terms of recombination in the base region [17–20]. Also, recombination in the emitter-base space-charge region cannot account for such a large β falloff at 7×10^{15} n/cm^2 [21]. In order to find the cause of the β falloff seen in Figure 12.7, the doping profile shown in Figure 12.5 was used together with a device-physics computer code [23]. The computer model showed the β fall-off to be due to the Kirk effect [24–26]. The Kirk effect occurs as a result of space-charge effects in the collector depletion region. Poisson's equation [Eq. (3.1)] for the collector depletion region may be written as [25]

$$\frac{dE}{dx} = \frac{e}{\epsilon}\left[N_D^+ - \frac{J_c}{ev_s}\right] \tag{12.16}$$

where J_c = collector current density ($J_c = nev_s$) and N_D^+ = collector epi

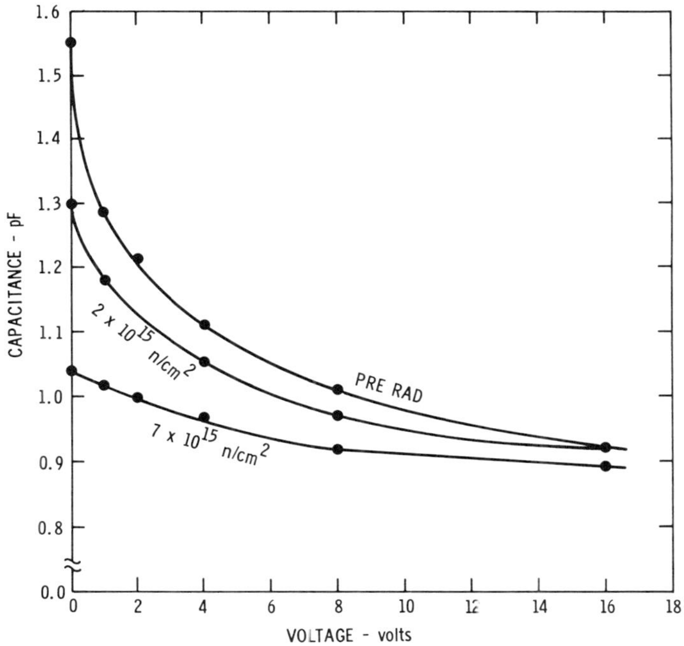

FIGURE 12.6
Capacitance-voltage characteristic of the collector-base junction of an irradiated 35821E transistor.

doping concentration. The solution to Eq. (12.16) may be obtained by integration as

$$E(x) = E(0) + \frac{e}{\epsilon}\left[N_D^+ - \frac{J_c}{ev_s}\right]\cdot x \tag{12.17}$$

where $x = 0$ at the collector-base junction. Equation (12.17) shows that if $J_c = 0$, the maximum magnitude of E will occur at $x = 0$ [$E(0)$ is a negative quantity]. As J_c increases, the slope of the electric field will change, becoming zero at

$$J_1 = ev_s N_D^+ \ . \tag{12.18}$$

Further increases in J_c will cause the electric field slope to reverse, and eventually the field will be suppressed to zero at the collector-base junction. When this occurs, there is an increase in the width of the region through which

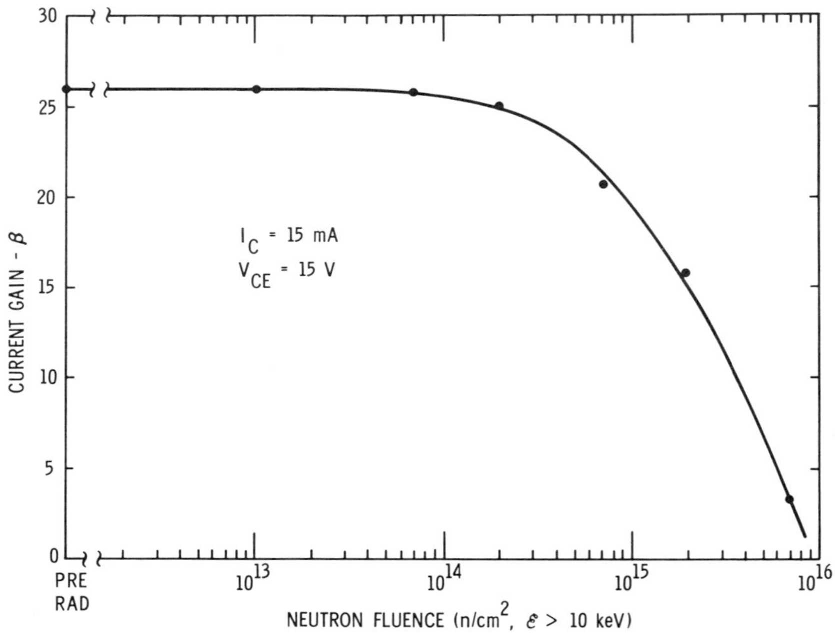

FIGURE 12.7
Low-frequency current gain versus neutron fluence for the 35821E microwave transistor (after Graham, Chaffin, and Gwyn [22]).

the carriers must move by diffusion. For this reason, it is often stated that the base width increases or base push-out occurs, and this is called the *Kirk effect*. The critical current for the onset base pushout is given by [27]

$$J_2 = e\mu N_D^+ (V_{c0} + |V_{cB}|)/L \tag{12.19}$$

where V_{c0} = "built-in" collector voltage and V_{cB} = applied voltage. For collector currents above this value a substantial increase in base diffusion time [see Eq. (12.8)] will occur, and thus recombination effects can become significant, causing β falloffs of the type shown in Figure 12.7.

The derivation represented by Eq. (12.16) to (12.19) is at best approximate. A more accurate picture of what is happening can be seen by examining the results of the device-physics computer model discussed earlier. Figures 12.8 to 12.11 show the calculated electric field profiles for the 35821E transistor for several different conditions. Figures 12.8 and 12.9 correspond to the prerad case.

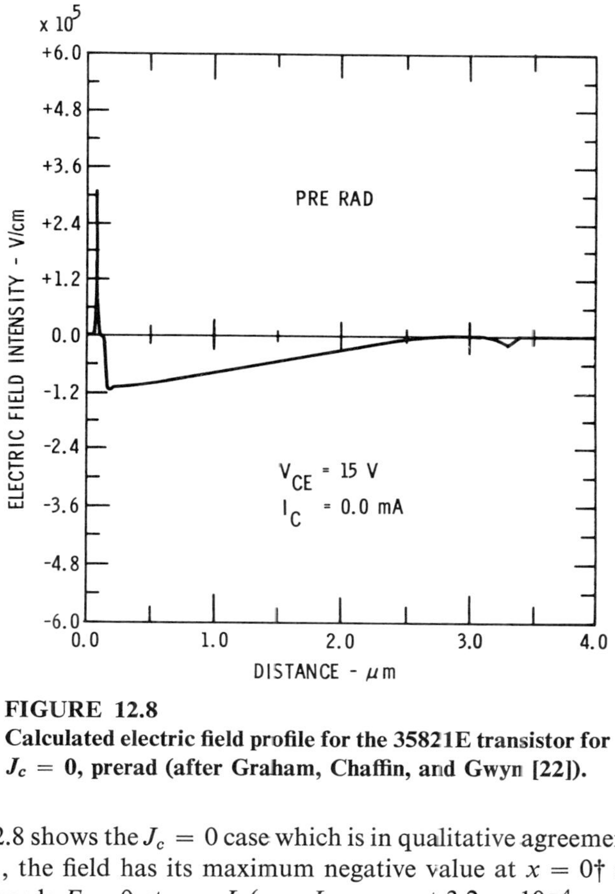

FIGURE 12.8
Calculated electric field profile for the 35821E transistor for $J_c = 0$, prerad (after Graham, Chaffin, and Gwyn [22]).

Figure 12.8 shows the $J_c = 0$ case which is in qualitative agreement with Eq. (12.17); i.e., the field has its maximum negative value at $x = 0$† and slopes upward towards $E = 0$ at $x = L$ ($x = L$ occurs at 3.2×10^{-4} cm in Figures 12.8 to 12.11).

Figure 12.9 shows the solution for the case when $I_c = 23$ mA, or $J_c = 3000$ A/cm^2,‡ which is very close to the J_1 value predicted by Eq. (12.18). From

† The C-B base junction is actually located at $x = 2 \times 10^{-5}$ cm in Figures 12.8 to 12.11.
‡ "Current crowding" effects (Figure 12.1) actually cause J_c to be somewhat greater than I/A.

Figure 12.9 it can be seen that the slope of the electric field is nearly zero throughout most of the collector region as predicted by Eq. (12.16).

Figure 12.10 shows the same bias conditions as Figure 12.9 for the postrad case. To obtain Figure 12.10, the collector epi doping concentration was reduced to 10^{13} per cm^3 in the computer solution. This is in keeping with Figure 12.6 which shows that the epi region in the post irradiated device

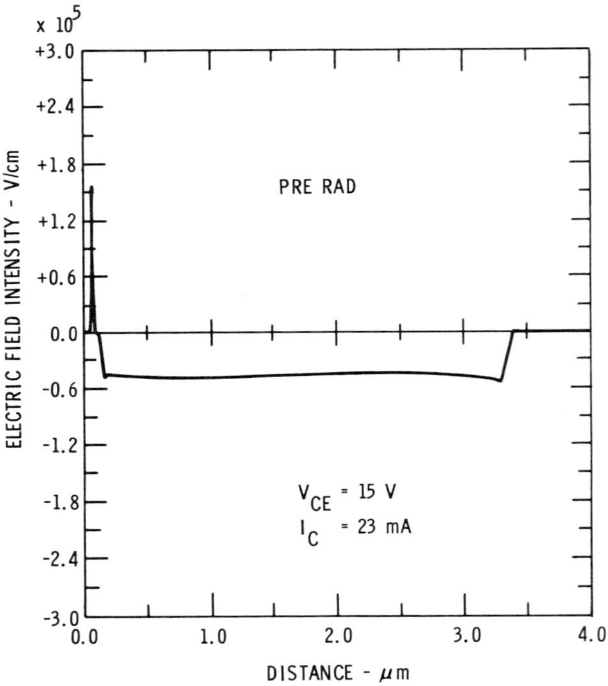

FIGURE 12.9
Calculated electric field profile for the 35821E transistor for J_c = 3000 A/cm^2 (23 mA) prerad (after Graham, Chaffin, and Gwyn [22]).

(7×10^{15} n/cm^2) is nearly punched-through at zero bias due to compensation of N_D^+ by trapped electrons. Figure 12.10 shows that base push-out occurs for $J_c = 3000$ A/cm^2 with the reduced effective collector doping concentration as is predicted by Eq. (12.19). Equation (12.19) also shows that for a fixed density the effective reduction in N_D^+ due to trapped electrons can be compensated for by raising the applied collector voltage.

Figure 12.11 shows that the base push-out effect can be eliminated by raising the applied collector voltage to 30 V for $J_c = 3000$ A/cm^2 postrad. Similar recovery in low-frequency current gain, β, is observed in the experimental case [22].

The conclusion of this work is that the original manufacturer-recommended bias point $V_{CE} = 15$ V, $I_c = 15$ mA approaches the base push-out value, and the radiation-induced change in the effective value of N_D^+ causes

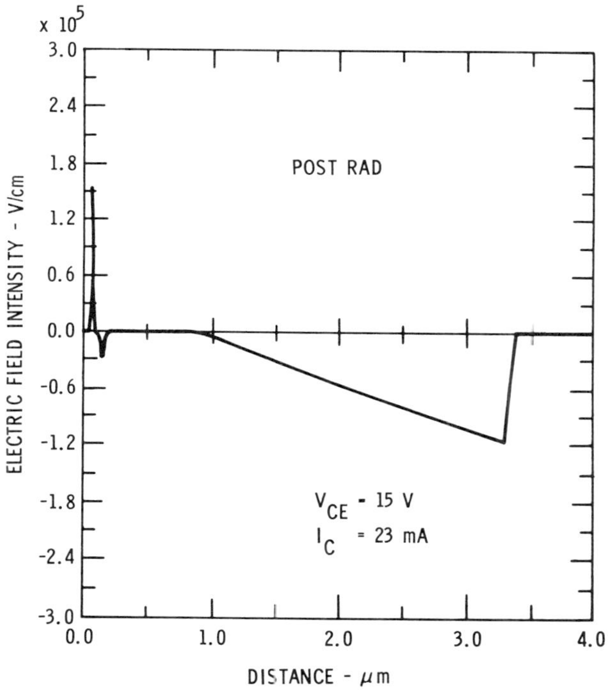

FIGURE 12.10
Calculated electric field profile for the 35821E transistor for $J_c = 3000$ A/cm^2 (23 mA) postrad (after Graham, Chaffin, and Gwyn [22]).

the transistor to go into base push-out after 7×10^{15} n/cm^2. The radiation tolerance of this transistor type can be improved by operating it at a higher collector voltage (eventually limited by breakdown) and at lower collector currents (eventually limited by low current β falloff [2]).

The 35821E transistors were also exposed to γ irradiation to a fluence of 10^7 rads from a Co^{60} source [22]. No significant changes in device operation at the $V_{CE} = 15$ V, $I_c = 15$ mA bias point were observed. Therefore it is

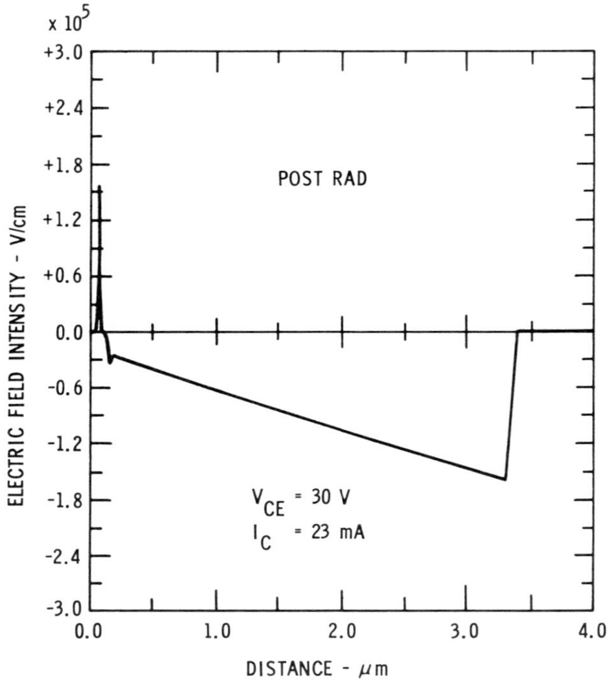

FIGURE 12.11
Calculated electric field profile for the 35821E transistor for increased value of V_{CE}, $J_c = 3000$ A/cm^2, postrad (after Graham, Chaffin, and Gwyn [22]).

possible to ignore the γ irradiation received in the ACPR reactor along with neutron fluence (10^7 rads for 10^{16} n/cm^2, $\mathscr{E} > 10$ keV) and attribute the damage solely to neutron-induced displacement effects.

12.4 Radiation-Induced Changes in Microwave Parameters

This section is devoted to a discussion of permanent radiation effects on the microwave properties of small-signal bipolar transistors. The discussion will be in two parts: (1) changes in the *S* parameters, and (2) changes in noise performance.

Radiation-Induced Changes in *S* Parameters. Very little has been published on the topic of radiation-induced changes in *S* parameters [22,28]. Graham et al. [22] have measured the *S* parameters of the 35821E transistor (discussed in the previous section) from 0.5 GHz to 8 GHz, before and after neutron irradiations ($\mathscr{E} > 10$ keV, Spectrum B). Typical results are shown in Figures 12.12 to 12.15. The most significant changes are seen in S_{21}. After 7×10^{15} n/cm^2, the forward transfer gain, S_{21}, drops below unity for all frequencies

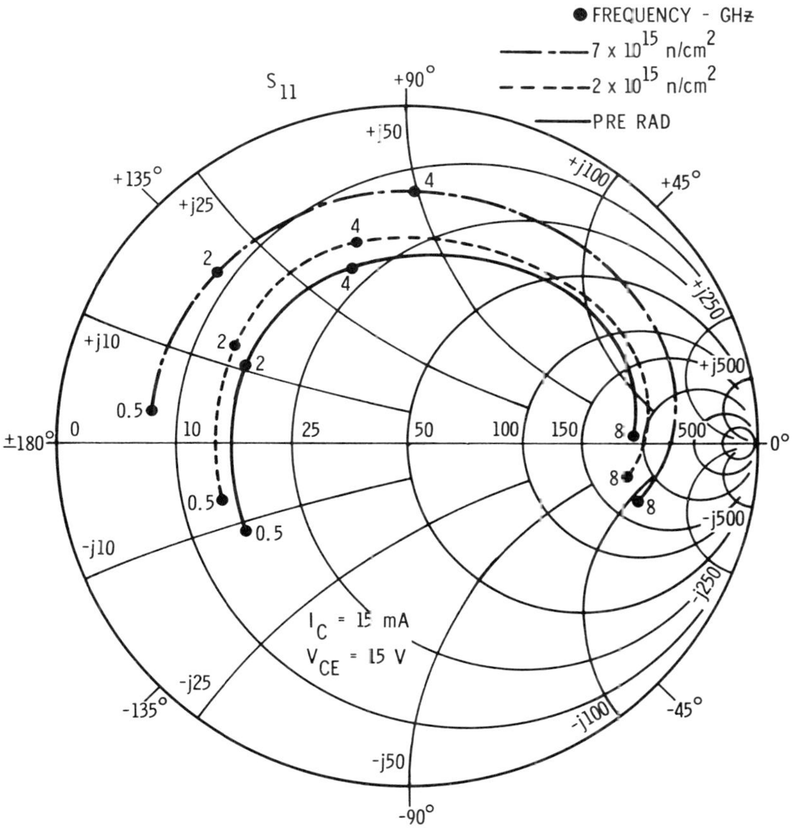

FIGURE 12.12
S_{11} versus frequency, pre- and postrad for the 35821E transistor (after Graham, Chaffin, and Gwyn [22]).

above 0.5 GHz. The other S parameters do not change appreciably with radiation since they are largely dependent on the package parasitics shown in Figure 12.13. The drop in $|S_{21}|$ can be seen more clearly from an examination of the $V_{CE} = 15$ V curves in Figures 12.16 and 12.17. It can be seen that f_s

(the frequency at which $|S_{21}| = 1$) drops from 4.5 GHz prerad to 540 MHz† after 7×10^{15} n/cm² at $V_{CE} = 15$ V. This drop in f_s is equivalent to a drop in the f_T of the device [$f_s \cong f_T$, see Eq. (12.15)]. From Eq. (12.5) and (12.6), we can write

$$f_T = \frac{1}{2\pi(\tau_e + \tau_b + \tau_c)} \cdot \tag{12.20}$$

The discussion in the previous section showed that there will be a drastic increase in τ_b for the $V_{CE} = 15$ V case at moderate collector currents after

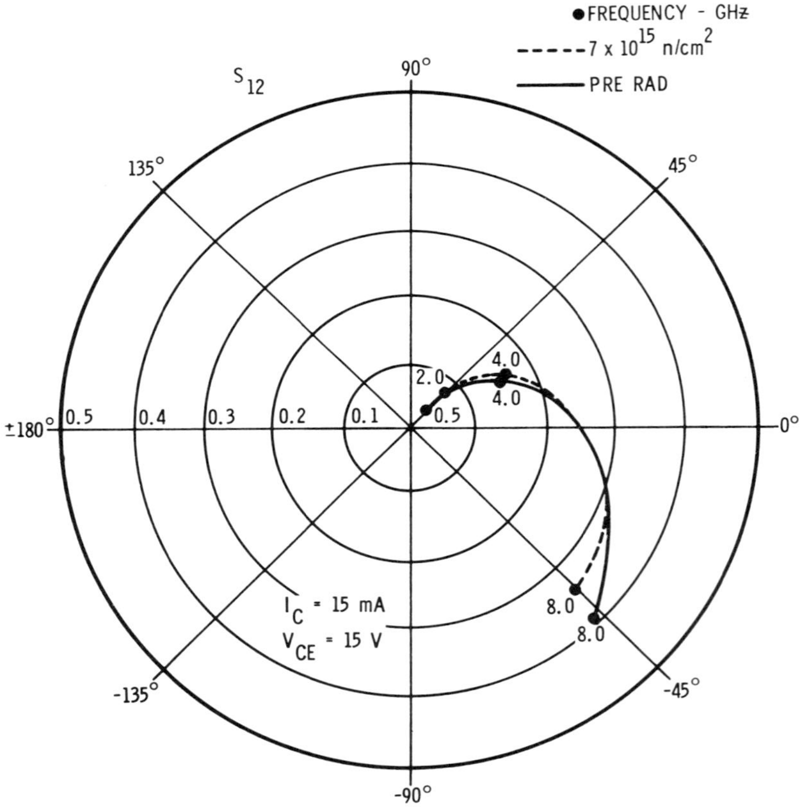

FIGURE 12.13
S_{12} versus frequency, pre- and postrad for the 35821E transistor (after Graham, Chaffin, and Gwyn [22]).

† The measurements in Figure 12.16 were taken several months after those shown in Figure 12.14, and a small amount of recovery in the value of f_T has occurred due to annealing.

7×10^{15} n/cm^2, due to the Kirk effect. Equation (12.19) and Figure 12.11 show that if the collector voltage is increased sufficiently, the base-push or Kirk effect can be suppressed. Figure 12.16 shows that if the collector bias on the irradiated transistor is increased from 15 V to 30 V, the f_s increases from 550 MHz to 3.0 GHz. Hence, most of the high-frequency gain degradation can be recovered by increasing collector bias voltage to suppress the Kirk

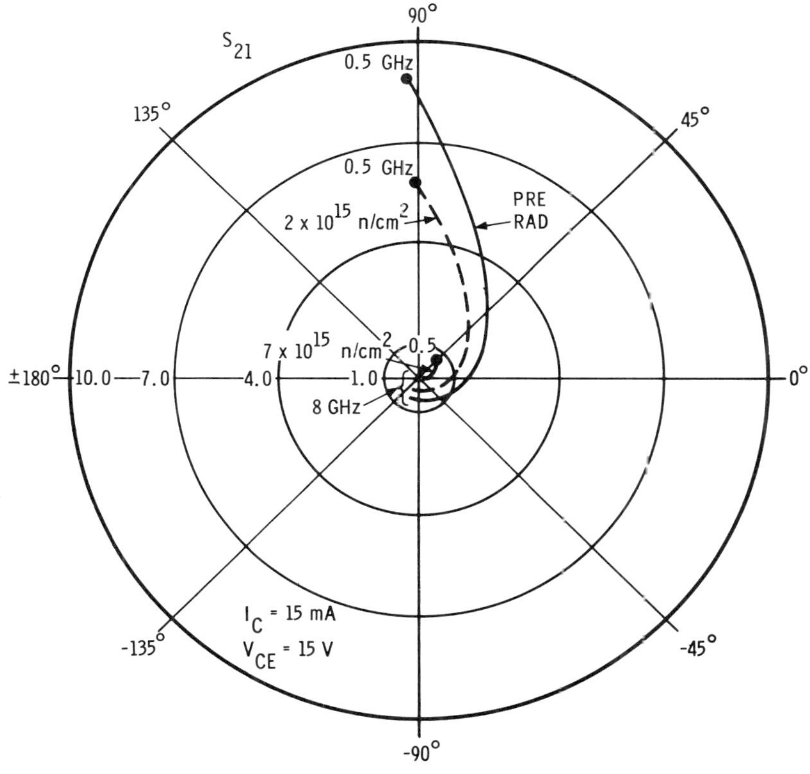

FIGURE 12.14
S_{21} versus frequency, pre- and postrad for the 35821E transistor (after Graham, Chaffin, and Gwyn [22]).

effect. Figure 12.17 shows that the change from 15 V to 30 V has very little effect on an unirradiated transistor (which still has N_5^+ high enough to prevent the Kirk effect from occurring).

The studies described here have been carried out using a second type of device. These devices were identical to the 35821E with the exception of the emitter dopant, which was changed from phosphorous to arsenic. The results obtained for the dc and rf parameters of the irradiated arsenic devices were virtually identical to those obtained with the phosphorous emitter devices. The reason for this behavior is that the arsenic and phosphorous emitter profiles are similar.

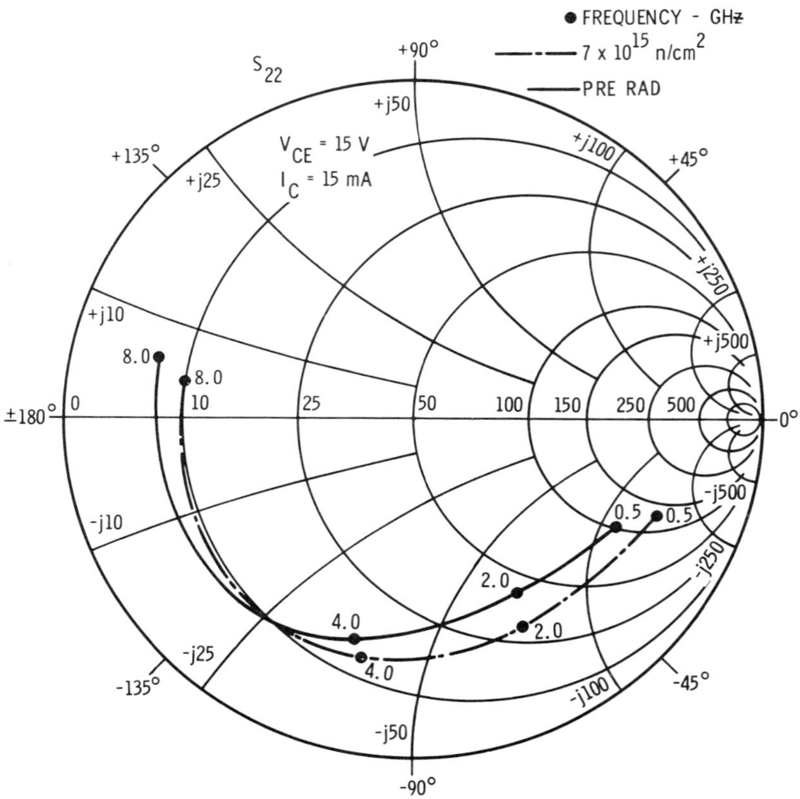

FIGURE 12.15
S_{22} versus frequency, pre- and postrad for the 35821E transistor (after Graham, Chaffin, and Gwyn [22]).

Radiation-Induced Changes in Noise Performance. One important application of the microwave bipolar transistor is in low-noise rf amplifiers used for the input stages of radar and communication systems. The noise performance of transistors used in these applications is extremely important since it is the input stage that usually determines the maximum sensitivity of the system.

This section investigates the effect of neutron fluence on the noise parameters of microwave bipolar transistor amplifiers.

Studies of changes in the noise performance of solid-state devices have been concentrated at lower frequencies [29–31]. Webb and Chaffin [32] have investigated the effects of radiation on the microwave noise performance of small-signal, bipolar transistors. They measured the noise figure of a 1.2 GHz transistor (35821E transistor) amplifier as a function of neutron fluence.

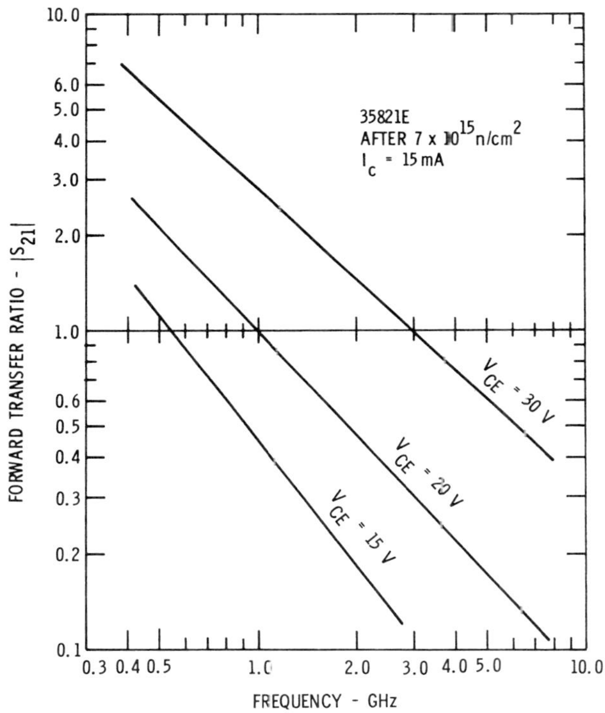

FIGURE 12.16
$|S_{21}|$ versus frequency at several bias voltages, postrad (after Graham, Chaffin, and Gwyn [22]).

Before discussing the experiment, a brief description of noise figure and its application to system parameters will be helpful. The usual figure of merit for a microwave receiver is called its noise figure. The noise figure, F, for a receiver is defined as

$$F \equiv \frac{S_i/N_i}{S_0/N_0} \tag{12.21}$$

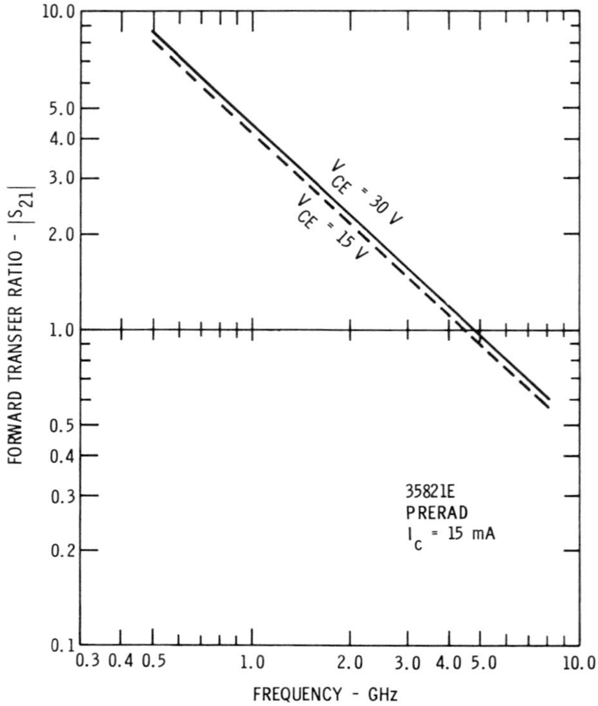

FIGURE 12.17
$|S_{21}|$ versus frequency at two bias voltages, prerad (after Graham, Chaffin, and Gwyn [22]).

where S_i and N_i are the input signal and input noise, and S_o and N_o are the output signal and output noise. Since noise free amplification is not possible, this requires $F > 1$. The closer F is to 1, the more sensitive the receiver will be. The minimum detectable signal for a receiver can be shown from Eq. (12.21) to be

$$S_i = FN_i \cdot \tag{12.22}$$

This is the input signal which results in a unity output signal-to-noise ratio. It can be seen that the receiver sensitivity is critically dependent on the noise figure. The noise figure of a single stage of the receiver can also be discussed using Eq. (12.21). For a single stage amplifier with gain, G_1, it can be shown that [33]

$$F_1 = \frac{S_i/N_i}{G_1 S_i/N_0} = \frac{N_o}{G_1 N_i} \cdot \tag{12.23}$$

Equation (12.23) shows that the noise figure of a single stage is the ratio of the output noise to amplified input noise. If the amplifier were perfect, the output noise would just equal the amplified input noise and F would be unity. However, since the amplifying device adds some noise of its own, $F > 1$. It can also be shown [33] that for two amplifiers in cascade the overall noise figure is

$$F_{12} = F_1 + \frac{F_2 - 1}{G_1} \tag{12.24}$$

where G_1 is the gain of the first stage. If G_1 is large, it can be seen from Eq. (12.24) that F_2 is relatively unimportant. This is the reason for the statement that the noise figure of a receiver is primarily determined by the input stage.

Flicker noise ($1/f$), surface noise, and generation-recombination noise are low-frequency effects and should not be important at microwave frequencies. Therefore, the radiation damage produced generation-recombination centers should not be of direct importance to the microwave transistor noise figure. Hence, the only degradation in microwave transistor noise performance due to irradiation should come about indirectly because of changes in transistor parameters (α, β, etc.) with radiation. The noise figure of high-frequency transistors has been studied by many authors [34–37]. The principal sources of noise within a microwave transistor are *shot noise*, due to the emitter and collector currents, and *thermal noise* (sometimes called resistor or Johnson noise) in the base resistance. These sources of noise are frequency independent below 1000 GHz. Fukui [36] has shown that the noise figure [Eq. (12.23)] of a microwave bipolar transistor can be expressed (ignoring header parasitic effects) as:

$$F = \left(1 + \frac{1}{\beta_0}\right)\left(1 + \frac{r_b}{R_g}\right) + \frac{kT}{2eI_cR_g} + \frac{eI_c}{2kTR_g}(r_b + R_g)^2\left[\frac{1}{\beta_{\text{dc}}} + \frac{1}{\beta_0^2} + \left(\frac{f}{F_T}\right)^2\right] \tag{12.25}$$

where β_0 = low-frequency current gain, $\beta_{\text{dc}} = I_c/I_B$, R_g = input source impedance, f_T = unity gain frequency, and f = operating frequency.

If no new high-frequency noise sources are introduced into the transistor by radiation, then the noise figure of a microwave transistor amplifier should still be predicted by Eq. (12.25) in the postirradiated case. To test this theory, a 1.2 GHz amplifier was constructed using a Hewlett-Packard 35821E microwave bipolar transistor. This transistor has an f_T of approximately 4.5 GHz. The amplifier had a prerad gain of 10 dB and a prerad noise figure of 3.9 dB. Adequate gain and noise figure were obtained without input impedance matching, thus $R_g = 50\ \Omega$ in Eq. (12.25). The noise figure was determined using the experimental setup shown in Figure 12.18. This was done by measuring the noise figure of the system (with an HP-340B automatic noise figure meter) with the first stage (test amplifier) removed (i.e., input at plane B-B' in Figure 12.18) and then measuring the noise figure with the first stage amplifier in place (i.e., input at plane A-A'). Then, from Eq. (12.24) and an independent measure of G_1, it was possible to calculate F_1, which is the noise figure of the first stage. The gain of the second amplifier, G_2, was 30 dB; consequently, the noise characteristics of the following portions of the receiver are unimportant. The first-stage amplifier was constructed so that the transistors could easily be removed and replaced. Ten transistors were used in

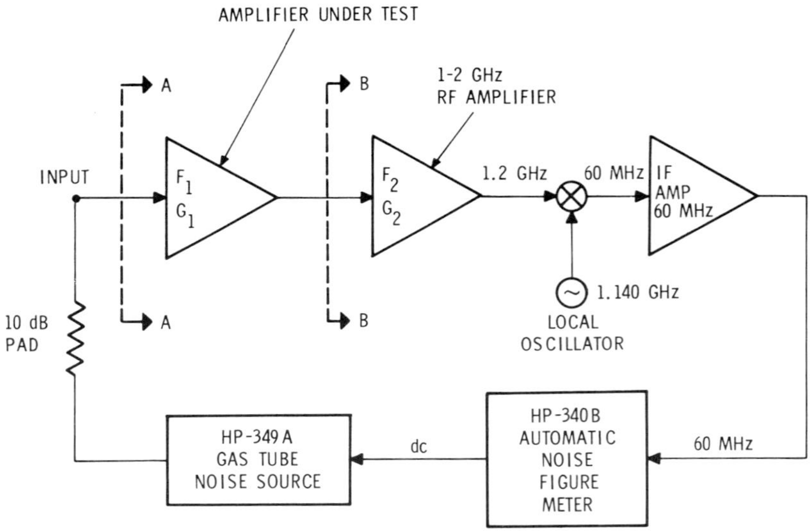

FIGURE 12.18
Experimental test setup for measuring the noise figure of a transistor amplifier (after Webb and Chaffin [32].

the test with two being held as controls. The transistors were neutron irradiated at the Sandia ACPR reactor (all fluences given are $\mathscr{E} > 10$ keV Spectrum B) and then replaced in the first-stage amplifier, and its gain and noise figure were remeasured. The collector current and voltage were held constant at 10 mA and 10 V, respectively, for all measurements. The experimental data obtained is shown in Figures 12.19 to 12.21. (These are eight sample averages, although the variance was less than 0.5 dB from sample to sample). Figure 12.19 shows that the gain, G_1, of the amplifier remains constant out to 2×10^{14} n/cm^2, falls off by 1.4 dB at 2×10^{15} n/cm^2, and fails completely (18 dB loss) at 7×10^{15} n/cm^2. Figure 12.20 shows the behavior of

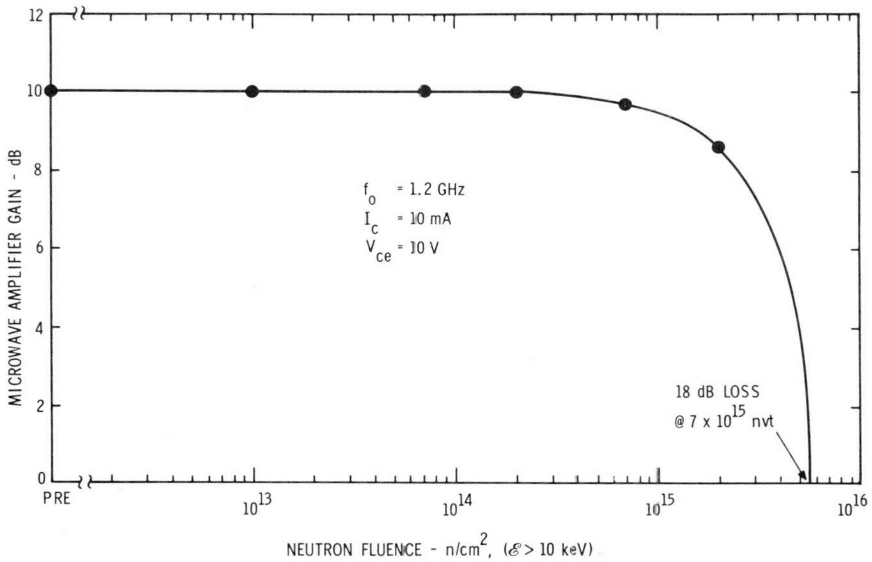

FIGURE 12.19
Amplifier gain versus neutron fluence (after Webb and Chaffin [32]).

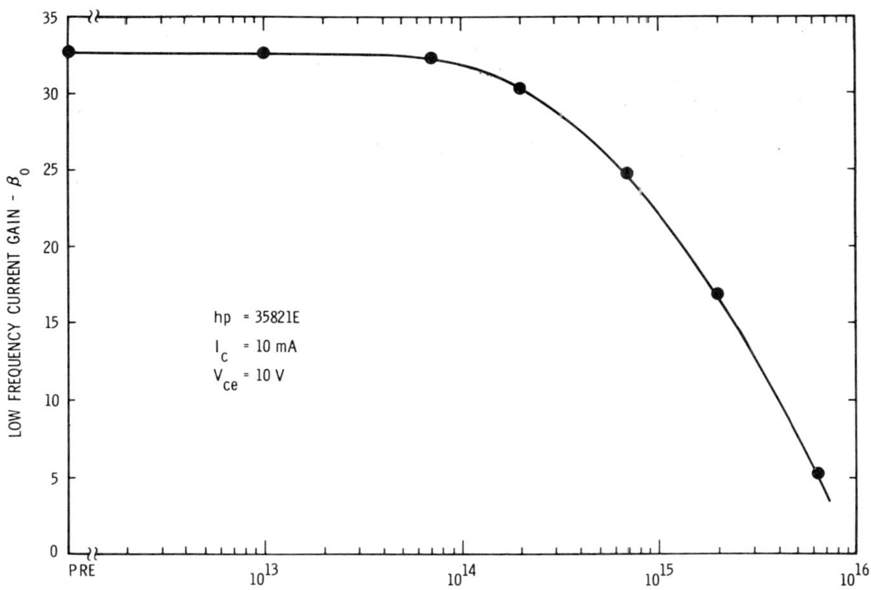

FIGURE 12.20
Low-frequency current gain versus neutron fluence (after Webb and Chaffin [32]).

the low-frequency current gain, β_0, versus neutron fluence. β_0 decreases steadily above 10^{14} n/cm^2 reaching a value of 4 at 7×10^{15} n/cm^2. The dc current gain, β_{dc}, was within a few percent of β_0 at all times (as expected for a small-signal device) and exhibited identical behavior with neutron fluence. Figure 12.21 shows the measured noise figure, F_1, of the amplifier stage, as well as the theoretical values of noise figure predicted by Eq. (12.25). The effect of neutron irradiation was brought into Eq. (12.25) through variation of

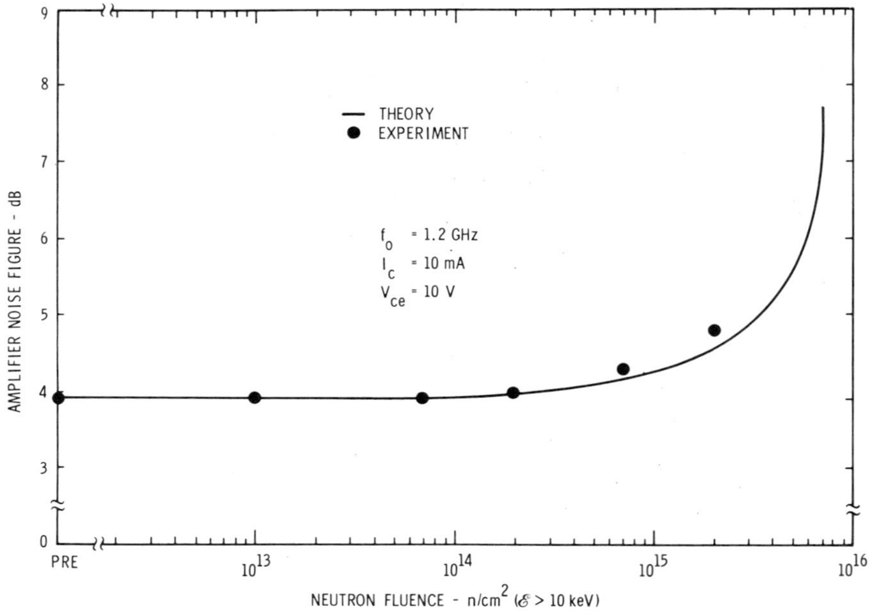

FIGURE 12.21
Theoretical and experimental noise figure versus neutron fluence (after Webb and Chaffin [32]).

β_0 (Figure 12.20) and β_{dc} with neutron fluence. Figure 12.21 shows that the experimental data agrees very well with the theoretical prediction of Eq. (12.25). The values used in Eq. (12.25) are $r_b = 4\ \Omega$, $f_T = 4.1$ GHz (typical for the 35821E transistor), $R_g = 50\ \Omega$, $I_c = 0.01$ A, and $T = 290°$K. The effects of neutron fluence on r_b and f_T were ignored. However, a slight decrease in f_T is seen at 2×10^{15} n/cm^2 [22]. This decrease would tend to improve the agreement between experiment and theory at high fluences (see Figure 12.21). The agreement between experiment and theory implies that the original hypothesis—that no new high-frequency noise sources are introduced by neutron irradiation of microwave bipolar transistors—is correct.

The conclusion to be drawn from this data is that the noise figure of bipolar microwave preamplifiers is fairly insensitive to neutron radiation. This is because the noise mechanisms introduced by neutron damage only contribute excess noise at low frequencies. The increase in noise figure which does occur can be attributed to the decreased current gain β_0.

12.5 Permanent Radiation Effects in Bipolar Microwave Power Transistors

Recent developments in bipolar transistor technology have resulted in transistors capable of producing several watts of output in the 0.5 GHz to 3 GHz range. These devices are somewhat similar to those previously discussed in Sections 12.2 to 12.4. The main differences are that they usually operate class "C," have much larger areas, and are usually operated at or near f_T. Webb and Chaffin [32] have presented data on neutron-irradiated rf power transistor operation.

The power gain of a microwave transistor power amplifier is given by [10]

$$G \cong h_{fe}^2 \frac{R_{LS}}{R_{IN}} \tag{12.26}$$

where h_{fe} (or β) is given by Eq. (12.10), R_{LS} is the series-equivalent load resistance, and R_{IN} is the series-equivalent input resistance. Quite often these amplifiers are operated at or slightly above f_T. From Eq. (12.10) this implies $\beta < 1$. However power gain is still achieved by virtue of the fact that usually, under matched conditions, $R_{IN} < R_{LS}$. For a power amplifier, the peak-to-peak rf voltage swing in the collector circuit determines the maximum output power. The maximum output power may be written as [38]

$$P_{\text{out}} = \frac{(V_{CC} - V_{SAT})^2}{2R_{LP}} \tag{12.27}$$

where V_{CC} = collector supply voltage and V_{SAT} = high-frequency collector-emitter saturation voltage.

Experimental measurements on a particular high-frequency power transistor have been carried out [32]. The device tested was a TRW-2N4430 transistor, which is a double-diffused, epitaxial silicon device similar in construction to the device shown in Figure 12.2. This is a power device, capable of dissipating 10 W at 25°C. The devices were measured under operating conditions in amplifiers which produced a nominal rf output of 8 W at a nominal frequency of 600 MHz. The devices were irradiated in the Sandia ACPR reactor (all fluences given are $\mathscr{E} > 10$ keV, Spectrum B). The most significant changes seen occurred in the values of V_{SAT} and h_{fe}. Figure 12.22

shows that V_{SAT} increases from approximately 1 V prerad to 13 V after 5×10^{15} n/cm^2. The effect of this increase was observed to degrade the maximum saturated rf output power available from the amplifier. This is in accordance with Eq. (12.27). The unsaturated gain of the amplifier was also observed to decrease† due to the decrease in high-frequency h_{fe} seen with irradiation (see

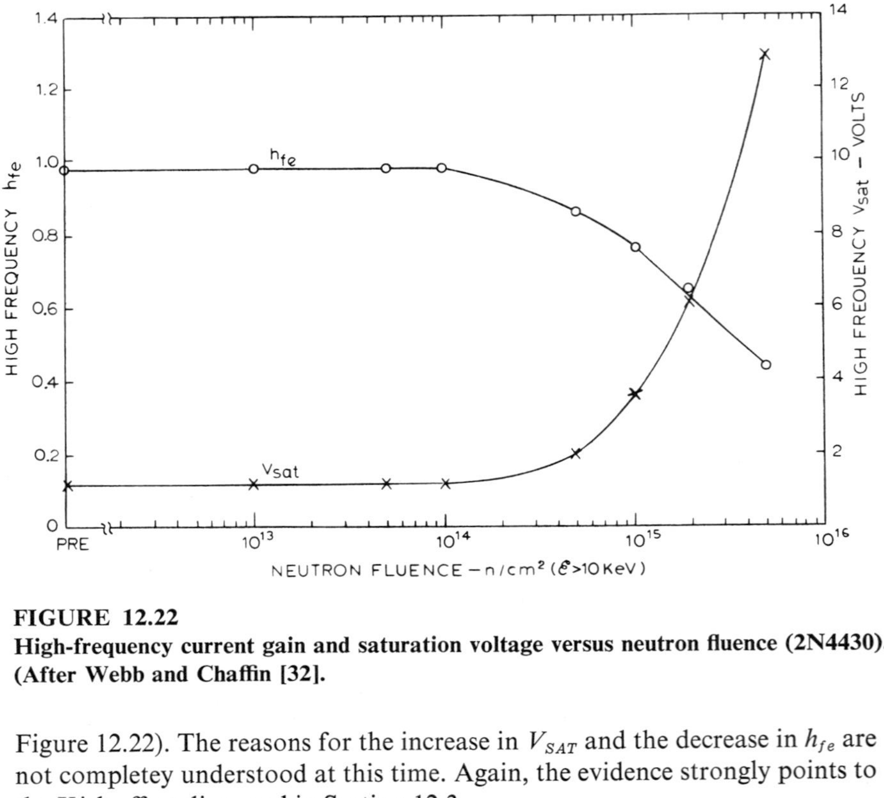

FIGURE 12.22
High-frequency current gain and saturation voltage versus neutron fluence (2N4430). (After Webb and Chaffin [32].

Figure 12.22). The reasons for the increase in V_{SAT} and the decrease in h_{fe} are not completey understood at this time. Again, the evidence strongly points to the Kirk effect discussed in Section 12.3.

12.6 Permanent Damage in Microwave FETs

Recent laboratory results [44] have shown that microwave operation of FETs (Field-Effect Transistors) is possible. These devices will probably find eventual applications as low-noise, small-signal, rf amplifiers in the 5 GHz to 20 GHz range. Unfortunately, no microwave performance data on irradiated devices has, as yet, been published. A great deal of work has been published

† Since R_{LS} and R_{IN} were observed to remain constant, see Eq. (12.26).

on radiation effects in low-frequency FETs [39–43]. Hypotheses can be made about what should occur in a microwave FET (junction or Schottky-barrier) on the basis of the low-frequency work [39–43]. However, it should be realized that these theories have not yet been confirmed by high-frequency measurements.

The basic changes produced by radiation fluence on FETs are attributed to carrier trapping effects in the channel of the device. The trapping effects lead to carrier removal effects, causing an increase in channel resistance and a corresponding decrease in channel current at a given V_{GD} (gate-to-drain voltage). Recall from Chapter 5 that these effects will become important when the density of the trapped charge approaches the original doping density. Nominally, four charges are trapped per n/cm^2 in bulk Si; therefore, if the channel-doping density is 10^{16} cm^{-3}, it would take 2.5×10^{15} n/cm^2 to strongly perturb the channel conductivity. The effective ionized doping density of the channel also changes due to compensation effects produced by the trapped charges. This effect can lead to frequency dependent changes in the transconductance of the devices [42,43]. These changes will become significant at about the same fluence that carrier removal effects become significant. The noise properties of a low-noise microwave FET are expected to have only a second-order dependence on irradiation fluence (i.e., due to changes in gain, bias current, etc.). The generation-recombination noise due to irradiation-induced trapping centers has been shown to be ineffective at microwave frequencies (see Section 12.4B).

In summary, it may be estimated that microwave FETs should operate without severe degradation to the radiation fluences at which up to 10% of the free carriers are trapped. (This corresponds to a neutron fluence per cm^2 of $\frac{1}{40}$ of the numerical value of the channel doping density, per cm^3; i.e., if $N_D = 10^{16}$ cm^{-3}, then $\Phi = 2.5 \times 10^{14}$ n/cm^2.) Severe degradation is to be expected at ten times this fluence level.

12.7 Transient Ionizing Radiation Effects

The effects of transient ionizing radiation on microwave bipolar transistors are similar to those seen in lower-frequency transistors. Larin [45] has provided an extensive discussion of the effects of ionizing radiation on transistors biased into their active region. The ionizing radiation produces a "primary" photocurrent due to the electron-hole pairs generated in the collector depletion region. If the ionizing radiation pulse lasts a long time in comparison with the transit time in the transistor (which is usually the case), then the primary photocurrent may be estimated from Eq. (5.36) as

$$J_{pp} \cong g \cdot e \cdot L \tag{12.28}$$

where L = collector depletion width (epi-layer thickness). This primary photocurrent flows backward through the transistor from the collector to the base. It can be shown [45] that the presence of these extra carriers in the base increases the emitter current and causes a secondary photocurrent to flow

$$J_{sp} \cong (1 + h_{fe})J_{pp} . \tag{12.29}$$

In other words, the primary photocurrent is multiplied by the common-emitter current gain of the transistor. Consequently, large collector currents can flow and temporarily interrupt high-frequency amplification (due to "current crowding" effects reducing f_T). This type of failure is only of a temporary nature. A more serious effect that can occur at high current levels is secondary breakdown [45]. Transistors usually have a "safe region" for high pulse power operation [2]. This rating is usually in the form of a maximum allowed (pulse power)·(time) product. If the transient ionizing radiation pulse forces the transistor operation to exceed this rating, hot-spotting and burnout may occur. The best solution to this problem is to provide current limiting in the collector supply if possible.

The effects described here apply to low-frequency transistors as well as microwave transistors. Generally microwave transistors might be expected to be slightly less sensitive to ionizing radiation than are low-frequency transistors. The reason for this is that microwave transistors tend to have smaller active volumes and higher operating current densities than do lower-frequency transistors.

REFERENCES

[1] H. A. Watson, ed., *Microwave Semiconductor Devices and Their Circuit Applications*, McGraw-Hill, New York, 1969, Chapter 17.

[2] S. M. Sze, *Physics of Semiconductor Devices*, Wiley-Interscience, New York, 1969, Chapter 6.

[3] H. V. Shurmer, *Microwave Semiconductor Devices*, Wiley-Interscience, New York, 1971, Chapter 7.

[4] M. H. White and M. O. Thurston, "Characterization of Microwave Transistors," *Sol. St. Elec.*, **13**, 523 (1970).

[5] J. Lange, "A Survey of the Present State of Microwave Transistor Modeling," *IEEE Trans. PGED*, **ED-18**, 1168 (1971).

[6] N. J. Gri, "Microwave Transistors—From Small Signal to High Power," *The Microwave Journal*, **14**, 45 (February 1971).

[7] H. F. Cooke, "Microwave Transistors, Theory and Design," *Proc. IEEE*, **59**, 1168 (1971).

[8] W. E. Beadle et al., "Design, Fabrication and Characterization of a Germanium Microwave Transistor," *IEEE Trans. PGED*, **ED-16**, 125 (1969).

[9] S. Kakihana and P. H. Wang, "Simple CAD Technique to Develop High-Frequency Transistors," *IEEE J. of Sol. St. Ckts.*, **SC-6**, 236 (1971).

[10] R. A. Battisti and K. O. Morey, "Amplifier and Oscillator Applications of the GHz Family of UHF Transistors," Applications Note, TRW Semiconductors, Lawndale, Calif.

[11] J. G. Tatum, "VHF, UHF Power Transistor Amplifier Design," Applications Note AN-1-2, ITT Semiconductor, West Palm Beach, Fla.

[12] J. A. Archer, "Design and Performance of Small Signal Microwave Transistors," *Sol. St. Elec.*, **15**, 249 (1972).

[13] H. F. Cooke, "Microwave Transistors," *Microwaves*, **8**, 46 (December 1969).

[14] M. K. Barnoski and D. Loper, "Ion Implanted Transistor Development," Air Force Report AFAL-TR-71 366 (November 1971).

[15] "*S*-Parameters, Circuit Analysis and Design," Application Note 95, Hewlett-Packard Co., Palo Alto, Calif.

[16] K. Hartman et al., "Equivalent Networks for Three Different Microwave Bipolar Transistor Packages," *Elec. Let.*, **1**, 510 (1971).

[17] J. A. Hood, "Predicting Current Gain Degradation in *n-p-n* Silicon Transistors after Irradiation by High Energy Neutrons," Report SC-TM-64-69, Sandia Corporation, Albuquerque, N.M. (1964).

[18] J. K. Loferski, "Analysis of the Effect of Nuclear Radiation on Transistors," *J. Appl. Phys.*, **29**, 35 (1958).

[19] C. A. Goben, F. M. Smits, and J. L. Wirth, "Neutron Radiation Damage in Silicon Transistors," *IEEE Trans. Nuc. Sci.*, **NS-14**, 6, 14 (1967).

[20] F. Larin, *Radiation Effects in Semiconductor Devices*, John Wiley & Sons, New York, 1968.

[21] C. W. Gwyn and B. L. Gregory, "Designing Ultrahard Bipolar Transistors," *IEEE Trans. Nuc. Sci.*, **NS-18**, 340 (December 1971).

[22] E. D. Graham, Jr., R. J. Chaffin, and C. W. Gwyn, "Radiation Effects in Microwave, Bipolar Transistors," *IEEE Trans. Nuc. Sci.*, **NS-19**, 335 (December 1972).

[23] C. W. Gwyn, D. L. Scharfetter, and J. L. Wirth, "The Analysis of Radiation Effects in Semiconductor Junction Devices," *IEEE Trans. Nuc. Sci.*, **NS-14**, 153 (December 1967).

[24] C. T. Kirk, Jr., "A Theory of Transistor Cutoff Frequency Falloff at High Current Densities," *IRE Trans. PGED*, **ED-9**, 164 (1962).

[25] R. J. Whittier and D. A. Tremere, "Current Gain and Cutoff Frequency Falloff at High Currents," *IEEE Trans. PGED*, **ED-16**, 39 (1969).

[26] H. C. Poon et al., "High Injection in Epitaxial Transistors," *IEEE Trans. PGED*, **ED-16**, 455 (1969).

[27] S. M. Sze, op. cit., p. 285.

[28] C. A. Goben, D. L. Gray, and Y. P. Han, "Nuclear Radiation Enhancement of Transistor Forward Gain at High Frequencies," *IEE Trans. Nuc. Sci.*, **NS-17**, 380 (December 1970).

[29] H. E. Kern and J. M. McKenzie, "Noise from Neutron Induced Defects in JFET's," *IEEE Trans. Nuc. Sci.*, **NS-17**, 256 (December 1970).

[30] J. L. Prince and R. A. Stehlin, "Effect of Co^{60} Gamma Radiation on the Noise Parameters of Bipolar Transistors," *IEEE Trans. Nuc. Sci.*, **NS-18**, 404 (December 1971).

[31] P. O. Lauritzen, "Effects of Radiation on the Noise Performance of Transistors," *IEEE Trans. Nuc. Sci.*, **NS-19**, 321 (April 1972).

[32] J. G. Webb and R. J. Chaffin, "Neutron Irradiation Effects on Microwave Transistor Amplifiers," *Proceedings of the IEEE Conference on Nuclear and Space Radiation Effects*, Seattle, Wash., July 1972.

[33] W. Mumford and E. Scheibe, *Noise Performance Factors in Communication Systems*, Horizon-House, Dedham, Mass., 1968.

[34] A. van der Ziel, "Theory of Shot Noise in Junction Diodes and Junction Transistors," *Proc. IRE*, **43**, 1639 (1955).

[35] E. Nielsen, "Behavior of Noise Figure in Junction Transistors," *Proc. IRE*, **45** 957 (1957).

[36] H. Fukui, "The Noise Performance of Microwave Transistors," *IEEE Trans. PGED*, **ED-13**, 329 (1966).

[37] H. F. Cooke, "Transistor Noise Figure," *Solid State Design*, **4**, 37 (February 1963).

[38] P. M. Chirlian, *Analysis and Design of Electronic Circuits*, McGraw-Hill, New York, 1968.

[39] W. L. George, "Optimization of Neutron Radiation Tolerance of JFET's," *IEEE Trans. Nuc. Sci.*, **NS-16**, 81 (December 1969).

[40] W. Shedd, B. Buchanan, and R. Dolan, "Radiation Effects on Junction Field Effect Transistors," *IEEE Trans. Nuc. Sci.*, **NS-16**, 87 (December 1969).

[41] A. F. Behle and R. Zulegg, "Fast Neutron Degradation of GaAs Field-Effect Transistors," *Proceedings of the IEEE Conference on Nuclear and Space Radiation Effects*, Durham, N.H., July 20, 1971, p. 184.

[42] B. L. Gregory, S. S. Naik, and W. G. Oldham, "Neutron Produced Trapping Centers in Junction Field Effect Transistors," *IEEE Trans. Nuc. Sci.*, **NS-18**, 50 (December 1971).

[43] S. S. Naik and W. G. Oldham, "Neutron Radiation Effects in Junction Field Effect Transistors," *IEEE Trans. Nuc. Sci.*, **NS-18**, 9 (October 1971).

[44] C. A. Liechti, E. Gowen, and J. Cohen, "GaAs Microwave Schottky-Gate FET," *Proc. IEEE ISSCC*, Philadelphia, Pa., February 1972, p. 158.

[45] F. Larin, *Radiation Effects in Semiconductor Devices*, John Wiley & Sons, New York, 1968, Chapter 10.

13

Radiation Damage in Tunnel Diode Devices

13.1 Introduction

The tunnel diode is a NDR (Negative-Differential Resistance) device which derives its NDR from junction rather than bulk conductivity effects. (See Chapter 9 for a discussion of bulk NDR devices.) The tunnel diode was discovered by L. Eskai [1,2] in 1958. He was the first to correctly interpret the anomalous forward V-I characteristic seen in very heavily doped (degenerate) *p-n* junctions. The tunnel diode produces a NDR up to very high frequencies (well above the microwave range), and therefore these devices find applications as low-power oscillators and amplifiers in the microwave range. The device has not gained wide acceptance due to its inherent low power and low impedance levels and also due to reliability problems associated with degenerately doped junctions.

Section 13.2 of this chapter will discuss the basic theory behind the operation of the tunnel diode device. Radiation effects on the device will be discussed in Section 13.3, and radiation effects on microwave circuits employing tunnel diodes will be discussed in Section 13.4. Transient radiation effects will be discussed in Section 13.5.

13.2 Tunnel Diode Theory

The theory of microwave tunnel diodes and circuits has been treated in detail in the literature [1–16], and, therefore, this discussion will be brief. Only those aspects of tunnel diode operation which are essential to an understanding of radiation effects in these devices will be discussed.

Figure 13.1 shows a typical V-I characteristic for a tunnel diode. This characteristic is frequency independent even at millimeter wave frequencies. From Figure 13.1 it can be seen that the V-I characteristic of a tunnel diode is considerably different from that of an ideal diode (see Figure 3.3). The reasons for this behavior will be discussed later. The forward-bias voltage corresponding to the valley current, I_v, is typically a few hundred millivolts. Due to the narrow voltage range (a few hundred mV) over which the NDR exists, these devices are limited to low power applications. Oscillator outputs of a few milliwatts or less are typical. Tunnel diodes may be made from any semiconductor material, although the most commonly used ones for microwave applications are Ge and GaAs.

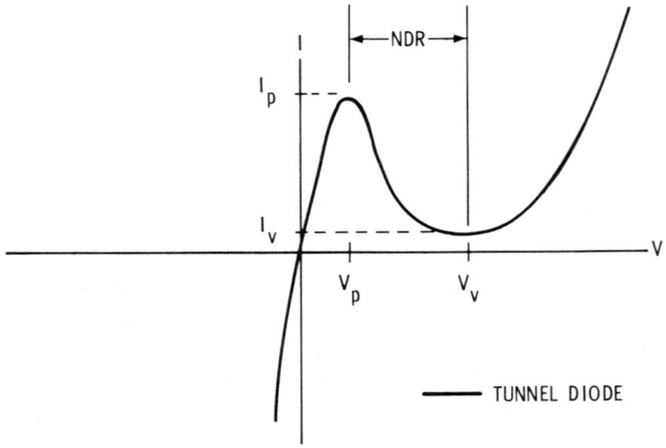

FIGURE 13.1
V-I characteristic of a typical tunnel diode.

The operation of the tunnel diode will be explained with the aid of Figure 13.2. The device structure consists of a very heavily doped p^+-n^+ junction formed by the alloy process (see Section 4.3). Figure 13.2 shows such a junction at several different applied biases. For very heavy (degenerate) doping concentrations Figure 13.2B shows that the Fermi level lies in the conduction band for the n^+ material and in the valence band for the p^+ material. Figure 13.2C shows that at a small forward bias, the electrons in the conduction band on the n^+ side of the junction are opposite available states in the valence band. Due to the finite probability of equi-energy quantum-mechanical tunneling, it is possible for an electron to tunnel through the forbidden gap and appear in the valence band (note both energy and momentum must be conserved, and therefore this process is more likely to occur in direct semiconductors such as GaAs [see Figure 2.4B]). It can be shown [5] that the probability of an electron tunneling through a triangular-shaped potential barrier is given by

$$P = \exp\left[-\frac{4}{3}\frac{\sqrt{2m_e}\mathscr{E}_g^{3/2}}{e\hbar E}\right] \tag{13.1}$$

where $\mathscr{E}_g$ = bandgap energy, $2\pi\hbar$ = Planck's constant, and E = electric field at the junction. From Eq. (13.1) it can be seen that the tunneling is enhanced for small effective masses. This is another reason why GaAs is used to make tunnel diodes (its effective mass is quite small, see Appendix A). Figure 13.2 shows an abrupt, degenerately doped p^+-n^+ diode at several different applied

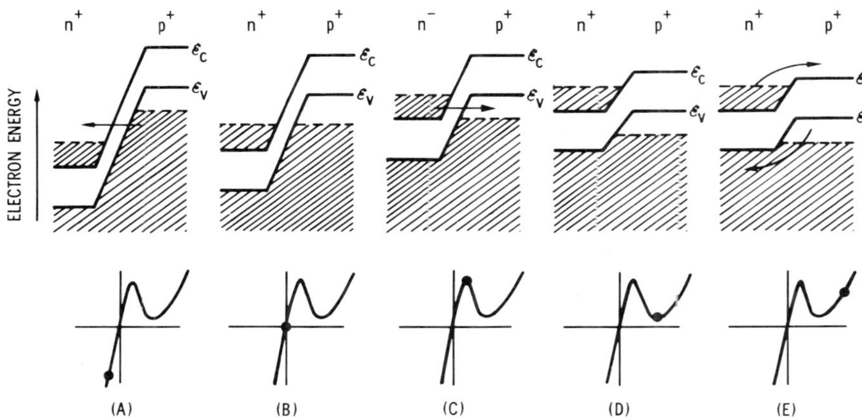

FIGURE 13.2
Explanation of V-I characteristic of a tunnel diode (after Hall [3]).

biases. The dot on the V-I characteristic below each energy band diagram shows the terminal voltage and current. The energy band diagrams are shown at the corresponding bias values. The cross-hatched areas represent states occupied by electrons, and the dashed lines indicate the Fermi levels. Figure 13.2B shows the zero-bias case. As forward bias is applied (Figure 13.2C), the energy of the electrons on the n^+ side is increased relative to the p^+ side. Figure 13.2C shows that some of the electrons on the n^+ side now lie directly opposite available states in the valence band, and tunneling occurs through the forbidden gap. This tunneling causes a current to flow as is shown in the lower portion of Figure 13.2C. As the forward bias is increased further, the electrons in the conduction band on the n^+ side are raised up (on the energy scale, see Figure 13.2D), and they now no longer lie opposite available states in the valence band. For this case the tunneling current drops essentially to zero, and the terminal current must drop as shown in Figure 13.2D. A further increase in forward voltage (Figure 13.2E) results in a sufficient lowering of the potential barrier so that the normal diode forward current region (see Section 3.3) is reached. For reverse applied bias, Figure 13.2A shows that considerable numbers of electrons can tunnel from the p^+ region to the n^+ region, and this number increases steadily with reverse bias. The reverse current therefore increases very rapidly with applied reverse bias. Hence, the V-I characteristic shown in Figure 13.1 can be explained on the basis of quantum-mechanical tunneling of electrons.

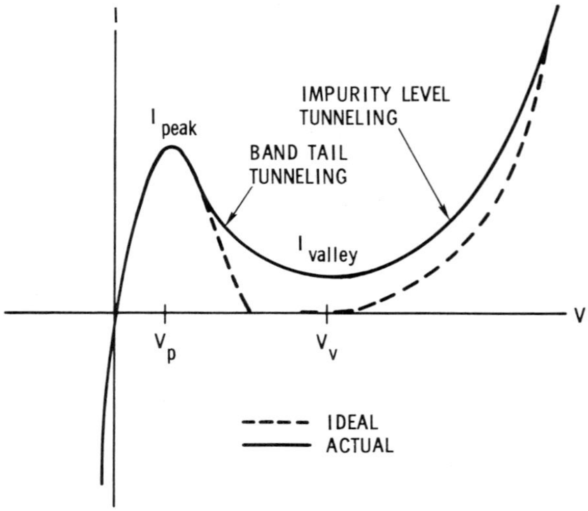

FIGURE 13.3
Ideal versus actual tunnel diode V-I characteristic.

In the real world, the V-I characteristic is found to deviate somewhat from that predicted by simple theories. Figure 13.3 compares measured and ideal tunnel diode V-I characteristics. From Figure 13.3 it can be seen that the valley current is not as low as would be expected from the simple model shown in Figure 13.2. The excess current that occurs for $V < V_v$ is usually attributed to band tail tunneling [4,5]; i.e., the conduction band and valence band edges become blurred, and the density-of-states function tails off exponentially into the forbidden gap in the degenerately doped case. For $V > V_v$, the excess current is attributed to impurity-produced energy levels deep within the forbidden gap [4,5]; i.e., electrons can fall from the conduction band into these deep levels and then tunnel into the valence band (more will be said on this later).

The high-frequency equivalent circuit of a packaged tunnel diode biased into the NDR region is shown in Figure 13.4. The R_s represents any series bulk or contact resistance; C_J is the junction capacitance (typically $2\mu f/\text{cm}^2$ [3]); $-R$ is the slope of the V-I characteristic; and L_s is the series inductance due to package parasitics. The effect of the parasitic elements L_s, R_s, and C_J is to limit the useful frequency range over which the tunnel diode can be used. The equivalent circuit shown in Figure 13.4 is valid for all frequencies.† Hence, it can easily be seen that microwave oscillators and amplifiers can be constructed by connecting the appropriate impedances across terminals

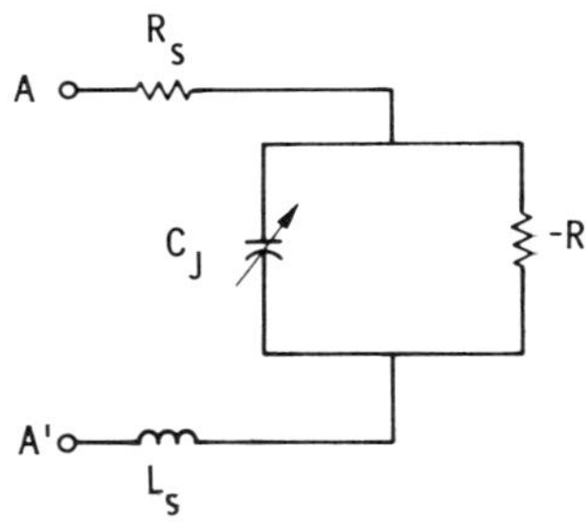

FIGURE 13.4
Equivalent circuit of a packaged tunnel diode.

A-A′ in Figure 13.4. Since the tunnel diode is a two-terminal device, amplifiers are usually fabricated with the aid of a circulator to isolate the input and output ports. Since tunnel diodes produce pure-shot noise, they have very low noise-figure capabilities. The minimum noise figure that can be obtained with a properly matched tunnel diode amplifier is given by [4]

$$F_{\min} = 1 + |R \cdot I|_{\min} \tag{13.2}$$

where $|R \cdot I|_{\min}$ = minimum value of the negative R times current product (see Figure 13.1). A tunnel diode oscillator can be formed by simply reactively terminating the terminals A-A' in Figure 13.4. The maximum power output that can be obtained from a tunnel diode oscillator is given by [6]

$$P_{\text{out}} \cong \tfrac{1}{8}(V_v - V_p) \cdot (I_p - I_v) \tag{13.3}$$

where the voltages and currents are defined in Figure 13.3. Since tunnel diodes do not suffer transit-time effects, the maximum usable frequency is

† Provided that the package dimensions are much less than a wavelength.

determined only by the parasitic elements shown in Figure 13.4. It can be shown [4] that the resistive cutoff frequency in the negative resistance region is given by

$$f_c = \frac{1}{2\pi|R|C_J} \cdot \sqrt{\frac{|R|}{R_s} - 1} \tag{13.4}$$

where C_J is measured at V_v.

13.3 Radiation Effects in Tunnel Diode Devices

This section is devoted to a discussion of permanent radiation effects in tunnel diode devices. The next section will consider circuit effects. Many researchers [7–17] have investigated permanent radiation effects in tunnel diode devices. Tunnel diodes made with Si, Ge, and GaAs have been tested in both electron and neutron radiation environments. One might expect tunnel diodes to be quite radiation tolerant. They are majority-carrier devices and do not depend on minority-carrier lifetime; also since they are quite heavily doped (10^{18}–10^{20} cm^{-3}), the carrier removal data given in Chapter 5 implies that the devices should operate to neutron fluences of 10^{18} n/cm^2. However, experimental data [7–17] shows that failure usually occurs in the 10^{15}–10^{17} n/cm^2 range. The primary failure mechanism for electron- or neutron-irradiated Si, Ge, and GaAs tunnel diodes is an increase in the valley current, I_v. Figure 13.5 shows the type of behavior usually seen. Curve A represents the prerad V-I characteristic, while curve B represents a low-temperature measurement following an electron irradiation, and curve C is a typical room-temperature result following a neutron irradiation. Curve B in Figure 13.5 will be discussed first.

Curve B shows a "hump" structure appearing in the V-I characteristic. The cause of the "hump" current in irradiated tunnel diodes has been attributed to the tunneling of electrons from a radiation-induced defect energy level in the forbidden gap to the valence band [10,12]. This is depicted in Figure 13.6. In Figure 13.6 it has been assumed that the radiation-induced defects produce energy levels which lie in a single narrow band (this is possible for low temperature irradiation and measurement, see Chapter 5). Since the Fermi level, $\mathscr{E}_f$, is well above the defect level, $\mathscr{E}_D$, the defect levels are expected to be fully occupied by electrons. Normally, the current would be quite low (near V_v) since the electrons in the conduction band cannot tunnel to the valence band. However, at some applied biases the defect energy level will be opposite the available states in the valence band (see Figure 13.6), and tunneling will occur. The defect energy band† lies across from the available

† If sufficient numbers of defects are introduced, the level may be expected to broaden slightly.

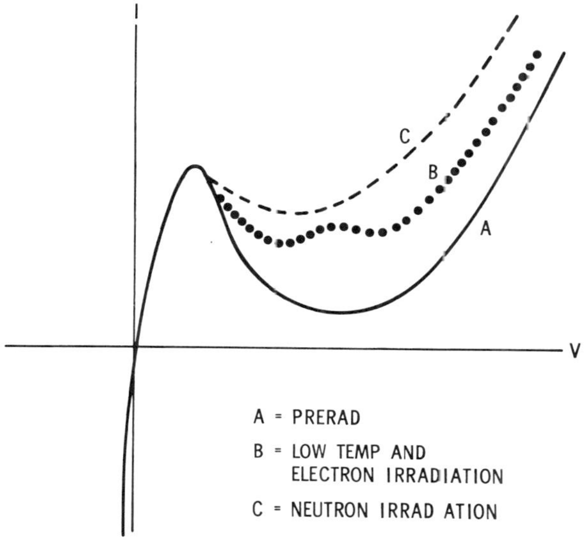

FIGURE 13.5
V-I characteristics of irradiated tunnel diodes.

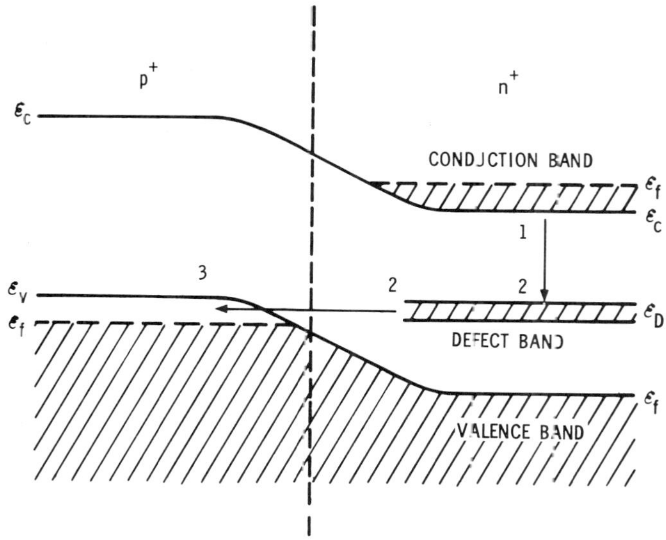

FIGURE 13.6
Tunneling from defect states to the valence band in an irradiated tunnel diode.

states in the valence band for only a fairly narrow range of applied biases. For this reason, the extra tunnel current produced by this mechanism will be quite voltage sensitive, and a "hump" in the V-I characteristic may occur (see Figure 13.5, trace B). If the Fermi levels in the n^+ and p^+ regions are known, it is possible to calculate the defect energy level position in the forbidden gap

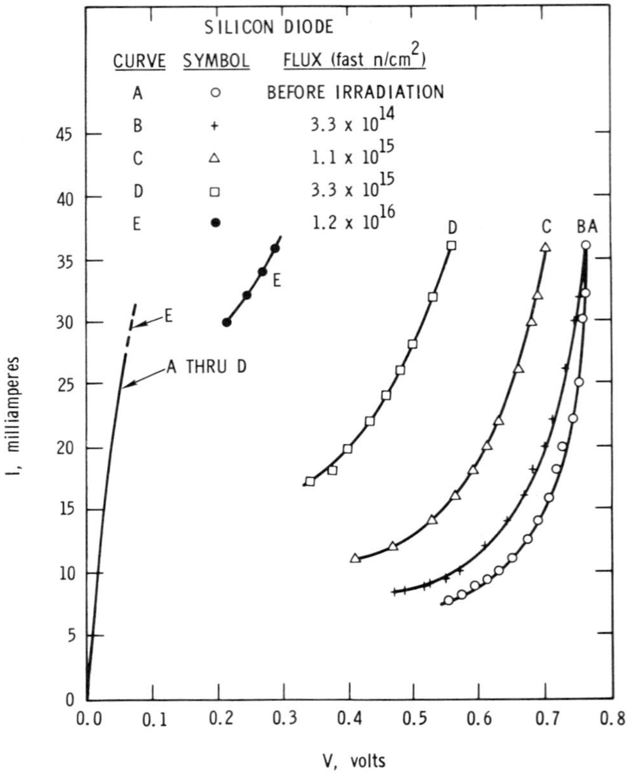

FIGURE 13.7
Characteristics of neutron-irradiated silicon tunnel diode (after Aukerman et al. [11]).

from the V-I characteristic [10]. If the diode is allowed to warm to room temperature (so that defect reordering occurs) or in the case of neutron bombardment, the V-I characteristic shown in Figure 13.5, trace C, occurs. The valley current is seen to increase in this case, but no "hump" or fine structure is seen. The reason for this behavior is that, in this case, the radiation produces many energy levels within the forbidden gap, and they lie close enough together so that they are essentially indistinguishable, and thus tunneling from these levels raises the current at all applied voltages in the valley region.

Experimental measurements on neutron-irradiated Si, Ge, and GaAs tunnel diodes are shown in Figures 13.7 to 13.10. The fluences given are for

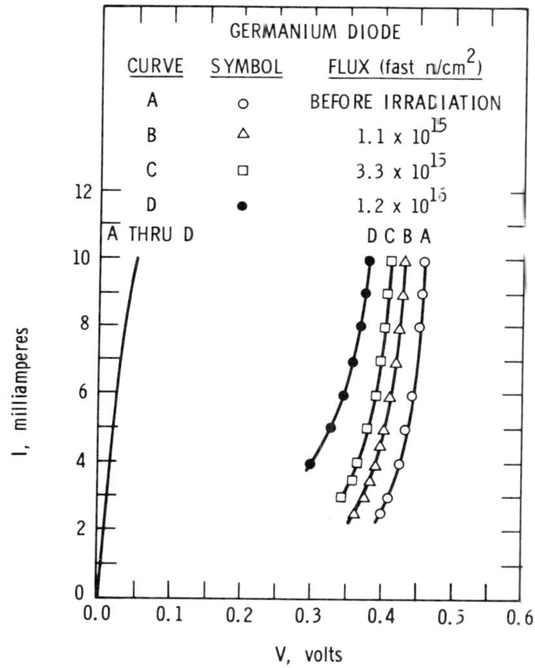

FIGURE 13.8
Characteristics of neutron-irradiated germanium tunnel diode (after Aukerman et al. [11]).

fast neutrons (exact cutoff is unknown [11]) Spectrum F. The discontinuous curves shown in Figures 13.7–13.9 occur when the tunnel diode is not loaded by a shunt resistance. In this case, as the voltage is increased, the device becomes unstable and "switches" from the peak to the valley; consequently these points correspond to the edges of the NDR region (see Figure 13.1). The data in Figures 13.7 to 13.9 show that the degradation with irradiation is primarily due to an increase in valley current, as was discussed earlier. The data in Figures 13.7 to 13.9 are summarized in Figure 13.10, which shows that

GaAs has the highest initial peak-to-valley ratio (discussed earlier) and the neutron fluence required to reduce the peak-to-valley ratio to one-half its initial value is approximately 2×10^{15} n/cm^2 for Si, Ge, and GaAs.

13.4 Radiation Effects in Tunnel Diode Circuits

This section is devoted to a discussion of radiation effects on microwave circuits employing tunnel diodes. It was mentioned earlier that the V-I

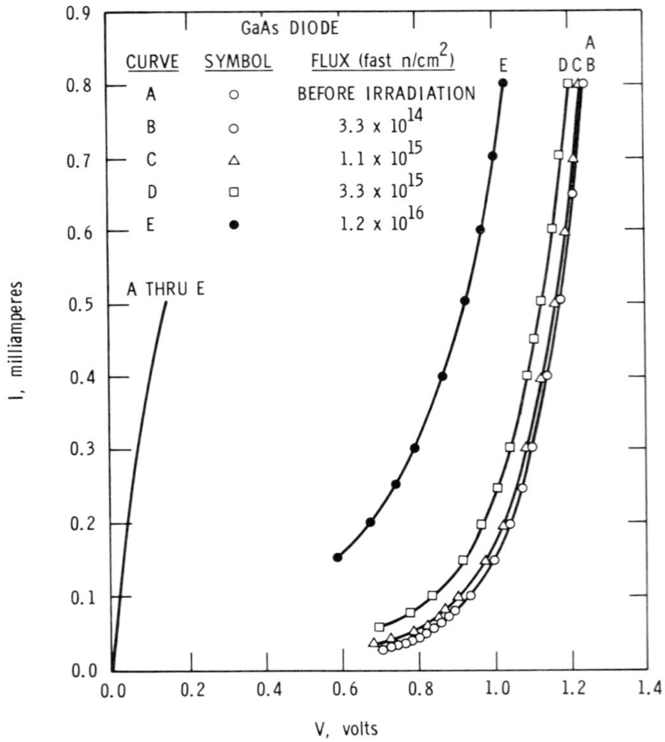

FIGURE 13.9
Characteristics of neutron-irradiated GaAs tunnel diode (after Aukerman et al. [11]).

characteristic for a tunnel diode is essentially frequency independent. This is also true in the post-irradiated case; therefore, the effects of radiation on microwave circuit performance may be calculated from changes in the V-I characteristics shown in Figures 13.7 to 13.9.

Microwave Oscillators. The effect of radiation on a Ge tunnel diode microwave oscillator has been examined by Easley and Blair [11]. Their results are shown in Figure 13.11. They observed a falloff in power output with gross degradation occurring at 9×10^{16} n/cm^2 ($\mathscr{E} > 0.1$ MeV, Spectrum G). A power output falloff of 3 dB and a frequency increase of 5 MHz (at 1.015 GHz) were observed at 3.5×10^{16} n/cm^2. The falloff in microwave power output can be understood from an inspection of Figure 13.8 and Eq. (13.3). From Figure 13.8 it can be seen that for a fluence of 1.2×10^{16} n/cm^2

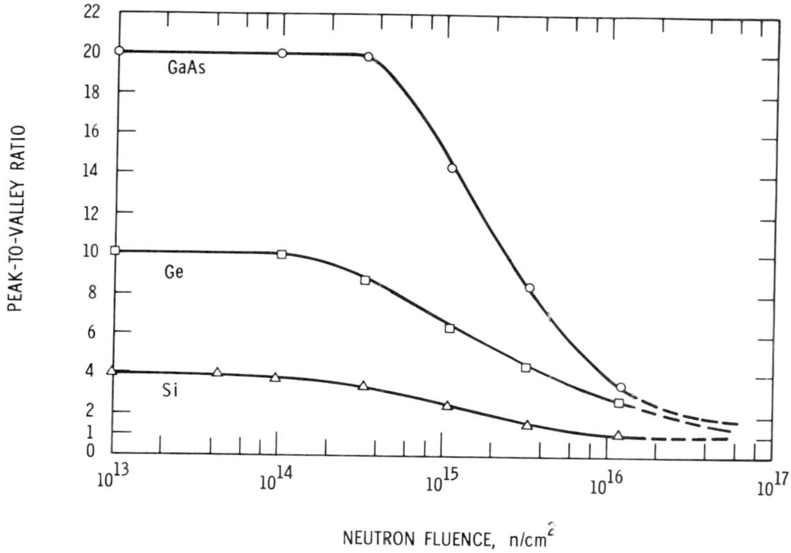

FIGURE 13.10
Peak-to-valley ratio of tunnel diodes as a function of neutron bombardment (after Aukerman et al. [11]).

(Spectrum F) the valley voltage decreases from 0.4 V to 0.3 V and the valley current increases from 2.5 mA to 4.0 mA, while the peak voltage and current do not change. If the values from Figure 13.8 (although it is not the same device) are inserted into Eq. (13.3), it can be seen that the power output should be reduced by 2.4 dB after 1.2×10^{16} n/cm^2 (Spectrum F). This is in reasonable agreement with the data shown in Figure 13.11. The frequency increase seen by Easley and Blair [7] can be explained in terms of a slight widening of the depletion width due to carrier removal effects. This reduces the value of C_J (see Figure 13.4) and raises the resonant frequency of the cavity-diode combination.

Microwave Amplifier. Published studies of radiation in low-noise microwave amplifiers using tunnel diodes are not available. However, in light of what already is known about radiation effects in general, and about tunnel diodes specifically, it is possible to hypothesize what should occur. Chapter 12 showed that the generation-recombination noise due to radiation-produced

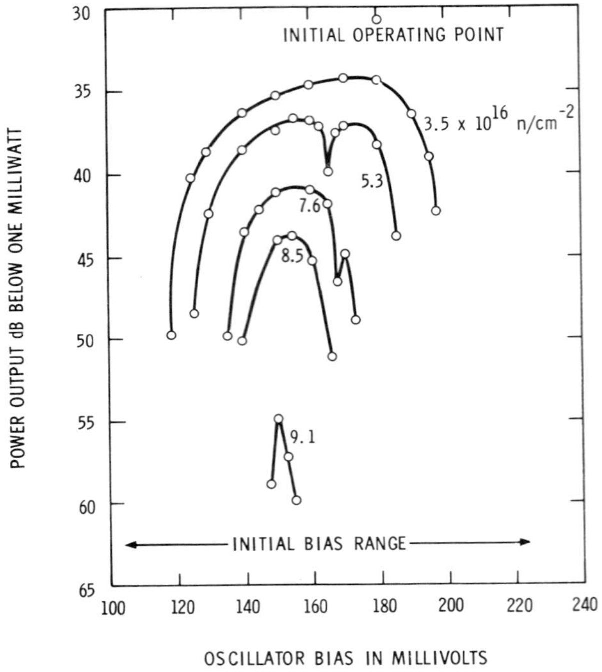

FIGURE 13.11
Power output versus diode voltage for a 1-GHz Ge tunnel diode oscillator at several neutron fluences. Dips at 170 mV due to cable reflections ($\mathscr{E}$ > 0.1 MeV, Spectrum G). (After Easley and Blair [7].)

trapping centers does not extend to the microwave range. Hence, no additional microwave noise components are expected, due to the introduction of trapping centers by irradiation. The changes that will occur are due to changes in the tunnel diode V-I characteristic. An inspection of Figures 13.5,13.7–13.9, and Eq. (13.2) shows that the minimum noise figure may be expected to increase (since $|R \cdot I|_{\min}$ increases with irradiation). Another consideration is impedance matching, i.e., the negative resistance of the diode is usually transformed to match a transmission line impedance, Z_0. A change in the negative resistance with irradiation will produce an impedance mismatch loss. One final consideration in radiation effects on tunnel diode amplifiers is the radiation-induced shift in the bias point. Figure 13.11 shows that in a fixed-bias amplifier the bias point may shift too far from the original optimum value. Considering all these possible effects, it can be concluded that a tunnel diode amplifier will be degraded whenever radiation produces a significant change in the V-I characteristic of the device. From Figures 13.7 to 13.9 it can be seen that this occurs at 1.1×10^{15} n/cm^2 (Spectrum F) for Si and Ge, and at times 1.2×10^{16} n/cm^2 for GaAs tunnel diodes.

Backward Diodes. Tunnel diodes operated at zero bias make good low-noise mixers and detectors (equal in performance to Schottky-barrier, devices), but since they have such a large nonlinearity near zero bias, they have an advantage in that they do not require a dc bias like the Schottky-barrier devices do. The V-I characteristic of a tunnel diode near zero bias is very steep (see Figure 13.1) and is dominated by tunneling in both the forward and reverse directions (see Figure 13.2). It is known [6] that the V-I characteristic is essentially independent of temperature in this region, which is another advantage of the backward diode over the Schottky-barrier. (Reliability problems and complex manufacturing techniques for backward diode devices keeps their applications to minimum.) Near zero bias tunneling of electrons from impurity to available states in the valence band, which is the failure mechanism in tunnel diode oscillators and amplifiers, does not occur since these energy levels do not line up. This can be seen from an inspection of Fig. 13.2B. Therefore, it is probable that backward diode detectors will be less susceptible to permanent radiation effects than are tunnel diode oscillators and amplifiers. The estimated failure level of a backward diode detector is 10^{17} n/cm^2, although no experimental data are available to support this conclusion.

13.5 Transient Ionizing Effects in Tunnel Diode Devices and Circuits

Tunnel diode devices have been exposed to neutron-gamma pulses from a fast-burst reactor [13] and to 17 MeV electron pulses and 2 MeV x-ray pulses [18]. Since tunnel diodes are very heavily doped, the generation of hole-electron pairs by ionizing radiation (see Chapter 5) is not expected to cause problems in the bulk material. Also, due to the heavy doping concentration and forward-bias operation, the depletion width is exceedingly narrow; consequently, generation of hole-electron pairs is not expected to be a problem here either. Bohan and Wagner [13] were unable to detect any changes in a GaAs tunnel diode V-I characteristic measured during a 80 μsec pulse from a fast-burst reactor. The flux during the pulse was 10^{17} neutrons cm^{-2} sec^{-1} and 10^7 rads(gamma) sec^{-1}. Tobolski et al. [18] were unable to make a tunnel diode logic device switch from the low-voltage state to the high-voltage state (like Figures 13.7 to 13.9) even when it was exposed to 17 MeV electron pulses at ionizing radiation rates of 10^9 rad·sec^{-1}. It can therefore be concluded that tunnel diode devices can survive very intense ionzing radiation pulses without interruption of operation. However, in the case of cavity oscillators, effects in the cavity must be considered (see Appendix B).

REFERENCES

[1] L. Esaki, "New Phenomena in Narrow Germanium *p-n* Junction," *Phys. Rev.*, **109**, 663 (1958).

[2] L. Esaki and Y. Miyahara, "New Device Using the Tunneling Process in Narrow *p-n* Junctions," *Sol. St. Elec.*, **1**, 13 (1960).

[3] R. N. Hall, "Tunnel Diodes," *IRE Trans. PGED*, **ED-7**, 1 (1960).

[4] H. A. Watson, *Microwave Semiconductor Devices and Their Circuit Applications*, McGraw-Hill, New York, 1969.

[5] S. M. Sze, *Physics of Semiconductor Devices*, Wiley-Interscience, New York, 1969.

[6] H. V. Shurmer, *Microwave Semiconductor Devices*, Wiley-Interscience, New York, 1971.

[7] J. W. Easley and R. R. Blair, "Fast Neutron Bombardment of Ge and Si Esaki Diodes," *J. Appl. Phys.*, **31**, 1772 (1960).

[8] A. G. Chynoweth et al., "Excess Tunnel Current in Silicon Esaki Junctions," *Phys. Rev.*, **121**, 684 (1961).

[9] R. A. Logan et al., "Electron Bombardment Damage in Si Esaki Diodes," *J. Appl. Phys.*, **32**, 1201 (1961).

[10] R. S. Claassen, "Excess and Hump Current in Esaki Diodes," *J. Appl. Phys.*, **32**, 2372 (1961).

[11] L. W. Aukerman et al., "Theoretical and Experimental Studies Concerning Radiation Damage in Selected Compound Semiconductors," Battelle Memorial Institute Report No. BMI-8064-65-5 (May 1962).

[12] C. B. Pierce et al., "Radiation Induced Hump Structure in the I-V Characteristic of Esaki Diodes," *J. Appl. Phys.* **33**, 3108 (1962).

[13] W. A. Bohan and A. J. Wagner, "The Effects of Steady State and Pulsed Nuclear Radiation on GaAs Tunnel Diodes," *IRE Trans. Nuc. Sci.*, **NS-9**, **1**, 346 (1962).

[14] C. B. Pierce and A. D. Kantz, "Defects Introduced in Neutron-Irradiated Esaki Diodes," *J. Appl. Phys.*, **34**, 1496 (1963).

[15] H. Roth et al., "Voltage Annealing of Radiation Damage in Tunnel Diodes," *J. Appl. Phys.*, **34**, 669 (1963).

[16] J. E. Dowdy and C. M. Travis, "An Analysis of Steady-State Nuclear Radiation Damage of Tunnel Diodes," *IEEE Trans. Nuc. Sci.*, **NS-11**, 55 (November 1964).

[17] P. Baruch, ed., *Radiation Damage in Semiconductors*, Academic Press, New York, 1964. See J. Easley, "The Semiconductor Device in Radiation Damage Research," p. 341.

[18] J. Tobolski, "Tunnel Diode Switching Induced by Transit Ionizing Radiation," *IEEE Trans. Nuc. Sci.*, **NS-14, 6,** 221 (1967).

Appendixes

Appendix A

Physical Constants of Semiconductors

Material Name	Effective Hole Mass (m^*/m_0)	Effective Electron Mass (m^*/m_0)	Bandgap (eV) at 300°K	Relative Dielectric Constant (ϵ_r)	Density $(\mathrm{gm/cm^3})$	Pair Formation Energy (eV)	Melting Point (°C)
Silicon (Si)	$m_{hh}^* = 0.5$ $m_{lh}^* = 0.16$ $m_{dh}^* = 0.55$ $m_{ch}^* = 0.45$	$m_{\perp}^* = 0.19$ $m_{\parallel}^* = 0.97$ $m_{de}^* = 0.33$ $m_{ce}^* = 0.26$	1.12	11.8	2.33	3.6	1420
Germanium (Ge)	$m_{hh}^* = 0.3$ $m_{lh}^* = 0.04$ $m_{dh}^* = 0.31$ $m_{ch}^* = 0.3$	$m_{\perp}^* = 0.082$ $m_{\parallel}^* = 1.6$ $m_{de}^* = 0.22$ $m_{ce}^* = 0.12$	0.80	16.0	5.33	2.6	937
Gallium Arsenide (GaAs)	$m_{hh}^* = 0.5$ $m_{lh}^* = 0.12$ $m_{dh}^* = 0.53$ $m_{ch}^* = 0.46$	$m_e^* = 0.068$ $m_{de}^* = m_e^*$ $m_{ce}^* = m_e^*$ $m_{2e}^* = 0.4$ (Upper Valley)	1.43	10.9	5.32	4.6	1238

Note: Some of the data in this table has been taken from: S. M. Sze, *Semiconductor Devices*, John Wiley & Sons, New York, 1969.

Appendix B

Microwave Circuit Boards and Cavities

The microwave semiconductor devices described in this book are usually mounted in a microwave circuit of some type. This microwave circuit typically consists of: (1) alumina microstripline; (2) Teflon-glass microstripline or stripline, or (3) a microwave cavity. In the course of studying radiation effects on devices, some useful information on radiation effects in microwave circuits was obtained. Some of this information will be presented here.

Microstripline and Stripline Circuits

A microstripline or stripline circuit usually consists of a narrow, flat metal transmission line separated from a ground plane (or planes in the case of stripline) by a dielectric substrate. The microwave properties of this configuration (impedance, wavelength, etc.) are well known [1] and depend on the mechanical dimensions of the conductors and the electrical properties of the dielectric substrate. Several past studies [2,8] have examined the low-frequency properties of irradiated dielectric materials, however measurements at microwave frequencies have only been made recently [3]. The complex dielectric constant may be expressed as:

$$\tilde{\epsilon} = \epsilon' - j\epsilon'' \tag{B.1}$$

and the loss tangent is defined as

$$\tan\delta = \frac{\epsilon''}{\epsilon'} \cong \frac{\epsilon''}{|\tilde{\epsilon}|}, \qquad \epsilon'' \ll \epsilon' \tag{B.2}$$

A small change in the real part of the substrate dielectric constant, ϵ', will cause a change in the circuit wavelength, λ, and the intrinsic impedance, Z_0 of the transmission line: the former will cause a frequency shift in resonant structures, and the latter will alter the circuit impedance level. Likewise a change in the imaginary part of the substrate dielectric constant, ϵ'', will affect circuit losses and resonator Q's.

A simple experiment to measure such changes with radiation will now be described. Improved measurement sensitivity to changes in the real and imaginary parts of the dielectric constant are obtained if one uses a resonant circuit. Figure B.1 depicts a resonant bandpass filter [4] with microstripline construction. Circuits of this type were fabricated from high purity alumina as well as Teflon-glass sub-

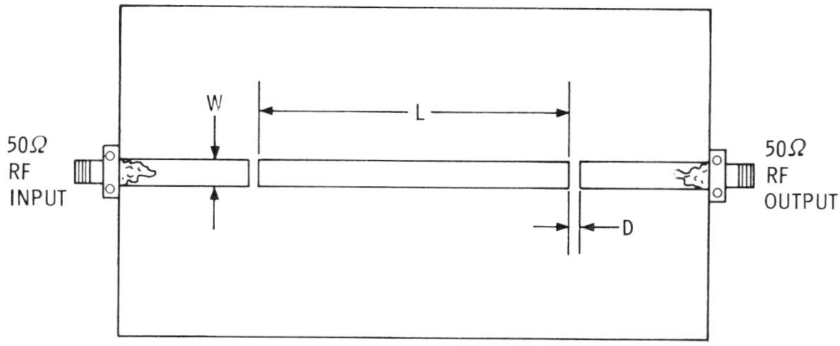

FIGURE B.1
Microstripline bandpass resonator.

strates. To first order, the resonant frequency of the filter shown in Figure B.1 is given by

$$f_r = \frac{2}{L\sqrt{\mu_0 \epsilon_{\text{eff}}}} \tag{B.3}$$

where $\epsilon_{\text{eff}} \cong \epsilon'$ for stripline and for microstriplines [5]

$$\epsilon_{\text{eff}} \cong \frac{\epsilon' + 1}{2} + \left(\frac{\epsilon' - 1}{2}\right) \cdot \left[1 + 10\frac{H}{W}\right]^{-1/2} \tag{B.4}$$

where H = substrate thickness. The unloaded Q, Q_u, of the resonator is related to the loss tangent by

$$\frac{1}{Q_u} = \frac{1}{Q_c} + \tan\delta \tag{B.5}$$

where Q_c is the Q if only conductor losses are considered. The loaded (measured) Q, Q_l, is related to the unloaded Q by

$$Q_l = Q_u(1 - \sqrt{T}) \tag{B.6}$$

where T = transmission coefficient at resonance. Changes in f_r, Q_l, and T can therefore be used to calculate changes in the real and imaginary parts of the dielectric constant of the substrate.

The alumina substrate resonators (Figure B.1) were exposed to both neutron and gamma fluences. The alumina resonator nominally had f_r = 5.3 GHz, Q_l =

100, and $T = -7$ dB. The only change in the alumina resonator that could be seen was a discoloration of the substrate. The color changed from white prerad to dark brown postrad. This was attributed to the excitation of trapped electron color centers by ionizing radiation [2]. This color change saturated at about 10^6 rads. No change in Q_l, f_r, or T of the resonator could be detected after 10^8 rads† γ (from a Co^{60} source) or a neutron fluence of 2×10^{15} n/cm^2 ($E > 10$ keV, Spectrum B). The measurement uncertainties in these parameters were $Q_l \pm 2\%$, $f_r \pm 2$ MHz, $T \pm 0.2$ dB. Therefore, it can be concluded that the complex dielectric constant of the alumina substrate is unchanged after large neutron and γ fluences.

The Teflon-glass substrate resonators were made from two commercially available substrate types: Budd dilecto 522 and Rogers's RT duroid 5870. The nominal parameters of the Teflon-glass substrate resonators were $Q_l = 100, f_r = 3$ GHz, $T = -6$ dB. The following changes were noted after γ-irradiation from the Co^{60} source: (1) decrease in f_{rs} (2) decrease in Q_l, and (3) decrease in T (increased insertion loss). No change in the physical appearance of the substrates was observed. The changes measured in the RT-5870 resonator versus γ fluence are plotted in Figure B.2. From this data it can be seen that the resonant frequency, f_r, and unloaded Q, both decrease with γ fluence. This corresponds to an increase in the real and imaginary parts of the dielectric constant [see Eq. (B.3) and (B.5)]. For example, at 10^8 rads the real part of ϵ_{eff} increases by 1.3%, and the loss tangent increases to 0.0015 (from its 0.0006 prerad value). Changes in the Bud 522 substrate resonator were similar to those seen in the RT-5870 resonator. Similar resonators were also exposed to fast-neutron fluence in the Sandia ACPR reactor to doses up to 2×10^{15} n/cm^2 ($E > 10$ keV, Spectrum B). The changes observed correlated with the γ dose (10^{-9} rad/nvt) obtained along with the neutrons from the reactor; consequently it is felt that the changes were due to the γ radiation rather than the neutron radiation. Careful measurements were made to insure that these changes were not due to mechanical changes (creepage) of the substrate. No dimensional changes to the limit of the measurement ability (1 part in 1000) were seen. This is not surprising since fiberglass reinforced Teflon is known to have reasonable mechanical properties to 10^8 rads [7]. The increase in ϵ' and ϵ'' with radiation for Teflon has been observed previously at lower frequencies [8]; therefore, it is not surprising that a similar result is seen at microwave frequencies.

Microwave Cavities

Quite often Gunn, BARITT, and IMPATT oscillators are mounted in microwave cavities (particularly at frequencies near X-band and above). This is done because a higher Q, and therefore lower noise and greater stability, can be achieved with a cavity than with stripline or microstripline. These cavities are usually air filled. There is very little or no permanent radiation damage to the cavities even at extreme fluences. However, in the presence of ionizing radiation, the air in the cavities is ionized and forms a very lossy plasma. This lossy plasma

† Radiations were done in air at temperatures between 20°C and 40°C.

will reduce the resonator Q substantially and can perturb the operation of the solid-state microwave oscillator at levels below which the semiconductor device itself is affected [9]. After the ionizing radiation ceases, the electrons and ions recombine, and the plasma disappears (this can take from tens of nanoseconds at atmospheric pressure to several microseconds at pressures around 1 torr). A detailed analysis follows.

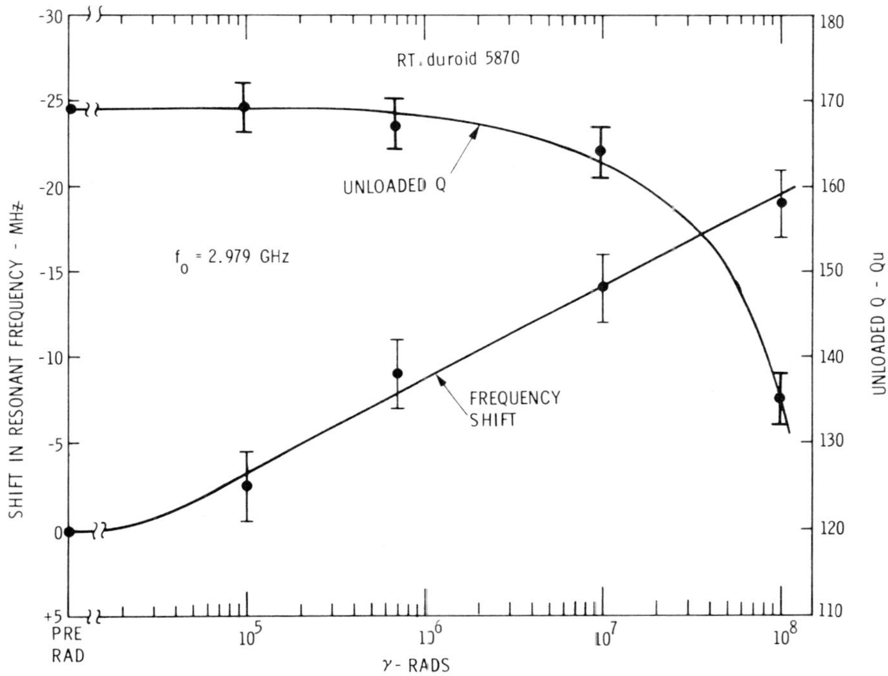

FIGURE B.2
Change in f_r and Q_u versus γ fluence for a Teflon-glass resonator.

Voth and Kenny [10] show that the uniform plasma formed in the cavity by the ionizing radiation causes a shift in the resonant frequency and a change in Q of the cavity which can be expressed as†

$$Q_1 \cong \frac{Q_0}{1 + Q_0 \dfrac{\omega_p^2 \nu}{\omega_0(\nu^2 + \omega_0^2)}}, \tag{B.7}$$

and

$$\frac{2\Delta\omega_0}{\omega_0} \cong \frac{\omega_p^2}{\nu^2 + \omega_0^2} \tag{B.8}$$

† This assumes the electron collision frequency is independent of velocity.

where

$$\begin{aligned} Q_1, Q_0 &= \text{cavity } Q \text{ with, without plasma} \\ \omega_0 &= \text{cavity resonant frequency} \\ \Delta\omega_0 &= \text{shift in resonant frequency with plasma present} \\ \nu &= \text{effective electron collision frequency} \\ \omega_p^2 &= ne^2/m\epsilon_0, \text{ plasma frequency} \\ n &= \text{electron density} \\ m, e &= \text{electron mass, charge} \\ \epsilon_0 &= \text{free space dielectric constant.} \end{aligned}$$

The case of interest here is when the cavity is filled with air at atmospheric pressure. If the electrons are assumed to have an energy on the order of 0.5 ev,†, then the collision frequency (at 760 mm) for momentum transfer may be found from Brown [11] as $\nu = 1.5 \times 10^{12}$ sec^{-1}. The electron-ion generation rate may be found from the known fact that it takes an average of 32.5 eV to produce an electron-ion pair in air at STP. Since the rad is a unit of energy deposition it can be shown that 1 rad per second produces 2.3×10^9 pair·cm^{-3}·sec^{-1}. To find the electron density, a differential equation must be solved relating the generation rate and loss rate (loss is due to several mechanisms). Instead of doing that here, assume that the energy is deposited in a time which is short compared to all loss mechanisms, thus an upper limit to the peak density may be calculated from the total dose (rather than the dose rate). This peak density will eventually decay back to zero due to the various loss mechanisms. For example, a dose rate of 4.8×10^8 rad/sec for 25 nsec results in a total dose of 12 rads, or a peak electron density of 2.75×10^{10} electrons/cm^3. Inserting the corresponding values of ν and ω_p into Eq. (B.7) and (B.8) we find (at 9 GHz and for $Q_0 = 800$)

$$Q_1 = 0.55\, Q_0 \tag{B.9}$$

and

$$\frac{2\Delta\omega_0}{\omega_0} = 4.2 \times 10^{-5}\,. \tag{B.10}$$

Therefore, these conditions (4.8×10^8 rad/sec for 25 nsec and a cavity filled with dry air at STP) result in a Q reduction of 45% and a resonant frequency shift of only 0.18 MHz. The reason for the large change in Q compared to the small frequency shift is the fact that $\nu^2 \gg \omega_0^2$, which makes the plasma much more resistive than reactive in nature. A Q reduction of 45% certainly cannot be neglected when the effect of ionizing radiation on a microwave diode oscillator is evaluated.

To verify the above calculations, an X-band microwave cavity was constructed. The cavity operated in the TM_{010} mode and had a resonant frequency of 9006 MHz and a loaded Q of 818. The cavity was a transmission type with input and output coupling loops. Ionizing radiation was obtained in the form of short (25 nsec), intense x-ray pulses from a Febetron flash x-ray generator. The setup used

† This is only a rough estimate of the kinetic energy of the electrons. Most of the electrons have energies between 0.025 eV (room temperature) and 16 eV, although the actual energy spectrum is unknown.

for the ionizing radiation experiment is shown in Figure B.3. The signal generator was tuned to the resonance peak, and the transmitted signal level was observed during and after the x-ray pulse with a crystal detector. Typical data obtained in this way are in Figure B.4, which shows the behavior of the signal transmitted through the cavity for doses of 12 rads, 137 rads, and 410 rads (25 nsec pulse width) while the cavity is filled with air at atmospheric pressure. The upper trace in each photograph corresponds to zero transmission. Figure B.4D shows the noise pickup in the signal lines. The changes shown in Figure B.4 are due to changes in Q only and not due to shifts in the resonant frequency. This was verified by setting the generator frequency slightly above and slightly below resonance and observing in each case that the transmitted signal decreased during

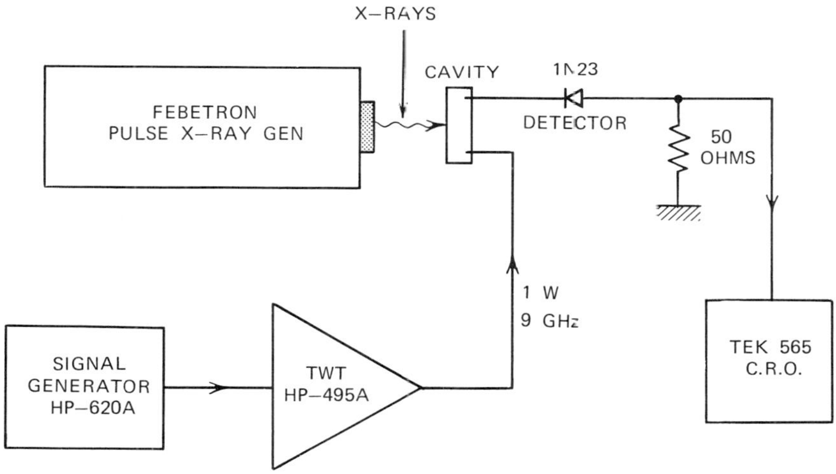

FIGURE B.3
Test setup for ionizing radiation in cavities experiment.

the radiation pulse. This is in agreement with Eq. (B.10). An evaluation of Figure B.4A shows the transmission drops by 7 dB at the peak. The change in transmission as a function of Q may be expressed [12] as:

$$\Delta L_{\text{db}} = 20 \log_{10} \left[\frac{Q_1}{Q_0} \right], \tag{B.11}$$

From Eq. (B.11) it can be seen that a 45% decrease in Q would cause the insertion loss to change by 5.52 dB. The change in insertion loss due to the resonance shift predicted by Eq. (B.10) is small compared to 5.2 dB and may be neglected. Hence, it can be seen that the change in transmission predicted by Eq. (B.1) (5.2 dB) is in close agreement with the measured value (7 dB) shown in Figure B.4A. The other changes shown in Figure B.4, for larger doses, cannot be analyzed in the same way

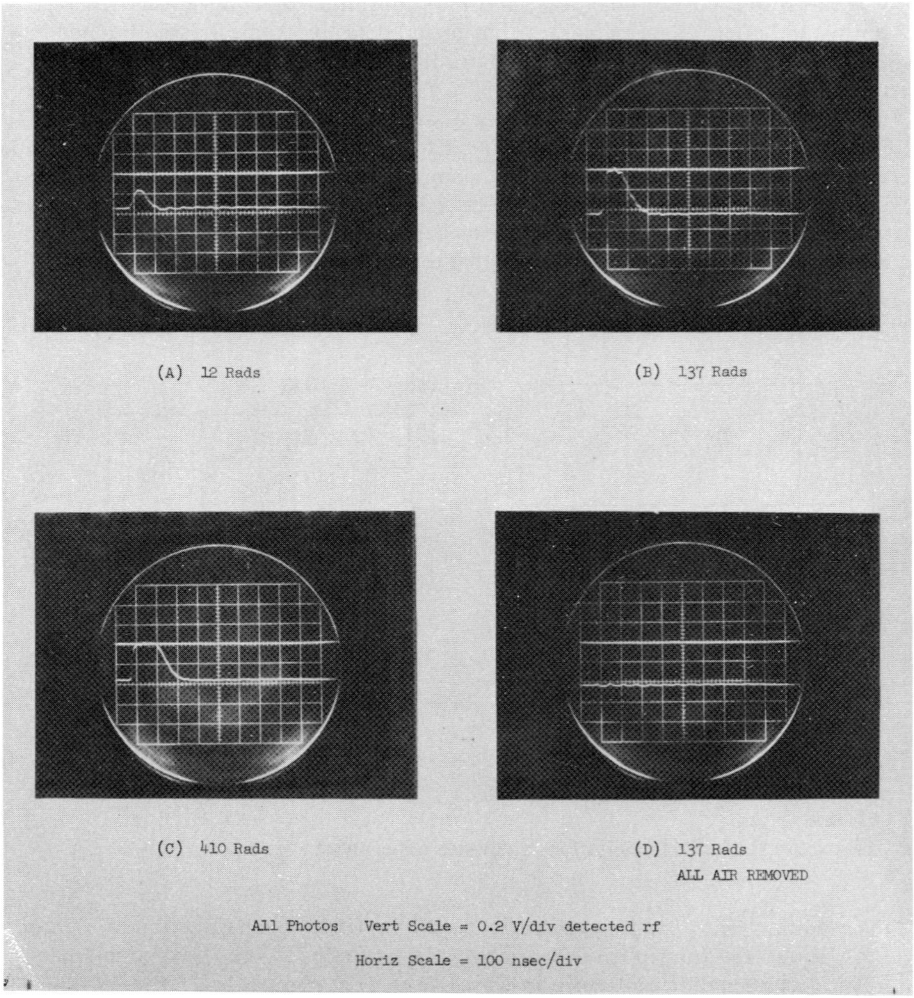

FIGURE B.4
Experimental data showing the effect of ionizing radiation pulses on air-filled microwave cavities: (A) 12 rads; (B) 137 rads; (C) 410 rads; and (D) 137 rads, all air removed. (For all photos: vertical scale = 0.2 V/div detected rf; horizontal scale = 100 nsec/div.)

because Eq. (B.7) to (B.11) are the results of perturbational analyses and as such are not valid for too large a Q change. It is safe to assume, however, that because of the nature of the high pressure plasma, the predominant cause of the drop in transmission shown in Figure B.4 is due to a decrease in Q. From Figure B.4 it can be seen that the period of the disturbance increases with increasing dose (or dose rate, assuming a constant x-ray pulse width). For example, the excursion beyond the 6dB point lasts: 50 nsec at 4.8×10^8 rad/sec; 140 nsec at 6.8×10^9 rad/sec; and 190 nsec at 2×10^{10} rad/sec. Therefore, if the drop in Q (50% for 6 dB) is enough to stop the oscillations, the delay effect will be a function of dose rate. This effect is seen to disappear if all of the air is pumped out of the cavity (see Figure B.4D). A similar effect should also occur in microstripline resonators; however, the time delay will not be as large since the recombination time of the electrons in the solid dielectric is much shorter than in a gas.

REFERENCES

[1] T. S. Saad, R. C. Hansen, and G. J. Wheeler, *Microwave Engineers' Handbook 1*, Artech House, Inc., Dedham, Mass., 1971.

[2] J. F. Kircher and R. E. Bowman, *Effects of Radiation on Materials and Components*, Reinhold Corp., New York, 1964.

[3] W. E. Wilson and R. J. Chaffin, "Radiation Damage Effects in Microwave Dielectric Substrate Materials," presented at IEEE Conference on Nuclear and Space Radiation Effects, July 1972.

[4] R. W. Peters, *Handbook of Triplicate Microwave Components*, Sanders Assoc., Nashua, N.H. 1956, Chapter 5.

[5] M. V. Schneider, "Microstriplines for Microwave and Integrated Circuits," *BSTJ*, 1421 (May 1969).

[6] C. G. Montgomery, R. H. Dicke, and E. M. Purcell, *Principles of Microwave Circuits 8*, MIT Radiation Lab Series, McGraw-Hill, New York, 1948, p. 238.

[7] J. B. Wattier et al., "Effects of Reactor Radiation on the Engineering Properties of Elastomers and Plastics," Battelle Research Center Report BRC-17829 (1962).

[8] J. J. Licari, *Plastic Coatings for Electronics*, McGraw-Hill, New York, 1970, Chapter 7.

[9] R. J. Chaffin, "Ionizing Radiation Effects in Microwave Cavities," *IEEE Sci.*, **NS-18**, 6, 436 (1971).

[10] M. Voth and E. Kenny, "Microwave Diagnosis of a Plasma Generated in a Reactor and Its Use as a Reactor Power Monitor," *Nuc. Instr. and Meth.* **63**, 45 (1968).

[11] S. C. Brown, *Basic Data of Plasma Physics*, MIT Press, Cambridge, Mass., 1966.

[12] R. J. Chaffin and J. Beyer, "Plasma Diagnostics with a Microwave Fabry-Perot Resonator," *IEEE Trans. MIT*, 37 (January 1968).

Appendix C

Spectrum Dependence of Neutron Damage

Most neutron damage studies are carried out using a nuclear reactor as a high-fluence neutron source. Chapter 5 discussed the energy dependence of neutron damage (see Figure 5.8) and stated that if a reactor is used as a neutron source, then a knowledge of the neutron energy spectrum (rather than just the total neutron fluence) is needed to calculate the density of initial Frenkel defects. This may be expressed as [see Eq. (5.7)]

$$N_F = N_0 \int_0^{\mathscr{E}_m} \sigma_d(\mathscr{E}) \cdot \Phi_n(\mathscr{E})\, d\mathscr{E} \tag{C.1}$$

where $\sigma_d(\mathscr{E})$ is given by Figure 5.8 and $\Phi_n(\mathscr{E})\, d\mathscr{E}$ is the number of neutrons with energies between $\mathscr{E}$ and $\mathscr{E} + d\mathscr{E}$.

One basic problem in attempting to compare the results of several different researchers in the scientific community is that different reactor types quite often are used to perform similar experiments. These different reactors do not have the same neutron energy spectra. As a result, the specification: neutron fluence = n/cm^2, $\mathscr{E} > 10$ keV does not tell the complete story when describing experimental results. This effect can be seen in Figure C.1 where the carrier removal rate for neutrons from several different reactor types is shown. Figure C.1 shows that the carrier removal rate seems to correlate well with the average neutron energy of the spectra. However, different results are obtained for materials with different resistivities [1]. The abbreviations used in Figure C.1 are: MTR is the Materials Test Reactor at Phillips Petroleum, Idaho Falls, Idaho. TRIGA is a water-moderated wide-pulse reactor at General Atomics, San Diego, California. X-10 is located at Oak Ridge National Laboratories, Oak Ridge, Tennessee. SPRF (SPR-I) is the Sandia Pulsed Reactor Facility, Sandia Laboratories, Albuquerque, New Mexico. BSF is the Bulk Shielding Facility at Oak Ridge National Laboratories.

Ideally, it should be possible, by using a curve like Figure 5.8 and Eq. (C.1), to derive an equivalent neutron fluence factor for various reactor spectra. Then reported neutron fluences could be converted to a 1 MeV or a 14 MeV (for example) equivalent neutron fluence which would allow a direct intercomparison of experiments done in different reactors. One of the basic problems with this approach is that the curve shown in Figure 5.8 is for primary defect density. It will be recalled from Chapter 5 that most of these initial defects reorder into defect complexes. The way in which the reordering occurs is strongly dependent on the sample history and on the neutron energy spectrum itself. Several researchers have discussed this problem [1–8]; but for the preceding and other reasons, no satis-

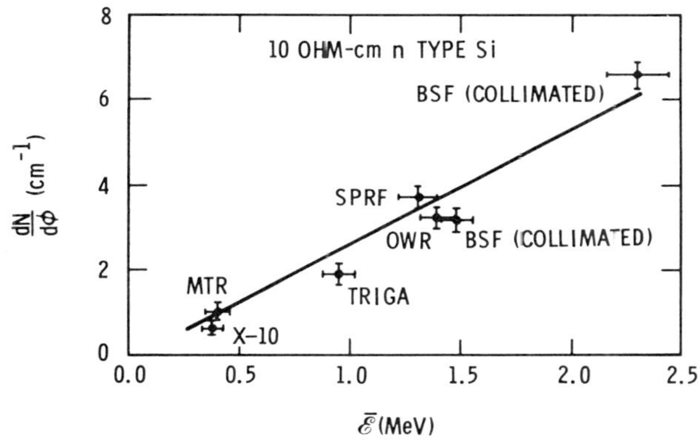

FIGURE C.1
Carrier removal rate versus average energy of the neutron experiment (after Kantz [1]).

factory neutron damage equivalence curve has as yet been accepted by the scientific community.

Rather than enter the current battle over neutron damage equivalence, this book has used the unambiguous neutron irradiation description: n/cm^2, $\mathscr{E} > 10$ keV, Spectrum X, specifying the reactor spectrum where it is known. The reactor spectra codes used are:

A = Sandia Laboratories SPR-II Rector (similar to unmoderated fission spectrum)
B = Sandria Laboratories ACPR Reactor (TRIGA-type reactor)
C = Oak Ridge Graphite Reactor, see Ref. [36], Chapter 5
D = Brookhaven National Laboratories, Research Reactor, see Ref. [61], Chapter 5
E = Sandia Laboratories SERF Reactor

F = Battelle Research Reactor with 0.02-inch cadmium shield, see Ref. [11], Chapter 13
G = Pennsylvania State University Reactor, see Ref. [7], Chapter 13
H = Ground Test Reactor, Air Force Nuclear and Aerospace Reactor Facility, Ft. Worth, Texas, see Ref. [41], Chapter 6

Two of the more commonly used spectra, A and B, are shown in Figures C.2 and C.3.

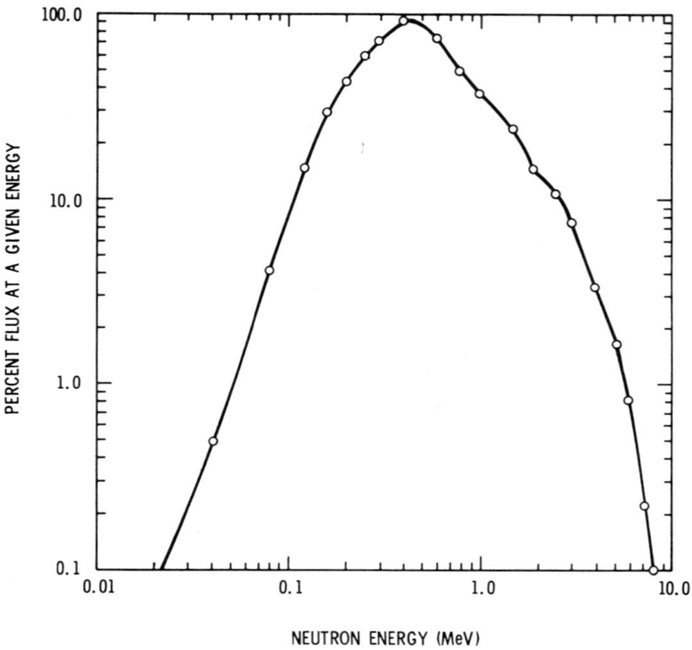

FIGURE C.2
Differential neutron energy spectrum in the SPR-II Reactor Room normalized to 100% at peak flux as determined by means of activation foils (after Morrison and Posey [9]).

REFERENCES

[1] A. D. Kantz, "Average Neutron Energy of Reactor Spectra and Its Influence on Displacement Damage," *J. Appl. Phys.* **34**, 1944 (1963).
[2] A. D. Rossin, "Dosimetry for Radiation Damage Studies," *IEEE Trans. Nuc. Sci.*, **NS-11**, 130 (November 1964).

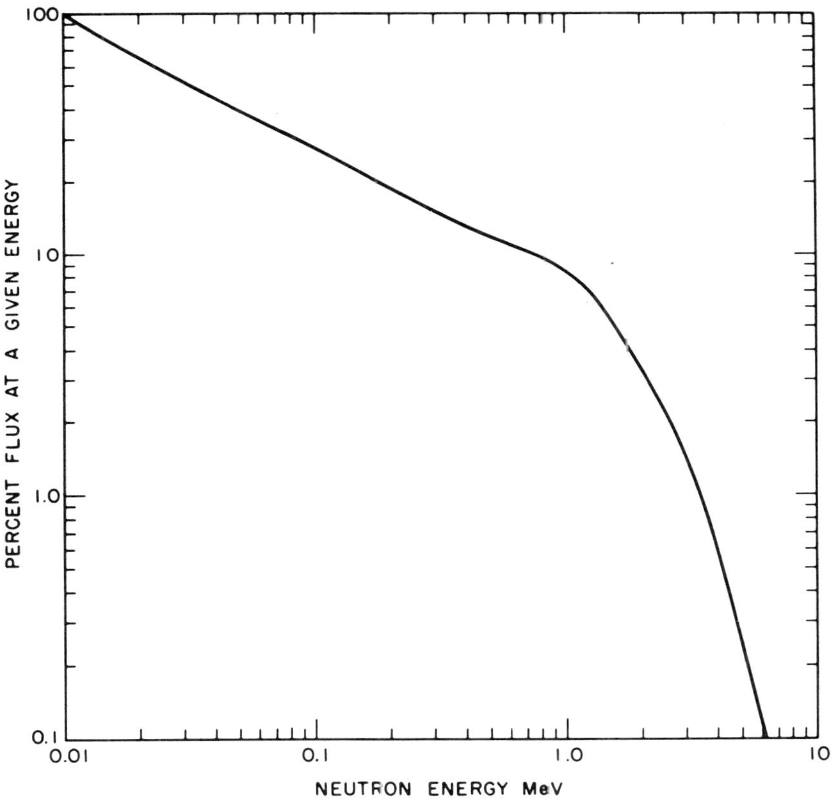

FIGURE C.3
Differential neutron energy spectrum of the ACPR Reactor normalized to 100% at peak flux (after Morris and Chodorow [10]).

[3] E. C. Smith et al., "Theoretical and Experimental Determination of Neutron Energy Deposition in Silicon," *IEEE Trans. Nuc. Sci.*, **NS-13,** 11 (December 1966).

[4] F. M. Smits, "On the Energy Dependence of Neutron Damage in Silicon Transistors," Report SC-R-64-196, Sandia Corporation, Albuquerque, N.M., 1964.

[5] H. J. Stein, "Energy Dependence of Neutron Damage in Silicon," *J. Appl. Phys.*, **38**, 204 (1966).

[6] F. Larin, *Radiation Effects in Semiconductor Devices*, John Wiley & Sons, New York, 1968.

[7] J. M. McKenzie, "Neutron Energy Dependence of Damage to Semiconductor Devices," *Proceedings of IEEE Conference on Nuclear and Space Radiation Effects*, Durham, N.H., July 1971, p. 97.

[8] E. E. Conrad, "Considerations in Establishing a Standard for Neutron Displacement Energy Effects in Semiconductors," *IEEE Trans. Nuc. Sci.*, **NS-18**, 200 (December 1971).

[9] L. W. Morrison and L. D. Posey, "Neutron and Gamma Measurements in the Sandia Pulsed Reactor Facility (SPR II)," Report SC-RR-68-85, Sandia Corporation, Albuquerque, N.M., 1968.

[10] F. M. Morris and A. M. Chodorow, "Annular Core Pulsed Reactor Experimenters Manual," Report SC-M-68-454, Sandia Corporation, Albuquerque, N.M., 1968.

List of Symbols

The following list of symbols contains the important symbols used throughout the book. In the case where a symbol is used only once, it is defined locally and is not given here. Since this book spans several different scientific areas, it was necessary to decide whether standard notation (which results in some cases of multiply defined symbols) should be used or unique nonstandard symbols should be used. It was felt that use of the standard definitions would make the book easier to use for a reference book, and for that reason standard definitions are used. In the case where a symbol is multiply defined, each definition is numbered in parentheses.

A	Device area
AF	Annealing factor
B	(1) Magnetic flux density (2) Susceptance
c	Velocity of propagation
C	Capacitance in general
C_d	Depletion capacitance
C_D	Diffusion capacitance
dB	Decibel
D_n, D_p	Electron, hole diffusion coefficients
e	Electronic charge
eV	Energy in electron-volts
E	Electric field ($\bar{E}$ = vector form)
$\mathscr{E}$	Energy in general
$\mathscr{E}_c$	Conduction band edge energy

$\mathscr{E}_d$	Lattice atom displacement energy
$\mathscr{E}_e$	Electron energy
$\mathscr{E}_f$	Fermi energy
$\mathscr{E}_g$	Bandgap energy
$\mathscr{E}_k$	Kinetic energy
$\mathscr{E}_n$	Neutron energy
$\mathscr{E}_p$	(1) Proton energy (2) Pair formation energy
$\mathscr{E}_T$	Trap energy Level
$\mathscr{E}_v$	Valence band edge energy
f	Frequency in general
f_c	Cutoff frequency
f_s	Frequency at which $\|s_{21}\| = 1$
f_T	Frequency at which $\|h_{fe}\| = 1$
$f(\mathscr{E})$	Fermi distribution function
F	Noise figure
g	(1) Pair generation rate (2) Conductivity
G	Conductance
h_{fe}	Common-emitter transistor current gain
$\hbar$	Planck's constant/2π
i	Intrinsic region carrier density
j	Particle current density
j_n, j_p	Electron, hole particle current density
J	Electric current density
J_n, J_p	Electron, hole electric current density
k	(1) Crystal momentum (2) Boltzman's constant
K_L	Diffusion length damage constant

K_τ	Lifetime damage constant
L	(1) Device length (2) Inductance (3) Transmission loss
L_n, L_p	Electron, hole diffusion length
m	Mass
m^*	Effective mass
m^*_{ce}	Conductivity effective electron mass
m^*_{ch}	Conductivity effective hole mass
m^*_{de}	Density-of-states effective electron mass
m^*_{dh}	Density-of-states effective hole mass
m^*_{hh}	Heavy-hole effective mass
m^*_{lh}	Light-hole effective mass
m^*_1	Lower-valley effective electron mass
m^*_2	Upper-valley effective electron mass
$m^*_\parallel$	Longitudinal effective mass
$m^*_\perp$	Transverse effective mass
M	(1) Carrier multiplication ratio in near avalanche conditions (2) Bias to threshold ratio for Gunn oscillators
n	(1) Electron density (2) Harmonic number
n_0	Equilibrium electron density
n_{n0}, n_{p0}	Equilibrium electron density in n, p regions
n/cm^2	Neutron fluence
N_A, N_D	Acceptor, donor doping density
$N_c(\mathscr{E})$, $N_v(\mathscr{E})$	Density-of-states distribution function in the conductance, valence bands
N_T	Trap density
p	Hole density
p_0	Equilibrium hole density

p_{n0}, p_{p0}	Equilibrium hole density in *n*, *p* regions
Q	(1) Total electronic charge (2) Microwave cavity quality factor
r	Recombination rate
R	(1) Abbreviation for rad (2) Resistance (3) Recombination rate
$\dot{\mathrm{R}}$	Rads per second
S_{ij}	*S* parameters
t_d, t_m, t_w	Excess diode, mixer, and white noise temperature
T	(1) Temperature (2) Transit time
$\overline{T}$, T_m	Average, maximum energy transferred during a collision
TSS	Tangential sensitivity
v	Carrier velocity
v_s	Saturated drift velocity
v_t	Thermal velocity
V	Voltage
V_A, V_D	Applied, diffusion voltage for *p-n* junction
W	Device width (same as L)
W_A, W_D	Width of depletion region on *p*, *n* side of junction
$W_T = W_A + W_D$	(Total depletion width)
x	Distance
X	Reactance
Y	Admittance
Z_0	Transmission line impedance
α	Common-base transistor current gain
$\alpha(E)$	Ionization coefficient
β	Common-emitter transistor current gain

γ	(1) Gamma radiation (2) Emitter injection ratio
δ	Collector collection factor
ϵ_0	8.85×10^{-12} Farad·m^{-1} (dielectric constant)
ϵ_s	$\epsilon_r \cdot \epsilon_0$ (semiconductor dielectric constant)
η	(1) Trap occupation probability (2) Base transport factor
λ	Wavelength
λ_n, λ_p	Electron, hole ac diffusion length
μ_n, μ_p	Electron, hole mobility
μ_H	Hall mobility
μ_1, μ_2	Lower-, upper-valley mobility in GaAs
ν	Collision frequency
ρ	(1) Charge density (2) Resistivity (3) Density gm/cm^3
σ	(1) Conductivity (2) Capture cross-section
σ_d	Total displacement cross-section
τ	Lifetime in general
τ_{ec}	Emitter-collector transit time
τ_n, τ_p	Electron, hole lifetime
Φ_e, Φ_n, Φ_p, Φ_γ	Electron, neutron, proton, gamma fluence
Φ_f	Failure fluence
Φ_m, Φ_s	Work function for metal, semiconductor
χ_s	Electron affinity potential
ω	Frequency in general
ω_α	Common-base cutoff frequency

Index